Local Structure
from Diffraction

FUNDAMENTAL MATERIALS RESEARCH

Series Editor: M. F. Thorpe, *Michigan State University*
East Lansing, Michigan

ACCESS IN NANOPOROUS MATERIALS
Edited by Thomas J. Pinnavaia and M. F. Thorpe

DYNAMICS OF CRYSTAL SURFACES AND INTERFACES
Edited by P. M. Duxbury and T. J. Pence

ELECTRONIC PROPERTIES OF SOLIDS USING CLUSTER METHODS
Edited by T. A. Kaplan and S. D. Mahanti

LOCAL STRUCTURE FROM DIFFRACTION
Edited by S. J. L. Billinge and M. F. Thorpe

Local Structure
from Diffraction

Edited by

S. J. L. Billinge and M. F. Thorpe

Michigan State University
East Lansing, Michigan

Plenum Press • New York and London

Library of Congress Cataloging in Publication Data

Local structure from diffraction / edited by S. J. L. Billinge and M. F. Thorpe.
     p.   cm.—(Fundamental materials research)
    "Proceedings of a Conference on Local Structure from Diffraction, held August 10–13,
1997, in Traverse City, Michigan"—T.p. verso.
    Includes bibliographical references and index.
    ISBN 0-306-45827-6
    1. Materials—Microscopy. 2. Diffraction patterns. 3. Long range order (Solid state physics) I. Billinge, S. J. L. II. Thorpe, M. F. III. Conference on Local Structure from Diffraction
(1997: Traverse City, Mich.) IV. Series.
    TA417.23.L63   1998
    620.1'1299—dc21   98-4348
                                                                       CIP

Proceedings of a Conference on Local Structure from Diffraction,
held August 10–13, 1997, in Traverse City, Michigan

ISBN 0-306-45827-6

© 1998 Plenum Press, New York
A Division of Plenum Publishing Corporation
233 Spring Street, New York, N.Y. 10013

http://www.plenum.com

10 9 8 7 6 5 4 3 2 1

SERIES PREFACE

This series of books, which is published at the rate of about one per year, addresses fundamental problems in materials science. The contents cover a broad range of topics from small clusters of atoms to engineering materials and involve *chemistry*, *physics*, *materials science* and *engineering*, with length scales ranging from Ångstroms up to millimeters. The emphasis is on basic science rather than on applications. Each book focuses on a single area of current interest and brings together leading experts to give an up-to-date discussion of their work and the work of others. Each article contains enough references that the interested reader can access the relevant literature. Thanks are given to the Center for Fundamental Materials Research at Michigan State University for supporting this series.

M.F. Thorpe, Series Editor
E-mail: thorpe@pa.msu.edu
East Lansing, Michigan

PREFACE

One of the most challenging problems in the study of structure is to characterize the atomic short-range order in materials. Long-range order can be determined with a high degree of accuracy by analyzing Bragg peak positions and intensities in data from single crystals or powders. However, information about short-range order is contained in the *diffuse scattering intensity*. This is difficult to analyze because it is low in absolute intensity (though the integrated intensity may be significant) and widely spread in reciprocal space.

The need to persevere and develop reliable techniques for analyzing diffuse scattering is becoming increasingly important. This is because many newly emerging materials, including some with potential technological applications, are quite disordered. These include materials such as semiconductor alloys, ferroelectric materials and other transition metal compounds, nanoporous and microporous materials such as zeolites and pyrolitic graphites, and molecular crystals, as discussed in these proceedings. In order to obtain a full solution of the structure, it is necessary to analyze both the long-range crystallographic order and short-range aperiodic deviations from the long-range order.

This workshop on *Local Structure from Diffraction* was organized to bring together leading researchers studying local structure using diffraction techniques. Surprisingly, there are few opportunities for the powder and single crystal diffuse scattering communities to come together in one place and discuss their common goals of local structure determination. This intimate and intensive workshop was held at the historic and picturesque Park Place Hotel in Traverse City, Michigan, USA from 10–14th August, 1997 and aimed at addressing that need. All the attendees were by invitation only; 25 of whom gave presentations at the meeting. There were also many animated and detailed private discussions. Seven different countries were represented with people coming from as far afield as Australia, Korea and Ukraine as well as a strong representation from Europe and North America.

The invited speakers were asked to produce a manuscript containing a pedagogical account of their work which would be of lasting value as a text for newcomers to the field and as a reference for established researchers. There is no other book currently available which has this scope, and our hope is that this volume fills a gap by bringing into one place descriptions of the various approaches which are used to collect, analyze and interpret diffuse scattering data. The authors responded conscientiously to our request, and the result is this present volume which contains 20 articles by many of the leading researchers in the field of local structure from diffraction studies.

We would like to thank Michigan State University for financing the meeting and the *Center for Fundamental Materials Research* at MSU for contributing to the cost of producing the proceedings. Also, the efforts of Lorie Neuman and Janet King, who organized the workshop and proceedings, are deeply appreciated as was the advice and efforts of the *Advisory Committee members*: W.I.F. David, T. Egami, S.C. Moss and D.L. Price.

Simon J.L. Billinge
Michael F. Thorpe
East Lansing, Michigan

CONTENTS

PDF Analysis Applied to Crystalline Materials...1
  Takeshi Egami

Anomalous X-Ray Scattering from Disordered Materials.....................................23
  David L. Price and Marie-Louise Saboungi

The Recording and Interpretation of Diffuse X-Ray Scattering.........................35
  T.R. Welberry

Recent Advances in Structure Refinement for Liquids
    and Disordered Materials...59
  A.K. Soper

Neutron Scattering and Monte Carlo Studies of Disorder
    in Oxides and Hydrides...85
  W. Schweika and M. Pionke

Reverse Monte Carlo Refinement of Disordered Silica Phases 101
  David A. Keen

Modelling Single Crystal Diffuse Neutron Scattering
    Using Reverse Monte Carlo..121
  V.M. Nield

Real Space Rietveld: Full Profile Structural Refinement
    of the Atomic Pair Distribution Function 137
  S.J.L. Billinge

Advances in Pair Distribution Profile Fitting in Alloys 157
  M.F. Thorpe, J.S. Chung, S.J.L. Billinge, and F. Mohiuddin-Jacobs

Local Atomic Arrangements in Binary Solid Solutions Studied
    by X-Ray and Neutron Diffuse Scattering from Single Crystals...........................175
  J.L. Robertson, C.J. Sparks, G.E. Ice, X. Jiang, S.C. Moss, and L. Reinhard

Fermi Surface Effects in the Diffuse Scattering
    from Alloys...189
  S.C. Moss and H. Reichert

Non-Mean-Field Theories of Short Range Order and
    Diffuse Scattering Anomalies in Disordered Alloys 207
  Igor Tsatskis

Diffuse Scattering by Crystals with Defects of Coulomb Displacement Field 233
  R.I. Barabash

Short-Range Disorder and Long-Range Order:
    Implications of the "Rigid Unit Mode" Picture 253
  Martin T. Dove, Volker Heine, Kenton D. Hammonds, Manoj Gambhir,
    and Alexandra K.A. Pryde

Vibrational Entropy and Local Structures of Solids 273
  Brent Fultz

Diffuse Scattering by Domain Structures 295
  Friedrich Frey

Recent "Local" Structural Studies: Metallic Alloys,
    Superconductors and Proteins 323
  George H. Kwei, Despina Louca, Simon J.L. Billinge, and H.D. Rosenfeld

Studies of Local Structure in Polymers Using X-Ray Scattering 337
  Michael J. Winokur

Diffuse X-Ray and Neutron Reflection from
    Surfaces and Interfaces 351
  Sunil K. Sinha and Roger Pynn

Studying Growth Kinetics of Metallic Multilayers
    Using Elastic X-Ray Diffuse Scattering 375
  Rogerio Paniago

List of Participants 391

Index 397

Local Structure
from Diffraction

PDF ANALYSIS APPLIED TO CRYSTALLINE MATERIALS

Takeshi Egami

Department of Materials Science and Engineering,
and Laboratory for Research on the Structure of Matter
University of Pennsylvania, Philadelphia, PA 19104-6272

INTRODUCTION

The method of atomic pair-density function (PDF) analysis has widely been used in the study of liquids, glasses and other amorphous materials.[1-3] The PDF describes the distribution of the atomic distances in the material, and can directly be determined from diffraction experiments. But it has been considered to be a method of last resort, when the usual crystallographic analysis fails and there is no other option. For this reason, while this method is beginning to be used for the study of crystalline materials,[4,5] it might not be easy to see the wisdom of applying it for crystals for which the structure can easily be determined with the aid of the Bragg's law. However, the PDF obtained by powder diffraction is simply a Fourier-transform of the structure function, $S(Q)$ ($Q = 4\pi\sin\theta/\lambda$), and thus the PDF carries no less information than the powder diffraction pattern. Therefore one can argue, in principle, that the real-space PDF analysis is at least equivalent to the methods of crystallographic powder diffraction analysis such as the Rietveld refinement that are carried out in the reciprocal space.

Moreover, if the material has some disorder in the structure, the shortcomings of the conventional crystallographic analysis and the advantage of the real-space PDF analysis become obvious. The crystallographic analysis that presumes periodicity takes only the Bragg peaks into account and ignores the diffuse scattering. Consequently it gives only a spatially averaged picture of the structure, and local variations in the structure are not correctly represented. In particular the information regarding the correlation among the local variations is lost as soon as the assumption of perfect periodicity is made. For instance if there are local atomic displacements away from the crystallographic sites, they are reflected in the crystal structure only in terms of the large thermal, or Debye-Waller, factor, and cannot be easily separated from the lattice vibration.

On the other hand, the real-space PDF method, that utilizes diffuse scattering as well as the Bragg scattering, can describe the disorder more accurately. In particular the correlation among the local variations is correctly represented, albeit in volume average. As shown in the examples below in many cases the local structure of a solid is different from the average structure. The PDF is capable of bringing this difference to light.

Strictly speaking few materials are perfectly periodic. Even when the crystal is without any lattice defects, lattice anharmonicity may introduce local collective deviations from perfect periodicity. Many materials of modern technological interest are complex in the structure, often containing internal disorder. For instance alloys, solid solutions and mixed ion crystals are inherently disordered at the atomistic level. The study of such materials would greatly benefit from the dual-space approach, or the parallel use of the real as well as reciprocal space analyses which are complementary to each other. In this article we will outline the real-space PDF method applied to the study of crystalline materials, discuss when in general the use of such a method is beneficial, summarize some of the recent studies in which the real-space or the dual-space method has proven their capability, including ferroelectric oxides, colossal magnetoresistance (CMR) materials, superconducting cuprates and MX polymers, and discuss future possibilities.

LIMITATIONS OF CRYSTALLOGRAPHY

The Bragg's law is the basis of crystallography, as we all know, and is indeed an extremely powerful law. By using this law it is possible to determine the lattice constant to an amazing accuracy of 10^{-4} Å or better. However, since the success of the Bragg's law is so pervasive and convincing that we tend to forget its premises and limitations. Let us for a moment try to extricate ourselves from the spell of the Bragg's law and see it objectively.

As a starter, it is strange that the lattice constant can be determined with the accuracy as good as 10^{-4} Å by using x-rays with the wavelength of 1 Å or so. Usually in order to access such a small lengthscale, because of the uncertainty principle $\Delta x \cdot \Delta k < 2\pi$, one ought to require a probe of a comparable wavelength, thus of an enormous energy. For instance γ-rays with $\lambda = 10^{-4}$ Å has the energy of 1.24 TeV. Indeed measurements with such a small lengthscale and high energy are being made in high energy physics, but not in crystallography. The answer is obviously that the crystallographic measurement does not directly measure the unit cell length a nor the atomic bond length. What is actually measured is the position of the Bragg peak, Q, with the accuracy of ΔQ. Now the group of density waves with the wavevector in the range from $Q - \Delta Q/2$ to $Q + \Delta Q/2$ will set up a wavepacket with the coherence length, $\ell = 2\pi/\Delta Q$, and the periodicity of $\lambda = 2\pi/Q$. When someone determines a lattice constant through the measurement of the Bragg peak with the Q resolution ΔQ, this person is really measuring the coherence length of the density wave, ℓ, and only by assuming perfect periodicity, relating it to a by,

$$\ell = Na. \tag{1}$$

The uncertainty in ℓ is of the order of λ, thus the accuracy in a is λ/N, which can be as small as 10^{-4} Å if N is as large as 10^4. Thus by utilizing the periodicity as a magnifying lens the small lengthscale of 10^{-4} Å can be indirectly accessed in crystallographic analysis without high energy probes. The critical assumption is the perfect periodicity. But how good is this assumption ?

We are so much used to thinking that a perfect crystal is perfectly periodic. However, that is not strictly true because of lattice vibrations, even at T = 0 K. Only when the lattice vibration is perfectly harmonic we can invoke the Debye-Waller approximation, and salvage the perfect periodicity in time average. The assumption of perfect periodicity breaks down when the structure contains some randomness, or when the lattice dynamics is very anharmonic. Any aperiodicity will result in the diffuse scattering which is excluded from the usual crystallographic analysis. Restating the obvious, the crystallographic structural analysis determines only the long range order of the lattice, the periodic component of the structure. If a particular property of interest depends on any other aspect of the structure, the crystal structure may not be able to provide information sufficient to explain that particular property. More often than not the properties of a material depend on the local atomic structure, rather than the long range lattice structure.[6] This is why a local probe to determine the local structure is so important.

THE PDF METHOD

The most commonly used local structural probe is the x-ray absorption fine structure (XAFS) method. By using the virtual photoelectron as the probe the XAFS method can determine the local distances to the neighboring atoms from the atom emitting the virtual photoelectron.[7] This method is credited for bringing in the local viewpoint of the atomic structure. Using this method new insights have been gained in the local structure in alloys,[8] co-operative phase transition,[9] and electron-lattice interaction,[10] just to name a few, and deeply influenced our views on these phenomena.

However, the method has limitations. The strong Coulomb interaction between the probe (photoelectron) and the scattering atom produces various complications including the phase shifts, inelastic scattering and multiple-scattering. These effects are theoretically corrected, so that the limitations of the theory became the limitations of the method itself. Also since the energy of the photoelectron is proportional to the square of the wavevector, the energy range in which the oscillation in the absorption coefficient has to be determined increases rapidly as the range of Q is increased. Thus in practice it is difficult to obtain data accurately beyond $k = 15$ Å^{-1}, or $Q = 2k = 30$ Å^{-1}. For these reasons the PDF of the XAFS is not reliable beyond the first peak, and even for the first peak the peak width cannot be accurately determined.

An alternative is the PDF obtained by x-ray or neutron diffraction measurement.[1-3] Since x-rays and neutrons interact more weakly with the matter than electrons do, their scattering can be treated kinematically by the Born approximation, making the analysis simpler and more reliable, compared to that of the XAFS. The PDF is equivalent to the Patterson correlation function of crystallography. However, in the PDF method the lattice periodicity is not assumed, or equivalently the size of the unit cell is assumed to be infinite. Therefore not only the Bragg peaks but the scattering in all the continuous Q-space, including the diffuse scattering, is considered. This enables the PDF to describe periodic as well as aperiodic structure. Indeed the PDF method has been used primarily for the study of liquids and glasses, while its application on crystals is commencing only recently.

In the case of powder diffraction the structure function S(Q) is determined from the measured diffraction intensity, I(Q), after a lengthy but straightforward procedure. At first it is important to measure, in addition to the scattering from the sample, the true background due to addenda, or anything other than the sample itself, such as the sample container and ambient gas (air, or preferably He). In many of the crystallographic methods such as the Rietveld refinement everything other than the Bragg peaks is treated as the background, fitted with a curve, and dropped. It thus becomes just another adjustable parameter in the refinement process. However, this type of "background" contains scattering coming from the sample such as diffuse scattering. In the PDF method, on the other hand, the diffuse scattering from the sample must be retained as it carries important information on disorder and local atomic correlation. Therefore the true background due to addenda has to be accurately measured, and subtracted from the data. Then the data have to be corrected for absorption, multiple scattering, and inelastic scattering (Compton scattering for x-rays and Plazcek correction for neutrons) to obtain S(Q).[11]

To calculate the PDF, S(Q) is then Fourier transformed by

$$\rho_0 g(r) = \rho_0 + \frac{1}{2\pi^2 r} \int_0^\infty [S(Q) - 1] \sin(Qr)Q\,dQ \qquad (2)$$

where ρ_0 is the number density of atoms. As an example the PDF of f.c.c. Ni powder at T = 10 K is shown in Fig. 1. This PDF was determined using pulsed neutron diffraction with the GLAD spectrometer of the Intense Pulsed Neutron Source (IPNS), the Argonne National Laboratory. It is compared with the PDF calculated for the f.c.c. structure. Excellent agreement of the measured PDF with the expected PDF demonstrates the accuracy of the technique. The PDF agrees with the model up to 40 Å or more.[11] The primary PDF calculated for the crystallographic structure is made of many δ-functions representing the distances between the atomic sites. In order to include the effect of the lattice vibrations these δ-functions are convoluted with a Gaussian function. The width of the Gaussian function corresponds to the Debye-Waller factor. In the case above the width ($\sigma = 0.064$ Å) is consistent with the zero-point Debye-Waller factor and the Debye temperature of Ni.

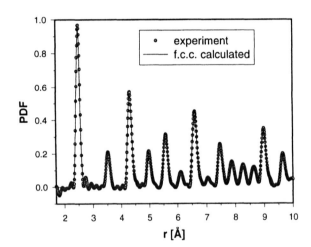

Fig. 1. Pulsed neutron PDF of f.c.c. Ni powder at T = 10 K, compared with the PDF calculated for the structure (solid line). The peak width is due to the zero-point lattice vibrations.

ACCURACY OF THE PDF METHOD

The main reason why the PDF method has not been used for crystalline materials for a long time is the termination error. While the Fourier-transform integral in eq. (2) requires S(Q) known up to Q infinity, the range of the Q value is limited by the wavelength of the scattering probe. If the integration is terminated prematurely, spurious oscillations called the termination errors are observed. If one uses a Mo-target sealed x-ray tube ($\lambda = 0.7107$

Å) the range of Q values is limited to 17 Å^{-1}, far short of a sufficient range, resulting in significant termination errors. The range of Q-space necessary for accurate Fourier-transformation depends upon the Debye-Waller factor. If the Debye-Waller envelope, $\exp(-Q^2 \langle u^2 \rangle)$, where u is the amplitude of lattice vibration and $\langle .... \rangle$ represents thermal average, becomes sufficiently small, and S(Q) converges to unity, it is unnecessary to collect the data beyond such a point.[11] For a reasonable amplitude of $\langle u^2 \rangle^{1/2} = 0.05$ Å, the 3σ of the Debye-Waller envelope is reached only at $Q = 42.4$ Å^{-1}. While this is a very high value of Q for conventional sources, it can be attained relatively easily using the synchrotron based radiation sources such as the spallation pulsed neutron sources (in the U.S. the IPNS of the Argonne National Laboratory and the LANSCE of the Los Alamos National Laboratory) and the synchrotron radiation sources (in the U.S. the NSLS of the Brookhaven National Laboratory, the CHESS of Cornell University, the SSRL of Stanford University and the APS of the Argonne National Laboratory), which provide high energy scattering probes. For most materials if a pulsed neutron source is used and S(Q) is determined up to about 40 Å^{-1}, the termination error is minimal, as shown in Fig. 1. For some strongly covalent bonds, however, termination at much higher Q values is required as we discuss later. In such a case high energy x-ray scattering and pulsed neutron scattering with a more intense source are the methods to be used.

The error in S(Q) can be propagated to the PDF by,[11]

$$\rho_0^2 \Delta g(r)^2 = \frac{1}{4\pi^4 r^2} \sum_v \left[\Delta S(Q_v) Q_v dQ_v sin(Q_v r) \right]^2 \qquad (3)$$

where v indexes the data points, and $\Delta S(Q_v)$ is the error in S(Q) for the v-th data point. Note that since the integration in (3) is only weakly dependent on r, the error actually decreases with r as $1/r$ as demonstrated later. Thus, contrary to the common perception the PDF is *more accurate at large distances*. Errors include statistical errors due to discrete photon counting ($\sqrt{N}$ error), and errors in corrections such as absorption correction and multiple-scattering correction. If the error is a slow function of Q, as many systematic errors are, then due to the Fourier-transformation in (2) their effect is contained only to the small r region, and the main part of the PDF remains accurate. This is why the PDF is remarkably robust and accurate in many cases. Moreover in the case of comparative studies, such as the study of changes due to temperature, fields, or sample processing conditions, the systematic errors are canceled, resulting in a higher accuracy of the measurement.

THE PDF IN HIGHER DIMENSIONS

Three-Dimensional PDF

The Patterson function is a 3-dimensional correlation function. Thus the PDF also originally is a 3-dimensional function, given by[11]

$$\rho_0 \left[g(\mathbf{r}) - 1 \right] = \frac{1}{8\pi^3} \int \left[S(\mathbf{Q}) - 1 \right] exp(-i\mathbf{Q} \cdot \mathbf{r}) d\mathbf{Q} \qquad (4)$$

where $S(\mathbf{Q})$ is the structure function determined in the 3-d Q-space, and $g(\mathbf{r})$ is the 3-d density correlation function,

$$g(\mathbf{r}) = \frac{1}{\rho_0^2 V} \int \rho(\mathbf{r}')\rho(\mathbf{r}' + \mathbf{r}) d\mathbf{r}' \qquad (5)$$

here V is the sample volume, and $\rho(\mathbf{r})$ is the single atom density function. In order to obtain this function the structure function $S(\mathbf{Q})$ has to be determined in the continuous 3-dimensional Q-space, but this is not an easy task. For instance if we use a relatively coarse mesh of $\Delta Q = 0.01$ Å^{-1}, in order to scan the entire three-dimensional Q-space of ± 40 Å^{-1}, scattering data have to be collected at as many as 5.12×10^{11} points in the Q-space. This is not impossible with the aid of a two-dimensional area detector, but it requires a long measurement time and a very large memory space, and has never been done over the size of the Q-space needed for accurate Fourier-transformation. An attempt to carry out such a measurement will be made in the near future using a pulsed neutron single crystal diffractometer with an area detector (SCD of LANSCE).

Since such a measurement cannot be normally done, the PDF is usually reduced to one-dimension by making the measurement on a powder sample and taking the orientational averaging, thus losing the angular information. This might appear as a very severe compromise, and one might argue that an accurate 3-dimensional single crystal study coupled with a measurement of the diffuse scattering over a limited Q-space as usually done is still better than the PDF study when some disorder is present. However, the merit of collecting all the diffuse scattering in the powder scattering should not be underestimated. Unless one has a very good idea where the diffuse scattering occurs in the 3-d Q-space one may miss important information in the study of diffuse scattering from a single crystal. The merit of being able to collect all the diffuse scattering by the powder measurement usually overweighs the disadvantage of losing angular information.

One-Dimensional PDF and layer correlation function

A mistake commonly made is to assume that if a one-dimensional scan of the single crystal scattering data is Fourier-transformed it produces the atomic PDF along that direction. For instance if one collects the scattering intensity from a single crystal along Q_z including the diffuse scattering and applied the Fourier-transform, what is produced by such a procedure is not the atomic PDF but the atomic layer-layer correlation function,

$$g_{layer}(z) = \frac{1}{L} \int \rho_2(z')\rho_2(z' + z) dz' \qquad (6)$$

where z is parallel to $\mathbf{Q}$, L is the length of the sample in the z-direction, and $\rho_2(z)$ is the layer-averaged single atom density function,

$$\rho_2(z) = \frac{1}{A} \int \rho(\mathbf{r}) dx dy \qquad (7)$$

where A is the area of the sample in x-y plane. The layer-layer correlation function (6) is often confused with the 1-dimensional correlation function averaged over the entire sample,

$$g_{1-d}(z) = \frac{1}{\rho_0^2 V} \int \rho(\mathbf{r}')\rho(\mathbf{r}' + \mathbf{z}) d\mathbf{r}' \qquad (8)$$

Note that in this case the correlation function is volume averaged, while in (6) and (7) the density function is averaged first, before evaluating the correlation function. In order to obtain the 1-d correlation function (8), one has to carry out an integration in the Q_x-Q_y space,

$$S_z(Q_z) = \frac{1}{A_Q} \int S(\mathbf{Q}) dQ_x dQ_y \qquad (9)$$

where A_Q is the area in the Q_x-Q_y space over which the integration is carried out, and then apply the Fourier-transform,

$$g_{1-d}(z) = \frac{1}{8\pi^3 L} \int \left[S_z(Q_z) - 1 \right] exp\left(-iQ_z z\right) dQ_z \qquad (10)$$

Thus $S(Q)$ has to be determined over the 3-d space anyway.

An example of this procedure is the one applied to multi-layered films. Normally the interfacial roughness is evaluated by measuring the specular reflectivity with the Q-vector along the normal of the surface. This, however, yields the layer-layer correlation as in (6), thus the effect of atomic diffusion (roughness at the atomic scale) and interfacial roughness cannot be separated. By measuring the diffuse scattering in the off-specular directions as shown in Fig. 2 and integrating the intensity in the Q_x-Q_y plane, these two can be separated.[12,13] In the case of the Pt/Co multilayered films with the <111> orientation the total thickness of the interface determined from the specular reflectivity is 8.7 Å, while the true local thickness due to diffusion is 6.9 Å.[12] Furthermore from the off-specular diffuse scattering it is possible to determine the interfacial height-height correlation function that characterizes the length scale of the interfacial roughness.[13]

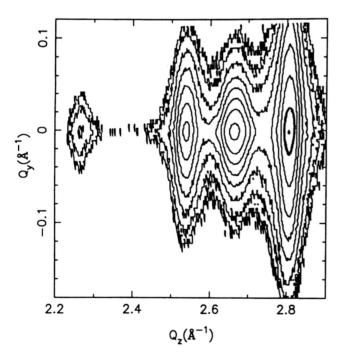

Fig. 2. Scattering intensity from a Pt/Co multilayered film as a function of Q_z, normal to the surface and Q_y, parallel to the surface.[12,13]

2-dimensional PDF

If the data such as those in Fig. 2 are averaged over the angle in the plane retaining the radial Q length $Q_r = \sqrt{Q_x^2 + Q_y^2}$,

$$S_2(Q_r, Q_z) = \frac{1}{2\pi Q_r} \int S(\mathbf{Q}) d\varphi \tag{11}$$

where φ is the angle in the Q_x-Q_y plane, the Fourier-transform of this function is the in-plane 2-d PDF,

$$\rho_0 g(R, Q_z) = \frac{1}{2\pi} \int [S_2(Q_r, Q_z) - 1] J_0(Q_r r) Q_r dQ_r \tag{12}$$

where $J_0(Q_r, Q_z)$ is the 0-th order spherical Bessel function. The averaging over the planar angle can easily be done by rotating the sample around an axis (z-axis) during the measurement.[14] This technique was successfully applied to the study of a decagonal quasicrystal $Al_{65}Cu_{15}Co_{20}$ which is quasiperiodic in the x-y directions, but is periodic in the z-direction. By using the PDF resolved for l the in-plane atomic structure of this compound was analyzed.

Partial wave analysis

Another possibility is to use a polar coordinate and decompose S(Q) and PDF into radial and angular functions,

$$S(\mathbf{Q}) = \sum_{\ell,m} S_\ell^m(Q) Y_\ell^m\left(\frac{\mathbf{Q}}{Q}\right)$$

$$g(\mathbf{r}) = \sum_{\ell,m} g_\ell^m(r) Y_\ell^m\left(\frac{\mathbf{r}}{r}\right) \tag{13}$$

where $Y_\ell^m(a)$ are the spherical harmonics. They are connected by

$$g_\ell^m(r) = \frac{i^\ell}{2\pi^2 \rho_0} \int S_\ell^m(Q) J_\ell(Qr) Q^2 dQ \tag{14}$$

where $J_\ell(x)$ is the ℓ-th order spherical Bessel function. This technique was applied to the analysis of anisotropic amorphous materials,[15,16] but it can also be applied to the study of crystalline materials.

X-RAY VS. NEUTRONS

We now compare the merits and demerits of the x-ray scattering and neutron scattering as far as the PDF method is concerned.

8

Q-Dependence of the atomic scattering factor

X-rays are primarily scattered by electrons. Thus the Fourier-transform of the scattered x-ray intensity, $I(Q)$, will produce the *electron-electron correlation function*, or the electronic PDF. This function is composed of broad peaks, reflecting the spatial extension of the electronic wavefunction. In order to extract the atom-atom correlation, the electronic PDF has to be deconvoluted by the electron density function. This process corresponds to dividing $I(Q)$ through the square of the x-ray atomic scattering factor, $f(Q)$. Since the Q dependence differs slightly among the elements, this division produces small inaccuracy for multi-element systems.[1-3] The effect of this approximation, however, is usually negligible, and if the differences are large it is possible to carry out the element specific deconvolution process in the real space.

On the other hand the scattering amplitude (length) of neutron, b, is approximately independent of Q due to the small size of the atomic nucleus, thus the Fourier-transform of the neutron scattering intensity, which is the nuclear PDF, is equivalent to the atomic PDF. This, however, is not true for the magnetic scattering of neutrons. Because of the spatial extension of the magnetic moment, or the density of the electrons carrying a magnetic moment, the magnetic form factor decreases rapidly with Q. Thus the direct Fourier-transform of $I(Q)$ produces the *spin-spin correlation function* which is spatially spread. For this reason it is very difficult to detect the magnetic contribution in the neutron PDF. In order to determine the atom-atom magnetic correlation, the magnetic component of the scattering intensity has to be separated, ideally by using spin-polarized neutrons, and then divided through the magnetic form factor,[17] just the same way as the x-ray intensity is divided through the x-ray scattering factor.

Q-Range

The spallation neutron sources provide high intensity of epithermal neutrons which are not fully thermalized, and thus have higher energies. The use of these epithermal neutrons greatly extends the Q-range. However, the spectrum of epithermal neutrons is nearly constant of λ, so that as a function of Q it decreases quickly as $1/Q^2$. As a result the statistical noise increases as Q, and the contribution of the noise to the integrand of eq. (2) as Q^2. Consequently the Q-range is limited by the noise. For instance the Q-range of the SEPD is practically limited to 40 Å^{-1}, even though it is possible to collect the data beyond.

It may appear that it is more difficult to extend the Q-range for x-ray scattering because of the rapid decrease of $f(Q)^2$ with Q. However, with the use of an intense high energy synchrotron source such as the wiggler beamline of CHESS, Cornell University, it is relatively easy to obtain sufficiently strong scattering at high values of Q, as shown below. In fact while the energy of the scattering probe is proportional to Q^2 for a given scattering angle for neutrons and electrons, it is proportional to Q for x-rays. Thus it is easier with x-rays to attain high Q values by escalating the probe energy. Moreover, the Q resolution is constant of Q for x-ray scattering, while it is proportional to Q for pulsed neutron scattering, so that x-ray scattering has advantages in obtaining high resolution data at high Q values.

Element specificity

The element specific PDF, or the PDF from a particular element, can be obtained with x-ray scattering by using anomalous dispersion near the absorption edge, or with neutrons by isotope substitution. The x-ray anomalous scattering is useful only when the energy of the absorption edge is sufficiently high to allow a large Q range, so that it can be applied only to relatively heavy elements. Isotopes are expensive, and not always available. Thus it is not always possible to determine the element specific PDF by diffraction methods. The XAFS method has a clear advantage when it comes to the element specific PDF. On the other hand, when the structure of the solid is approximately known it is usually possible to refine the structure without the element specific PDF.

Lattice dynamics

It is often stated that x-rays give a snap-shot picture of the structure, and neutrons yield the average structure, since at the same wavelength x-rays have a much higher energy than neutrons. While this is often true it needs more comments. The dynamics of the measurement is dictated by the energy resolution, not by the energy of the probe itself. Whether it is x-ray or neutrons, for the scattering process with the momentum transfer Q and the energy transfer ω, the theoretical scattering intensity by an assembly of moving atoms is proportional to the dynamic structure factor,[18]

$$S(Q,\omega) = \frac{1}{NT} \int_0^T \sum_{i,j} \exp\left(i\mathbf{Q} \cdot \left[\mathbf{r_i}(0) - \mathbf{r_i}(t)\right] - i\omega t\right) dt \qquad (15)$$

where $r_i(t)$ denotes the position of the i-th atom at time t, and N is the number of atoms. During the actual measurements, however, both Q and ω can be specified only within a non-zero resolution, ΔQ and $\Delta \omega$. Thus with respect to energy the integration of (4) over the resolution window, $\Delta \omega$,

$$S_m(Q,\omega,\Delta\omega) = \int_{-\Delta\omega/2}^{\Delta\omega/2} S(Q,\omega') d\omega' \qquad (16)$$

gives the actual photon or neutron counts. This integration sets the time scale of observation. For instance if the energy window is wide open ($\Delta\omega \to \infty$), then the integration removes all the correlation but t = 0, and thus the same time correlation function, or the snap shot picture, is obtained. On the other hand if $\Delta\omega = 0$, all the temporal correlations are included, so that the correlation among the average density is obtained. Since a typical energy resolution of x-ray scattering (~ 10 eV) is much higher in energy than the energy scale for phonons (~ 0.1 eV), phonons look completely frozen for x-rays. However, by improving the energy resolution of the x-ray scattering it is possible to observe the inelastic scattering by phonons. In the case of pulsed neutron scattering with the TOF spectrometer, even though there is no energy discriminator in the system, due to the Placzek shift[19] the effective energy window is about 20 meV.[20] Lattice dynamics slower than this would look static, while those faster than this would be averaged out.

REAL SPACE MODELING

Since the PDF is a one-dimensional quantity it requires modeling to recover a three-dimensional structure from the data. Such modeling can be done in the similar way as the regular structural refinement, but done in the real space rather than in the reciprocal space.[21] The procedure is the following:

1. Build a model of the atomic positions.
2. Calculate the distances among the atoms and obtain the PDF made of many δ-functions.
3. Convolute the δ-functions with a Gaussian function representing the lattice vibrations to obtain $\rho_0 g_{mod}(r)$.
4. Calculate the agreement (A) factor,

$$A^2 = \frac{\int_{r_1}^{r_2} \left[g_{exp}(r) - g_{mod}(r) \right]^2 dr}{\int_{r_1}^{r_2} dr} \tag{17}$$

where $g_{exp}(r)$ is the experimentally determined PDF. This A-factor is similar to the R factor in the crystallographic analysis.
5. Improve the model by the Monte-Carlo simulated annealing process to minimize the agreement factor.

For a perfect crystal at low temperatures the real-space modeling and the usual reciprocal space modeling such as the Rietveld refinement were shown to produce almost identical results as shown below. When some disorder is present the advantage of the real-space method becomes clear. In particular a step-wise local refinement, in which one starts with modeling the immediate locality, such as the nearest neighbors, and gradually extends the range of modeling, proved to be a powerful method of modeling the local structure.[21] By this approach the local structure was often found to be quite different from the average structure, including symmetry, as illustrated in the examples below.

EXAMPLES

Structure of PZ and PZT

$Pb(Zr_{1-x}Ti_x)O_3$ (PZT) is a widely used ferroelectric/piezoelectric oxide system, and its end member, $PbZrO_3$ (PZ), is well known for antiferroelectricity. Nevertheless some controversies remain regarding the structure of both PZ and PZT. Their structure was recently studied using pulsed neutron diffraction, with the data analyzed both in the reciprocal space by the Rietveld refinement and in the real space by the PDF method.[22] At T = 10 K the structures refined by the two methods agreed quite well, typically within 0.01 Å. The PDF A-factor (8.11 %) was comparable to the Rietveld R-factor (7.99 %). The experimental and model PDF's are compared in Fig. 3. The only notable difference was found in the thermal factor. The thermal factor refined by the PDF method was appreciably smaller and closer to what is expected from the Debye temperature, while that obtained by the Rietveld method was larger and appeared to overestimate the vibrational magnitude.

At higher temperatures anharmonic local displacements of Pb atoms along the c-axis were better characterized by the PDF method, and the structure of the intermediate phase (503 - 510 K) was for the first time found to be a different type of antiferroelectric structure by the real space method. An interesting finding is that the PDF of the ferroelectric phase just below the Curie temperature and that of the paraelectric phase just above are very similar, as shown in Fig. 4. This shows that the local structure does not change so much through T_C, and local lattice distortions already exist above T_C. Similar findings were made for various ferroelectric solids by the XAFS and PDF studies.[9,23]

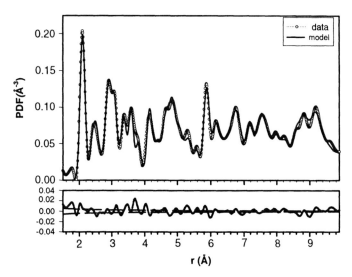

Fig. 3. Pulsed neutron PDF of PbZrO$_3$ at T = 20 K (circles) compared with the model PDF (solid line) (above), and the difference between the two (below). The dashed line (below) shows the statistical error, eq. (3) which decreases with r approximately as 1/r.[22]

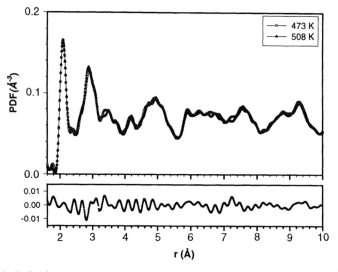

Fig. 4. Pulsed neutron PDF of PbZrO$_3$ at T = 473 K (anti-ferroelectric phase) and at T = 508 K (paraelectric phase), showing little difference (below, expanded scale) between them in spite of the phase transition.[22]

12

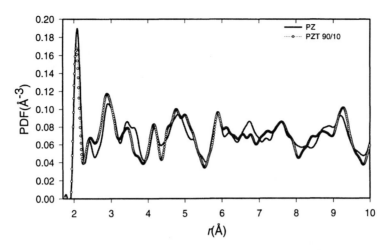

Fig. 5. Pulsed neutron PDF's of $PbZrO_3$ (PZ) and $Pb(Zr_{0.9}Ti_{0.1})O_3$ (PZT 90/10).[24]

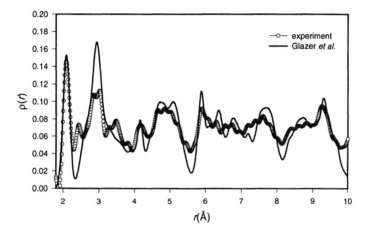

Fig. 6. Pulsed neutron PDF of PZT 90/10 compared with the PDF calculated for the model structure refined by Rietveld method by Glazer et al.[25]

When a small amount (5 %) of Zr atoms are replaced by Ti atoms the displacive order changes from antiferroelectric to ferroelectric, and the symmetry changes from tetragonal to orthorhombic. It is rather strange that such a small amount of Ti can induce such major changes in the structure. As it happens the local structure described by the PDF again does not change much in spite of the changes in the crystal structure, as shown in Fig. 5. The PDF of the rhombohedral phase is significantly different from the one calculated from the structure refined by the Rietveld method (Fig. 6). The real space refinement shows that the local structure is similar to that in the orthorhombic phase, but the axes are randomly rotated around the [111] axis, producing the rhombohedral symmetry in average.[24]

The common feature in these phase changes is that the phase transition is brought about not by uniform changes in the structure, but by the rearrangement of the local structural units which themselves remain little changed through the transition. This is presumably because the energy scale to change the local atomic bonding (usually of the order of 1 eV) is larger than the energy scale for phase transition (of the order of 0.1 eV). It would be of great interest to see how generally this conclusion applies.

Local structure of CMR manganites

Manganites such as $(La_{1-x}A_x)MnO_3$ (A = Ca, Sr, Ba, Pb) that show colossal magnetoresistance (CMR) represent a spectacular case of the electron-lattice interaction governing the properties. One of the issues here was whether or not the doped charges form lattice polarons as predicted by Millis et al.[26] The PDF studies offered convincing evidence of polaron formation and their role in phase transitions.[27,28]

$LaMnO_3$ has a distorted perovskite structure, with Mn^{3+} ion occupying the center of MnO_6 octahedron. Mn^{3+} is in d^4 configuration. The cubic component of the crystal field splits the d-level into a triplet t_{2g} and doublet e_g levels. These 4 d-electrons are spin polarized due to strong Hund's coupling, and fill up the spin polarized t_{2g} level, and singly occupy the e_g level. This leads to the Jahn-Teller (JT) distortion of the MnO_6 octahedron which becomes elongated to accommodate the e_g electron in the d_z^2 orbital. The PDF of $LaMnO_3$ (Fig. 7) shows a well-split Mn-O first peak into 4 short (about 1.98 Å) and two long (about 2.23 Å) Mn-O bonds. This peak is negative because the neutron scattering length b of Mn is negative. The PDF calculated for the Rietveld refined structure[28] shows excellent agreement.

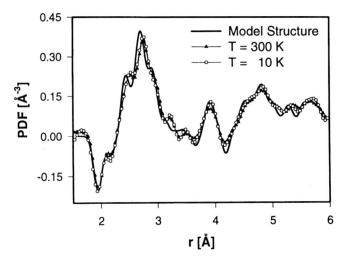

Fig. 7. Pulsed neutron PDF of $LaMnO_3$ at T = 10 K and 300 K, compared to the PDF of the crystallographic structure.[28]

If LaMnO$_3$ is doped with divalent ions (hole doping) the macroscopic JT lattice distortion decreases quickly and disappears, for instance at 16 % Sr. However, the PDF clearly shows that locally the JT distortion does not go away so quickly. The local JT distortion can be detected by the presence of long Mn-O bonds and conversely the reduced number of short Mn-O bonds. Fig. 8 shows the position of the Mn-O peaks as a function of Sr concentration. Even though the Mn-O bonds deduced from the crystal structure converge rapidly into one bondlength beyond 16 % Sr, locally the Mn-O bond lengths remain unchanged, at 1.98 Å and 2.23 Å.

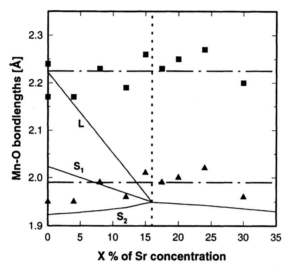

Fig. 8. The positions of the PDF Mn-O peaks as a function of the Sr concentration, compared to those for the crystal structure (solid lines).[28]

The number of short Mn-O bonds, N_{Mn-O}, can be determined by integrating the first Mn-O subpeak. For the JT distorted LaMnO$_3$ it is four, while in SrMnO$_3$ the JT distortion is gone, the MnO$_6$ octahedron is undistorted, so that N_{Mn-O} is six. At room temperature N_{Mn-O} changes linearly from 4 to 6 as Sr concentration is increased (Fig. 9), indicating that the system is made of a mixture of JT distorted Mn^{3+} and undistorted Mn^{4+}. In other words the doped hole is localized on one Mn site, forming a polaronic Mn^{4+} state. This is a direct demonstration that the lattice polaron related to the JT distortion is formed in this system.

At T = 10 K, the dependence of N_{Mn-O} on Sr concentration becomes stronger, with the slope about three times of the room temperature data, as shown in Fig. 9. This implies that the holes are slightly more delocalized, and spread over three Mn sites. Inside this three-site polaron the Mn-Mn exchange interaction must be changed from negative to positive due to the double-exchange interaction within the three-site polaron. Now it is of great interest that N_{Mn-O} shows no discontinuity across the metal-insulator (M-I) transition at x = 0.17.[29] Usually once the metallic state is reached polarons disappear because of high dielectric constant in the metal. However, probably because of the spatial spin-charge separation the charges are not uniformly distributed even in the metallic state, and some local

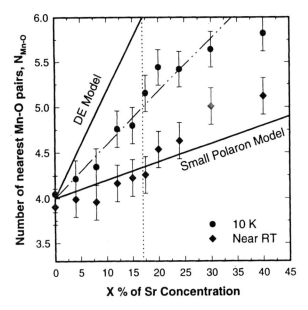

Fig. 9. The number of short Mn-O bonds per Mn ion as a function of Sr concentration, at T = 10 K and near RT. See text for the explanation of the two solid lines.[28]

distortions indicating locally varying charge density and JT distortion remain. It is possible that the polarons self-organize into stripes, and the charges flow freely inside the stripes.[28,30] This state may strongly resemble that of the superconducting cuprates.[31]

Local structure of HTSC cuprates

The pulsed neutron PDF analysis was applied to superconducting oxide immediately after they were discovered.[32] Notable deviations from the crystallographic structure were found, and it was observed that the local structure remains unchanged through the tetragonal/orthorhombic phase transition. Small local displacements of oxygen ions were found in all the high-T_C oxides studied by the PDF method so far.[5,33] In particular strong change in the dynamics of displacement was detected near T_C for $Tl_2Ba_2CaCu_2O_8$[34] as shown in Fig. 10.

It is possible that these local oxygen displacements are related to the dynamic spin-charge stripes[35] postulated in these materials. A recent high energy phonon measurement offers a strong evidence that fragmented dynamic stripes do exist in cuprates.[36] The size of the charge domain, 8 × 20 Å, as suggested by the phonon measurement, agrees well with the size of the polarization domain suggested by the PDF measurement of $YBa_2Cu_4O_8$.[37] Details aside the existence of some local structure in the Cu-O plane with the size of 2a × 5a appears to be certain. What implication this observation has to the superconductivity remains to be seen.[31]

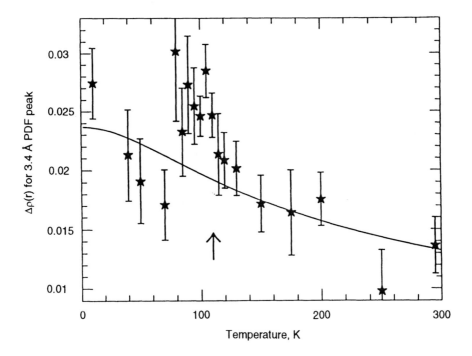

Fig. 10. Anomalous temperature dependence of the PDF peak height at 3.4 Å for $Tl_2Ba_2CaCu_2O_8$, compared to the expected dependence calculated from the phonon density of states (solid curve).[34] The arrow indicates superconducting T_C.

Charge density wave in the MX chain compounds

High-energy x-ray scattering measurements were recently carried out to determine the atomic structure of an organic 1-dimensional chain compound, $[Pt(en)_2I_2][Pt(en)_2](ClO_4)$, where *en* stands for 1,2-diaminoethane, with the incident x-ray energy of 60 keV.[38] This compound is one of the family of the so-called MX chain compounds that show strong charge density waves (CDW) due to the Peierls distortion.[39] Figure 11 shows the integrand of eq. (2), $i(Q) = Q[S(Q) - 1]$, determined at T = 30 K. At such high Q values the scattering intensity is dominated by Compton scattering. By using a solid state Ge detector the Compton contribution was separated, thus attaining excellent statistical accuracy even at a such high Q range.

In Fig. 11 it is seen that the oscillations persist at least up to 40 Å^{-1} and perhaps above, due to the strong covalent bond between Pt and I. Furthermore the oscillation has a slight beat, with the node around 22 Å^{-1}, indicating the presence of two bondlengths. Indeed the PDF ($G(r) = 4\pi\rho_0 r[g(r) - 1]$) shown in Fig. 12 clearly shows two strong peaks due to Pt-I correlation, at 2.7 Å and 3.1 Å, indicating the split of the Pt-I correlation due to the CDW. This result demonstrates the capability of the high-energy x-ray scattering in determining S(Q) up to very high values of Q, and obtaining the PDF with a high spatial resolution.

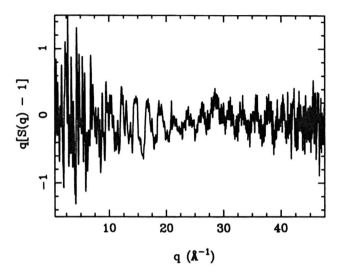

Fig. 11. $i(q) = q[S(q) - 1]$ of Pt-I chain compound $[Pt(en)_2I_2][Pt(en)_2](ClO_4)$ determined by high-energy x-ray scattering measurement.[38]

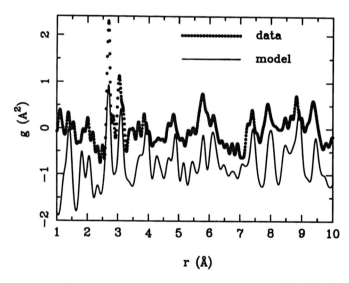

Fig. 12.   PDF, G(r), of the Pt-I chain compound determined by the high-energy x-ray scattering, compared with the PDF calculated for the crystal structure.[38]

APPLICATION OF THE PDF METHOD IN OTHER FIELDS

Other examples of ordered solids for which the PDF method was applied and proved effective in determining the local structure and relating it to properties include quasicrystals,[4,14,40,41] fullerenes,[42-44] relaxor ferroelectrics,[45,46] ferroelectric phase transition,[23,47] automotive catalytic support oxides,[48] and semiconductor alloys.[49] Other fields the PDF method potentially can impact are, martensitic transformation including the shape memory alloys, piezoelectric materials, orientational glasses, surfaces and interfaces (in particular the use of 2-dimensional PDF), and strong type II superconductors such as the chevrel phase. With the upcoming LANSCE expansion and the start of the APS operation high energy radiation sources will become even more readily available. It is hoped that the PDF method will be practiced by a larger number of researchers, and eventually will be regarded just as one of the standard methods. In particular the coupled use of both the real-space and reciprocal-space methods, the dual-space method, may well prove to be the best tool in characterizing complex materials.

CONCLUSIONS

Crystallographic methods are powerful techniques to determine the atomic structure of perfect crystals with amazing accuracy. However, if the structure is not perfectly periodic, due to dynamic or static local deviations, the limitations of the crystallographic methods become apparent. The PDF analysis is emerging as a most useful and convenient alternative for such cases. For a perfect crystal the PDF method is essentially equivalent to the crystallographic powder diffraction methods. When the system is partially disordered, the PDF method provides a much more accurate picture of the structure than the crystallographic methods do. While the PDF describes the local atomic correlation, the crystal structure only signals the presence of disorder. This is because the PDF analysis includes the diffuse scattering which contains the information regarding the aperiodic atomic correlation, while the diffuse scattering is neglected in the usual crystallographic analysis. By modeling the PDF in separate ranges of atomic distances, for instance in short, intermediate, and long ranges, the difference between the local structure and the average structure can be brought out to light. When they differ, it is most likely that it is the short range structure that influences the properties the most. The use of the both techniques, the dual-space method in which refinement is made both in the reciprocal and real spaces, is ideal when the presence of structural disorder is suspected. This method holds a great promise for the study of complex materials in general.

ACKNOWLEDGMENT

The advances in the PDF analysis technique and its application to crystalline solids were achieved through the tireless effort by the current and former students and postdoctoral fellows in my group, in particular, W. Dmowski, B. H. Toby, S. Nanao, X. Yan, W. Yu, D. D. Kofalt, Y. Suzuki, Wendy Spronson (Frydrych), S. J. L. Billinge, Ruizhong Hu, H. D. Rosenfeld, T. R. Sendyka, R. J. McQueeney, S. Teslic and Despina Louca. Their enthusiasm, energy and ingenuity have been marvelous, and are warmly appreciated. We also owe to D. L. Price for the design of the IPNS PDF software and the development of GLAD, and to many other collaborators and colleagues who helped the development of this technique. The author is extremely grateful for the steady support by the National Science

Foundation over many years, at present through DMR96-28134, which has been essential for the success of this work. He also thanks the Office of Naval Research for the support through N00014-91-J-1036.

REFERENCES

1. B. E. Warren, *X-ray Diffraction*, Dover, New York (1969, 1990).
2. H. P. Klug and L. E. Alexander, *X-Ray Diffraction Procedures for Polycrystalline and Amorphous Materials*, 2nd ed., John Wiley and Sons (1968).
3. G. S. Cargill, III, *Solid State Physics*, **30**, 227 (1975).
4. D.D. Kofalt, S. Nano, T. Egami, K.M. Wong and S.J. Poon, *Phys. Rev. Lett.* **57**, 114 (1986).
5. T. Egami and S. J. L. Billinge, in *Physical Properties of High Temperature Super-conductors V*, ed. D. M. Ginsberg (World Scientific, Singapore, 1996) p. 265.
6. A. P. Sutton, *Electronic Structure of Materials* (Clarendon Press, Oxford, 1992).
7. E. A. Stern, D. E. Sayers and F. W. Lytle, *Phys. Rev. B* **11**, 4836 (1975).
8. A. Erbil, W. Weber, G. S. Cargill III and R. F. Boehme, *Phys. Rev. B* **34**, 1392 (1986).
9. E. A. Stern and Y. Yacoby, *J. Phys. Chem. Solids* **57**, 1449 (1996).
10. S. D. Conradson and I. D. Raistrick, *Science* **243**, 1340 (1989).
11. B. H. Toby, and T. Egami, *Acta Cryst. A* **48**, 336 (1992).
12. X. Yan, T. Egami, E. E. Marinero, R. F. C. Farrow and C. H. Lee, *J. Mater. Res.* **7**, 1309 (1992).
13. X. Yan and T. Egami, *Phys. Rev. B* **47**, 2362 (1993).
14. Y. He, R. Hu, T. Egami, S. J. Poon and G. J. Shiflet, *Phys. Rev. Lett.* **70**, 2411 (1993).
15. Y. Suzuki, J. Haimovich and T. Egami, *Phys. Rev. B* **35**, 2162 (1987).
16. T. Egami, W. Dmowski, P. Kosmetatos, M. Boord, T. Tomida, E. Oikawa and A. Inoue, *J. Non-Cryst. Solids* **192&193**, 591 (1995).
17. Y. Wu, W. Dmowski, T. Egami, M. E. Chen and J. D. Axe, *J. Appl. Phys.* **61**, 3219 (1987).
18. S. W. Lovesey, *Theory of Neutron Scattering from Condensed Matter* (Clarendon Press, Oxford, 1984).
19. G. Placzek, *Phys. Rev.* **86**, 377 (1952).
20. R. J. McQueeney, unpublished.
21. T. Egami, *J. Phys. Chem. Solids* **56**, 1407 (1995).
22. S. Teslic and T. Egami, *Acta Cryst. B*, in press.
23. G. H. Kwei, S. J. L. Billinge, S.-W. Cheong and J. G. Saxton, *Ferroelectrics* **164**, 57 (1995).
24. S. Teslic, T. Egami and D. Viehland, *Ferroelectrics* **194**, 271 (1997).
25. A. M. Glazer, A. A. Mabud and R. Clarke, *Acta Cryst. B* **34**, 1060 (1978).
26. A. J. Millis, P. B. Littlewood and B. I. Shairman, *Phys. Rev. Lett.* **74**, 5144 (1995).
27. S. J. L. Billinge, R. G. DiFrancesco, G. H. Kwei, J. J. Neumeier and J. D. Thompson, *Phys. Rev. Lett.* **77**, 715 (1996).
28. D. Louca, T. Egami, E. Brosha, H. Röder and A. R. Bishop, *Phys. Rev. B* **56**, R8475 (1997).
29. A. Urushibara, Y. Moritomo, T. Arima, A. Asamitsu, G. Kido and Y. Tokura, *Phys. Rev. B* **51**, 14103 (1995).

30. D. Louca and T. Egami, unpublished.

31. T. Egami, D. Louca and R. J. McQueeney, *J. Superconductivity*, in press.

32. T. Egami, W. Dmowski, J.D. Jorgensen, D.G. Hinks, D.W. Capone, II, C.U. Segre and K. Zhang, in *"High-T_c Superconductors III"*, ed. Bose and Tyagi (World Scientific, Singapore, 1988), p. 101; *Rev. Solid St. Sci.* **1**, 247 (1987).

33. T. Egami and S. J. L. Billinge, *Prog. Mater. Sci.* **38**, 359 (1994).

34. B. H. Toby, T. Egami, J. D. Jorgensen and M. A. Subramanian, *Phys. Rev. Lett.* **64**, 2414 (1990).

35. J. N. Tranquada, B. J. Sternlieb, J. D. Axe, Y. Nakamura, and S. Uchida, *Nature* **375**, 561 (1995).

36. T. Egami, R. J. McQueeney, Y. Petrov, G. Shirane and Y. Endoh, unpublished.

37. T. R. Sendyka, W. Dmowski, T. Egami, N. Seiji, H. Yamauchi and S. Tanaka, *Phys. Rev. B* **51**, 6747 (1995).

38. T. Egami, S. J. L. Billinge, S. Kycia, W. Dmowski and A. S. Eberhardt, *Nucl. Instrum. Methods,* in press.

39. B. Scott, S. P. Love, G. S. Kanner, S. R. Johnson, M. P. Wilkerson, M. Berkey, B. I. Swanson, A. Saxena, X. Z. Huang and A. R. Bishop, *J. Molecular Structure*, **356,** 207 (1995).

40. R. Hu, T. Egami, A.-P. Tsai, A. Inoue and T. Masumoto, *Phys. Rev. B* **46**, 6105 (1992).

41. Y. Shen, W. Dmowski, T. Egami, S. J. Poon and G. J. Shiflet, *Phys. Rev. B* **37**, 1146 (1988).

42. F. Li and J. S. Lannin, *Phys. Rev. Lett.* **65**, 1905 (1990).

43. R. Hu, T. Egami, F. Li and S. J. Lannin, *Phys. Rev. B* **45**, 9517 (1992).

44. S. Teslic, T. Egami and J. E. Fischer, *Phys. Rev. B* **51**, 5973 (1995).

45. T. Egami, H. D. Rosenfeld and Ruizhong Hu, *Ferroelectrics* **136**, 15 (1992).

46. H. D. Rosenfeld and T. Egami, *Ferroelectrics* **150**, 183 (1993).

47. S. J. L. Billinge and G. H. Kwei, *J. Phys. Chem. Solids* **57**, 1457 (1996).

48. T. Egami, W. Dmowski and R. Brezny, *SAE Publication* **970461** (1997).

49. F. Mohiuddin-Jacobs and S. J. L. Billinge, unpublished.

ANOMALOUS X-RAY SCATTERING FROM DISORDERED MATERIALS

David L. Price and Marie-Louise Saboungi

Argonne National Laboratory, Argonne, IL 60439

INTRODUCTION

Many structural issues in disordered systems, such as liquids, polymers and glasses, are still unresolved and controversial. Part of the problem stems from the fact that a single diffraction measurement of an n-component system yields a weighted average structure factor of the $n(n+1)/2$ separate partial structure factors:

$$S(Q) = \sum_{ab} W_a(Q) W_b(Q) S_{ab}(Q)$$

(1)

where $W_a(Q)$ is the weighting factor for element a and $S_{ab}(Q)$ is the partial structure factor for the element pair (a,b).[1] For a multicomponent system it is therefore difficult to extract reliable information about a specific atom pair from a single experiment. However, in recent years improvements in the radiation sources available have led to the exploitation of complementary techniques which allow a more detailed investigation of the structure. Spallation neutron sources and instruments dedicated to amorphous and liquid systems make it possible to carry out neutron diffraction (ND) measurements with sufficient resolution to resolve peaks in the radial distribution function from different atom pairs; in favorable cases, isotope substitution can also be used to vary the weighting factors in Eq. (1).[2] An especially important advance has been the advent of high-powered synchrotron x-ray sources which make it possible to obtain reliable difference measurements near the absorption edge of a particular element with anomalous x-ray scattering (AXS), varying the weighting factor for that element in Eq. (1) by changing the x-ray energy.[3]

In this review we present the formalism of AXS as applied to partial structure analysis of disordered materials. We give first the theoretical formalism, and then some examples of how the technique has been applied in specific problems of current interest.

FORMALISM

The primary experimental quantity in an x-ray diffraction measurement may be taken as the dead-time corrected counts in a detector due to single scattering normalized by the dead-time corrected counts in a beam monitor. This gives the *intensity of single scattering*

$$I^s = \int \rho \frac{d\sigma^s}{d\Omega} \frac{\eta_d}{\eta_m} \Delta\Omega \exp\left(-\frac{2\mu x}{\sin\theta}\right) \frac{dx}{\sin\theta}$$

(2)

where ρ is the total number density, $d\sigma^s/d\Omega$ is the *average scattering cross section* per atom, η_d and η_m are the efficiencies of detector and beam monitor, μ the linear attenuation coefficient in the sample at the incident energy, x the depth of the scattering point in the sample and 2θ the scattering angle. The specific form given above for Eq. (2) assumes a symmetric reflection geometry and a slab sample, but it may be readily generalized to other geometries. To obtain I^s as given in Eq.(2), it is necessary to correct the measured beam intensity for fluorescence, resonant Raman scattering, Compton scattering, and attenuation and multiple scattering in the sample. The form of these corrections depends on the instrumental configuration; standard procedures are generally available for taking care of them. For a uniform slab of thickness d, Eq. (2) may be readily integrated to give

$$I^s = \frac{\rho A}{2\mu} T(d) \frac{d\sigma^s}{d\Omega}$$

(3)

where $A = \Delta\Omega\eta_d/\eta_m$ is the normalization constant, usually determined *a posteriori*, and $T(d) = [1-\exp(-2\mu d/\sin\theta)]$ is a thin sample correction, equal to unity for thick samples.

The scattering cross section is given by a sum of products of scattering factors and phase factors over all pairs of atoms of the system:

$$\frac{d\sigma^s}{d\Omega} = \frac{1}{N} \sum_{ij} f_i(\mathbf{Q}) f_j^*(\mathbf{Q}) \exp\left[i\mathbf{Q}\cdot(\mathbf{r}_i - \mathbf{r}_j)\right]$$

(4)

where $f_i(\mathbf{Q})$ is the form factor of atom i evaluated at the scattering vector $\mathbf{Q}$ and $\mathbf{r}_i$ its instantaneous position. It is convenient to reformulate Eq. (4) as the sum of three terms which, in the Faber-Ziman formulation[4], becomes

$$\frac{d\sigma^s}{d\Omega} = I(\mathbf{Q}) + \left[\left\langle|f(\mathbf{Q})|^2\right\rangle - |\langle f(\mathbf{Q})\rangle|^2\right] + N\delta_{Q0}|f(0)|^2$$

(5)

where the angular brackets represent averages over all atoms in the sample. The first term in Eq. (5) is the *interference scattering* which contains the details of the atomic structure, the second is the *Laue diffuse scattering* which varies slowly with $\mathbf{Q}$, and the third is the *forward scattering* which is singular in the small-angle limit for a homogeneous system. Clearly the first term is the important one in the present context. In a multicomponent system, it can be expressed as a weighted sum over element pairs:

$$I(\mathbf{Q}) = \sum_{ab} c_a c_b f_a(\mathbf{Q}) f_b^*(\mathbf{Q}) S_{ab}(\mathbf{Q})$$

(6)

where the *partial structure factor* $S_{ab}(\mathbf{Q})$ representing the structure associated with a given pair of elements (a,b) is given by

$$S_{ab}(\mathbf{Q}) = \frac{N}{N_a N_b} \sum_{i\in a, j\in b} \exp\left[i\mathbf{Q}\cdot(\mathbf{r}_i - \mathbf{r}_j)\right] - \frac{N}{N_a}\delta_{ab} + 1 - N\delta_{Q0}$$

(7)

It is often convenient to define an *average structure factor* $S(\mathbf{Q})$:

$$S(\mathbf{Q}) = \frac{I(\mathbf{Q})}{|\langle f(\mathbf{Q})\rangle|^2}$$

$$= \sum_{ab} c_a c_b \frac{f_a(\mathbf{Q}) f_b^*(\mathbf{Q})}{|\langle f(\mathbf{Q}) \rangle|^2} S_{ab}(\mathbf{Q})$$

(8)

In the Faber-Ziman formalism used here, both $S(\mathbf{Q})$ and $S_{ab}(\mathbf{Q}) \to 1$ as $Q \to \infty$.

To obtain real-space information, pair correlation functions are obtained by Fourier transformation of the appropriate structure factors, *e.g.* :

$$g(\mathbf{r}) - 1 = \frac{1}{8\pi^3 \rho} \int [S(\mathbf{Q}) - 1] \exp(-i\mathbf{Q} \cdot \mathbf{r}) d\mathbf{Q}$$

(9)

where $g(\mathbf{r})d\mathbf{r}$ has a direct physical interpretation as the number of atoms in a volume element $d\mathbf{r}$ at a distance $\mathbf{r}$ from a reference atom at the origin. Often we are interested in a particular structural feature, *e.g.*, a shell of atoms at a certain distance from a reference atom at the origin, that can be associated with a particular region of $g(\mathbf{r})$ — call it $g^n(\mathbf{r})$. Then the integrated volume of this region,

$$A^n = \rho \int g^n(\mathbf{r}) d\mathbf{r}$$

(10)

is a linear combination of the coordination numbers $C_a^n(b)$ of b atoms in the shell about an average a atom:

$$A^n = \sum_{ab} \frac{f_a f_b^*}{|\langle f \rangle|^2} c_a C_a^n(b)$$

(11)

where the form factors are evaluated at $Q = 0$.[5] Normally such a feature is associated with a single (a,b) pair, in which case the corresponding $C_a^n(b)$ is given uniquely by Eq. (11).

We now treat the case of special case of anomalous scattering. Suppose that the incident energy is near an absorption edge for element A and we make a small change ΔE in the incident energy, which we assume has a significant effect on f_A but not on the other f_a. From Eq. (6), the corresponding change in $I(\mathbf{Q})$ is given by

$$\Delta I(\mathbf{Q}) = c_A \Delta f_A \sum_b c_b f_b^*(\mathbf{Q}) S_{Ab}(\mathbf{Q}) + c.c.$$

(12)

In general, and especially near an absorption edge, the form factor contains anomalous energy-dependent terms as well as the regular term:

$$f(\mathbf{Q}) = f_0(\mathbf{Q}) + f'(E) + if''(E)$$

(13)

Substitution of Eq. (13) into Eq.(12) gives, after some rearrangement,

$$\Delta I(\mathbf{Q}) = 2c_A \Delta f_A' \sum_b c_b \left[\left(f_{b0}(\mathbf{Q}) + f_b' \right) S_{Ab}' + f_b'' S_{Ab}'' \right] + 2c_A \Delta f_a'' \sum_b c_b \left[f_b' S_{Ab}' - \left(f_{b0}(\mathbf{Q}) + f_b' \right) S_{Ab}'' \right]$$

(14)

By analogy with Eq. (8), we can define a *difference structure factor* for element A:

$$S_A(\mathbf{Q}) = \frac{\Delta I(\mathbf{Q})}{2c_A \Delta f_A' \langle f_0(\mathbf{Q}) + f' \rangle + 2c_A \Delta f_a' \langle f'' \rangle}$$

$$= \frac{2c_A \Delta f_A' \sum_b c_b \left[(f_{b0}(\mathbf{Q}) + f_b')S_{Ab}' + f_b' S_{Ab}'' \right] + 2c_A \Delta f_a'' \sum_b c_b \left[f_b' S_{Ab}' - (f_{b0}(\mathbf{Q}) + f_b')S_{Ab}'' \right]}{2c_A \Delta f_A' \langle f_0(\mathbf{Q}) + f' \rangle + 2c_A \Delta f_a' \langle f'' \rangle} \tag{15}$$

which also $\to 1$ as $Q \to \infty$. Notice that $S_A(\mathbf{Q})$ as defined in Eq. (15) implicitly depends on both E and ΔE. The ΔE dependence could be removed by taking the limit of Eq. (15) as $\Delta E \to 0$. However, the error in the measured value of $S_A(\mathbf{Q})$ obviously becomes very large as ΔE is made small. In practice we try to choose a compromise value for ΔE that is large enough to reduce this error and small enough so that we can neglect the changes in f_b, $b \neq A$. The values of the f_b in Eq. (15) are then evaluated at $E + 1/2\,\Delta E$. The choice of the distance of E from the absorption edge of element A is also a compromise between large values for the anomalous terms in f_A versus an accurate knowledge of Δf_A.

Eq. (15) can be simplified with the help of two approximations. First, a good approximation below the edge of element A is

$$\Delta f_A'' \ll \Delta f_A'. \tag{16}$$

E.g., for Ge at E_{Ge}-17 eV, $f_{Ge}' = -6.169$, $f_{Ge}'' = 0.494$; at E_{Ge}-200 eV, $f_{Ge}' = -3.987$, $f_{Ge}'' = 0.504$. With this approximation Eq. (15) simplifies to

$$S_A(\mathbf{Q}) = \frac{\sum_b c_b \left[(f_{b0}(\mathbf{Q}) + f_b')S_{Ab}' + f_b' S_{Ab}'' \right]}{\langle f_0(\mathbf{Q}) + f' \rangle} \tag{17}$$

Second, if the incident energies are far removed from the absorption edges of the elements $b \neq A$, a reasonable approximation is

$$f_b'' \ll f_{b0}, \, b \neq A. \tag{18}$$

In this case, taking account of the fact that S_{AA} has to be real, Eq. (17) simplifies finally to

$$S_A(\mathbf{Q}) = \sum_b c_b \frac{(f_{b0}(\mathbf{Q}) + f_b')}{\langle f_0(\mathbf{Q}) + f' \rangle} S_{Ab}'. \tag{19}$$

To obtain the corresponding real-space information, we follow the procedure of Eqs. (8-10) above and obtain the volume of a particular feature in $g_A(\mathbf{r})$:

$$A_A^n = \rho \int g_A^n(\mathbf{r}) d\mathbf{r} \tag{20}$$

giving the coordination numbers of the elements associated with this feature about A :

$$A_A^n = \sum_b \frac{\left(f_{b0} + f_b' \right)}{\langle f_0 + f' \rangle} C_A^n(b) \tag{21}$$

where the form factors are evaluated at $Q = 0$. Again, normally only a single (A, b) pair is involved.

SPECIFIC EXAMPLES

We now illustrate these principles with some results of AXS measurements carried out at the X-7A beam line at the National Synchrotron Light Source (NSLS) at Brookhaven National Laboratory. In all the measurements described here, the monochromator was set to produce a monochromatic beam of x-rays with energies near the K edges of the elements of interest. Usually data were collected at two energies, one just below the edge (typically by about $2\delta E$ where δE is the FWHM energy spread of the incident beam) and one further below (typically by about $0.02E$). AXS measurements are generally made on the low-energy side of an absorption edge to remove the fluorescence, although a certain degree of resonant Raman scattering is inevitable, especially in the near-edge measurement.

Data were collected in an energy-dispersive intrinsic Ge detector, with detector and sample rotated in a $(\theta, 2\theta)$ mode. A multi-channel analyzer was used to give integrated signal counts over specific scattered energy ranges. Normally three ranges were selected, corresponding to elastic scattering, K_α resonant Raman scattering and total scattering; additional ranges were used in some cases where fluorescence from other elements was involved. Integrated counts in these ranges were recorded independently as a function of scattering angle 2θ, making it possible to correct the counts in the elastic scattering channel for the K_β resonant Raman scattering contribution (the K_β/K_α ratio was determined in a separate measurement with an x-ray energy above the absorption edge) and dead-time. Air scattering, appreciable only at small angle, was subtracted. Further, the elastic scattering was corrected for the effects of multiple scattering and attenuation in the sample to give finally the single scattering intensity I^s defined in Eq. (2). The data at each energy were then reduced to $S(Q)$ [Eq. (8)] and $g(r)$ [Eq. (9) (since these measurements involve isotropic disordered materials, Q and r can be represented by scalar quantities). The normalization constant A in Eq.(2) was determined at each energy from the condition that $S(Q) \to 1$ as $Q \to \infty$, averaging $S(Q)$ over a suitable region at high Q. Then the difference was taken to derive $S_A(Q)$ [Eq. (15)] and $g_A(r)$.

Germania and germanate glasses

While the short-range structure of network glasses is generally well understood, usually in terms of structural units identified thorough comparison with analogous crystalline compounds, the intermediate-range structure — the manner in which these units are organized to form a large random network — remains a controversial issue.[6] Further, the modification of such glasses by the addition of metal oxides produces an order on an extended length scale that is also not well understood.[7] We have addressed these issues in a series of combined ND and AXS experiments on experiments on germania and rubidium germanate glasses.[8] Germania- rather than silica-based glasses, and rubidium as the modifier element, were chosen because the energies of the Ge and Rb edges were suitable for AXS at the X-7A beamline.

The glasses were prepared in solid form and polished to give smooth (~50 μ roughness) flat surfaces toward the x-ray beam. A series of AXS measurements were carried out at the Ge and Rb K edges using the procedures described above. The Rb edge measurements in the rubidium germanate glass were complicated by the high level of Ge fluorescence.

Average structure factors $S(Q)$ for GeO_2 and $(Rb_2O)_{0.2}(GeO_2)_{0.8}$ are shown in Fig. 1, germanium difference factors $S_{Ge}(Q)$ in Fig. 2, and the rubidium difference factor $S_{Rb}(Q)$ for the $(Rb_2O)_{0.2}(GeO_2)_{0.8}$ glass in Fig. 3. Putting the appropriate weighting factors into Eqs. (8) and (19), we get for GeO_2:

$$S = 0.404S_{GeGe} + 0.463S_{GeO} + 0.133S_{OO}, \qquad (22a)$$
$$S_{Ge} = 0.626S_{GeGe} + 0.373S_{GeO.} \qquad (22b)$$

27

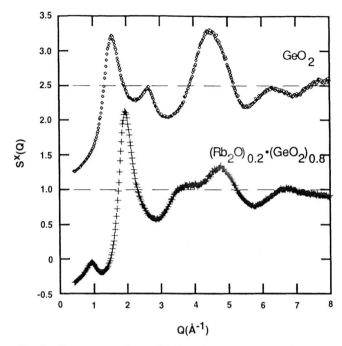

Fig. 1. X-ray structure factor of GeO$_2$ and (Rb$_2$O)$_{0.2}$(GeO$_2$)$_{0.8}$ glasses.

and for (Rb$_2$O)$_{0.2}$(GeO$_2$)$_{0.8}$:

$$S = 0.077S_{RbRb} + 0.245S_{RbGe} + 0.195S_{GeGe} + 0.156S_{RbO} + 0.248S_{GeO} + 0.079S_{OO}, \quad (23a)$$
$$S_{Ge} = \qquad\quad 0.284S_{RbGe} + 0.429S_{GeGe} \qquad\qquad + 0.287S_{GeO} \qquad\qquad (23b)$$
$$S_{Rb} = 0.243S_{RbRb} + 0.483S_{RbGe} \qquad\qquad\qquad + 0.274S_{RbO} \qquad\qquad (23c)$$

where the x-ray scattering factors are calculated at $Q = 0$.

For GeO$_2$, $S_{Ge}(Q)$ has the same features as $S(Q)$, the main difference being that the peak at $Q_2 = 2.8$ Å^{-1}, attributed to Coulomb ordering of Ge and O, is considerably enhanced. In molecular dynamics (MD) simulations of SiO$_2$ glass, S_{SiSi} and S_{OO} make positive contributions to this peak while S_{SiO} makes a negative contribution of almost identical magnitude.[9] In all three partials, the peak arises from Coulomb oscillations in the partial pair correlation functions $g_{ab}(r)$ with period $L_2 \sim 2.2$ Å, in phase with $sin(Q_2 r)$ for g_{SiSi} and g_{OO} but out of phase for g_{SiO}. Taking into account the appropriate weighing factors for GeO$_2$, one would expect this peak to be considerably enhanced in S_{Ge}, as indeed observed.

For (Rb$_2$O)$_{0.2}$(GeO$_2$)$_{0.8}$, the intensity of the first peak at $Q_0 = 0.95$ Å^{-1}, associated with extended-range order,[7] is positively correlated with the scattering amplitude of Ge (Fig. 2), but *negatively* correlated with that of Rb (Fig. 3). This behavior can be explained in terms of a chemical ordering of Rb and Ge with a period $L_0 \sim 6.6$ Å, S_{RbRb} and S_{GeGe} being positive and S_{RbGe} negative at Q_0. The second peak at $Q_1 = 2.0$ Å^{-1} is associated with intermediate-range order.[6] It has a strong positive correlation with both Rb and Ge scattering amplitudes, consistent with the crucial role of the cation-cation correlations generally found in features characteristic of intermediate-range order such as the "first sharp diffraction peak" (seen here as the first peak in the GeO$_2$ structure factor at $Q_1 = 1.54$ Å^{-1}). In molecular dynamics simulations[10] on the analogous silicate system (Rb$_2$O)$_{0.2}$(SiO$_2$)$_{0.8}$, the Rb-Rb partial structure factor shows positive peaks at 1.1 and 1.9 Å^{-1}, in agreement with this interpretation.

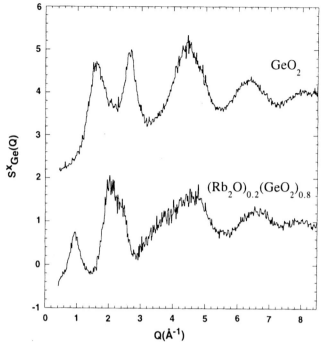

Fig. 2. Germanium difference structure factor of GeO_2 and $(Rb_2O)_{0.2}(GeO_2)_{0.8}$ glasses.

Nanoclusters in zeolite

Porous structures such as zeolites, which are about 30-50% void space after dehydration, provide a mechanism for confining materials in a controlled fashion, leading to a wide range of applications including molecular filters, electronic devices and nuclear waste storage. We have formed nanoclusters of Se and RbSe and CsSe compounds inside zeolite cages 1.3 nm wide and investigated the local surroundings of the Se and Rb atoms with AXS.[10] This technique is extremely powerful for solving the structure of nanoclusters contained as a minority species in a macroscopic host. In contrast to the more traditional semiconductors (e.g., CdSe, CdTe), the materials studied have not been synthesized in nanocrystalline form. Furthermore, their melting point is low, they can be readily incorporated in host porous structures, and their elements have x-ray absorption edges in a favorable energy region for X-7A.

The semiconducting materials were introduced into the host (Nd-substituted Y zeolite) by vapor diffusion . TEM measurements revealed a uniform cage structure with a periodicity of ~1 nm, as expected.[12] There was no sign of external Se particles, while energy-dispersive analysis confirmed the presence of Se in the host crystal. Neutron powder diffraction measurements, analyzed by Rietveld profile refinement, showed that the Nd-exchanged zeolite has a typical Y-zeolite structure with Nd^{3+} ions residing in the sodalite cages.[13] These results indicated that (a) the structure of the zeolite host is preserved upon incorporation of the semiconducting materials and (b) the incorporated atoms are largely out of registry with the zeolite host. For the AXS measurements, powder material was loaded into aluminum holders with slab-shaped cavities sealed with thin kapton foil.

The real-space correlation function $T_{Se}(r) = 4\pi\rho r^2 g_{Se}(r)$ for the Se-loaded zeolite is shown in Fig. 4, and those for the Rb-Se loaded zeolite at the Se and Rb K edges in Fig. 5. The first three peaks are fitted by Gaussians in Fig. 4, and the first four for each edge in Fig. 5. For

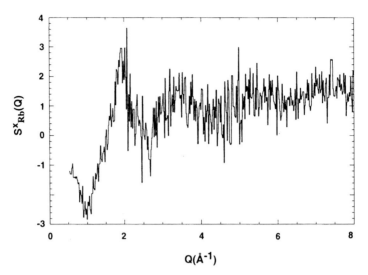

Fig. 3. Rubidium difference structure factor of $(Rb_2O)_{0.2}(GeO_2)_{0.8}$ glass.

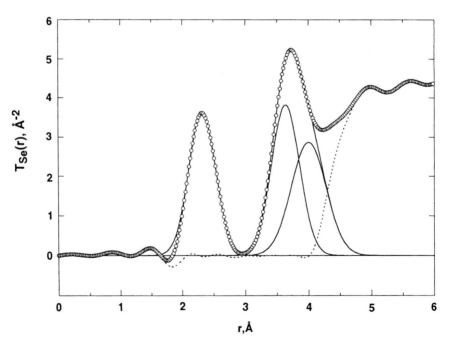

Fig. 4. Selenium difference pair correlation function for Se in Nd-Y zeolite, with fitted Gaussian peaks.

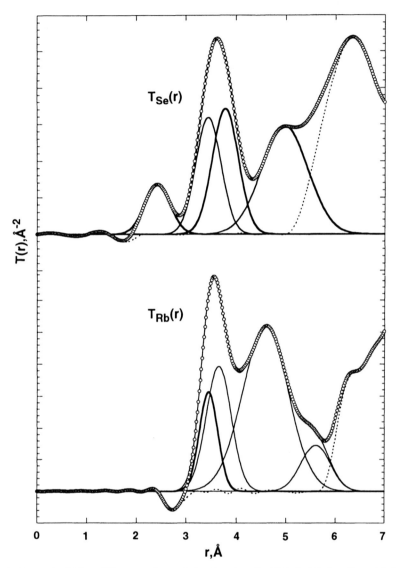

Fig. 5. Selenium and rubidium difference pair correlation functions for RbSe in Nd-Y zeolite, with fitted Gaussian peaks.

the Se-loaded zeolite, comparison with bulk crystalline and amorphous Se leads to the assignment of the two peaks at 2.32 and 3.66 Å with Se-Se correlations and the third at 4.00 Å with Se-host atom correlations. Comparison of these results with data for the various crystalline, glassy and liquid phases of bulk Se leads to several important conclusions. First, the nearest-neighbor bond length and coordination number in the Se-loaded zeolite is similar to those in the bulk phases. Second, the second-neighbor bond length is also similar, while its coordination number is significantly lower than in bulk crystalline Se. A disordered chain model for liquid Se (and by implication for amorphous Se) was proposed in which trans- and cis- configurations are adopted randomly with statistical weights that depend on temperature.[14] The short-range structure observed here points to a similar disordered chain structure for the Se nanoclusters.

For the alkali-Se loaded zeolite, chemical analysis confirmed the formation of $Rb_{0.52}Se_{0.48}$. According to the Zintl-Klemm rule in which electron transfer from the alkali moves the chemical properties of the Group-VI atom into line with those of Group VII, Se_2^{2-} polyanions will be present in the crystal and by analogy with KTe^{15} are likely to persist in the disordered state. It is seen immediately from Fig. 5 that the peak at 2.42 Å appears only in $T_{Se}(r)$ and not in $T_{Rb}(r)$, indicating that it is due to Se-Se nearest neighbors. Its coordination number of about one shows that Se_2^{2-} Zintl ions are present in the RbSe clusters. The second peak in $T_{Se}(r)$ — the first in $T_{Rb}(r)$ — is a broad structure between 3 and 4 Å, partially resolved from the high-r part of the pair correlation function. The Gaussian fits to these peaks give unresolved doublets. It is reasonable to associate the first peak of the doublet in both functions with the Rb-Se nearest-neighbor correlation, and the second with subsequent correlations involving both Rb and Se atoms for the Se edge function and Se atoms for the Rb edge function, with distances similar to those observed in the RbSe crystal. It is evident that the RbSe forms clusters in the zeolite that mimic the structure of the bulk compound, despite the fact that only a small number of atoms are involved.

These examples make it clear that AXS is a powerful technique for obtaining partial structure information in complex disordered materials, and in particular can probe the structure of minority species in a majority host. With the more powerful sources now available, and a corresponding development in experimental methods and analysis techniques, we can expect that before too long full partial structure analysis, with complete derivation of all partial structure factors, will become routine.

Acknowledgments. We thank Dr. D. E. Cox and Prof. T. Egami for helpful discussions and guidance in the experiments, and Drs. P. Armand and K. Suzuya for the analysis of the x-ray data. The work was supported by the U.S. Department of Energy (DOE), Division of Materials Sciences, Office of Basic Energy Sciences, under Contract W-31-109-ENG-38. NSLS is supported by the Divisions of Materials and Chemical Sciences (DOE) at Brookhaven National Laboratory.

REFERENCES

1. Y. Waseda, *The Structure of Non-Crystalline Materials*. (McGraw-Hill, New York, 1980); *Methods in the Determination of Partial Structure Factors of Disordered Matter by Neutron and Anomalous X-Ray Diffraction*, Ed. J. B. Suck, P. Chieux, D. Raoux and C. Riekel (World Scientific, Singapore, 1993).
2. D. L. Price, Ed., *Research Opportunities In Amorphous Solids With Pulsed Neutron Sources*, J. Non-Cryst. Solids 76, No. 1 (1985)
3. P. H. Fuoss, P. Eisenberger, W. K. Warburton and A. Bienenstock, Phys. Rev. Lett. 46, 1537 (1991).
4. T. E. Faber and J. M. Ziman, Phil. Mag. 11, 153 (1965).
5. D. L.Price, unpublished work.
6. S. C. Moss and D. L. Price, in *Physics of Disordered Materials*, ed. D. Adler, H. Fritzsche and S. R. Ovshinsky (Plenum, New York, 1985), p. 77; D. L. Price, S. C. Moss, R. Reijers, M.-L. Saboungi and S. Susman, J. Phys.: Condens. Matter 1, 1005 (1989); S. R. Elliott, Nature 354, 445 (1991).

7. P. Armand, M. Beno, A. J. G. Ellison, G. S. Knapp, D. L. Price and M.-L. Saboungi, Europhys. Lett. 29, 549 (1995).

8. D. L. Price, A. J. G. Ellison, M.-L. Saboungi, R.-Z. Hu, T. Egami and W. S. Howells, Phys. Rev. B 55, 11249 (1997).

9. P. Vashishta, R. K. Kalia, J. P. Rino and I. Ebbsjö, Phys. Rev. B41,12197 (1990).

10. J. Kieffer and D. Nekhayev, unpublished work.

11. P. Armand, M.-L. Saboungi, D. L. Price, L. E. Iton. C. Cramer and M. Grimsditch, Phys. Rev. Letters 79, 2061 (1997).

12. P. Armand, A. Goldbach, C. Cramer, R.Csencsits, L.E. Iton, D. L. Price and M.-L. Saboungi, J. Non-Cryst. Solids 205-207, 797 (1996).

13. P. Armand, L. Iton , J. W. Richardson and M.-L. Saboungi, private communication

14. M. Misawa and K. Suzuki, Trans. J. Phys. Soc. Jpn. 44, 1612 (1978)

15. J. Fortner, M.-L. Saboungi and J. E. Enderby, Phys. Rev. Lett. 69, 1415 (1992)

THE RECORDING AND INTERPRETATION
OF DIFFUSE X-RAY SCATTERING

T.R. Welberry

Research School of Chemistry
Australian National University
Canberra, ACT 0200
Australia

INTRODUCTION

Crystal structure analysis based on Bragg diffraction data reveals only information about the *average* crystal structure, such as atomic positions, temperature factors and site occupancies. Additional information about static and thermal disorder within the crystal can be obtained by analysing the diffuse scattering. A number of reviews of diffuse scattering have appeared in recent years [1-8]. Unlike the analysis of Bragg data, for which the same basic methods are applicable for structures containing only a few or as many as thousands of atoms, the interpretation and analysis of diffuse scattering requires markedly different treatment according to the complexity of the system. For relatively simple disordered systems such as alloys or metal oxides, methods have evolved for the quantitative extraction of correlation parameters which describe the local environment of the different atomic species in the material. At the other extreme, in systems containing larger (even relatively modest) numbers of atoms, extraction of the same level of detail is simply not possible. Here it is often the case that the primary problem is to determine just what is causing the diffuse scattering. Only if the problem can then be simplified in some way (*e.g.* by assuming part of the system does not contribute to the diffuse scattering, or treating groups of atoms as single entities) is it likely that further progress can be made.

Our approach to this general problem of trying to interpret and analyse diffuse scattering has been to use computer simulation as an aid. We attempt to construct a computer model, which captures the essential features of the particular disordered system. The model, based on well established physical and chemical knowledge, can be as detailed as the computer resources allow. Generally it is preferable to keep the model as simple as possible and usually it involves only a few parameters which can be iteratively adjusted until satisfactory agreement with the observed diffraction pattern is obtained. Although the model that is obtained by this method does not generally give a quantitative fit to the observed data it is at least physically and chemically plausible, and allows incorrect possibilities to be readily investigated and rejected. Quite often we find that distinctive qualitative features of a diffraction pattern which can be reproduced by a simple model are more convincing proof of the validity of the model than any quantitative measure of overall agreement.

In developing a suitable model for any real system an understanding of the basic diffraction equations and how different terms contribute to the diffraction pattern, is of paramount importance. Consequently in the following section we give a brief account of these equations, for later reference. In the next section we outline the experimental methods we use for recording diffuse scattering patterns in the laboratory, and this is followed by a description of the specifications necessary to produce diffraction patterns, of a comparable quality to the experimental ones, from computer simulations. We then describe the various simulation techniques which may be used to build a model crystal containing occupational and displacement disorder. Finally we describe in some detail aspects of our ongoing research into cubic stabilised zirconias (CSZ's) which provide excellent pedagogical examples of how these various techniques, used in conjunction, can give meaningful insight into the origins of very complex diffuse diffraction patterns.

THE DIFFRACTION EQUATIONS

A general description of diffuse scattering that allows for both short-range compositional order and local atomic displacements can be obtained by expanding the exponential in the kinematic scattering equation in powers of displacement.

$$
\begin{aligned}
I &= \sum_{n=1}^{N} \sum_{m=1}^{M} f_m f_n \exp\left[ik \cdot \left(R_m + u_m - R_n - u_n\right)\right] \\
&\approx \sum_{n=1}^{N} \sum_{m=1}^{M} f_m f_n \exp\left[ik \cdot \left(R_m - R_n\right)\right] \\
&\times \left\{1 + ik \cdot \left(u_m - u_n\right) - \frac{1}{2}\left[k \cdot \left(u_m - u_n\right)\right]^2 - \frac{i}{6}\left[k \cdot \left(u_m - u_n\right)\right]^3 + \ldots\right\}
\end{aligned}
\tag{1}
$$

Here I is the scattered intensity and f_m is the scattering factor of the atom m associated with the lattice site at the location R_m and which is displaced from its site by a small amount u_m. The scattering vector is $k = h_1 a^* + h_2 b^* + h_3 c^*$. Equ. (1) expresses the fact that the intensity distribution may be written as the sum of component intensities: the first term being independent of the displacements, the second term dependent on the first moment of displacements, the third term on the second moment etc. After removal of the Bragg peak component due to the average structure the remaining diffuse intensity can similarly be expressed as the sum of corresponding components,

$$
I_{\text{Diffuse}} \approx I_0 + I_1 + I_2 + I_3 + \ldots
\tag{2}
$$

It is usual in analyses of alloys and simple oxides to truncate this Taylor expansion at second order, although recently we have shown that in some cases the third and higher order terms are also important. Table 1 summarises the properties of these different components. Further details can be found in Welberry and Butler[7].

The first term I_0 is the intensity component due to short-range order and is not dependent on displacements. There is one term in this summation for every different interatomic vector lmn along which significant correlation may be present. Each term in the sum involves a short-range order parameter, α_{lmn}^{ij}, defined by,

$$\alpha^{ij}_{lmn} = 1 - P^{ij}_{lmn}\big/c_j \qquad (3)$$

where P^{ij}_{lmn} is the conditional probability of finding an atom with label j at the end of a vector $\mathbf{r}_{lmn}$ given that there is an atom with label i at its origin. c_j is the site occupancy or concentration of atom type j. $\mathbf{I}_1$, $\mathbf{I}_2$ and $\mathbf{I}_3$, which involve the displacement-parameter components, X^{ij}, Y^{ij}, Z^{ij}, similarly also have terms for every different interatomic vector. For a simple binary system $\mathbf{I}_1$ has 6 terms for each interatomic vector, $\mathbf{I}_2$ has 18 and $\mathbf{I}_3$ has 30 terms. Fitting equations of this form with such large numbers of parameters is a formidable task even for simple oxide systems, but for systems containing more than one or two atoms per asymmetric unit it is quite prohibitive.

Table 1. Summary of the properties of the different components of the diffuse intensity appearing in equation 2.

Term	$\mathbf{I}_0$	$\mathbf{I}_1$	$\mathbf{I}_2$	$\mathbf{I}_3$
Description	Short-range order (SRO) term	Warren Size-effect	Huang Scattering 1st order TDS	3rd order size term
Lattice averages involved	SRO parameters α^{ij}	$\langle X^{ij}\rangle, \langle Y^{ij}\rangle$ etc.	$\langle (X^{ij})^2 \rangle$, $\langle X^{ij}Y^{ij}\rangle$ etc.	$\langle (X^{ij})^3 \rangle$, $\langle (X^{ij})^2 Y^{ij}\rangle$ etc.
Type of Summation	cosine	sine	cosine	sine
Symmetry	symmetric	anti-symmetric	symmetric	anti-symmetric
Variation in k-space	nil	linear, i.e. with h_1, h_2 etc.	quadratic, i.e. with $h_1^2, h_1 h_2$ etc.	cubic, i.e. with $h_1^3, h_1^2 h_2$ etc.
Dependence on f_A, f_B for binary	$(f_A - f_B)^2$	$f_A(f_A - f_B)$, $f_B(f_A - f_B)$	$f_A^2, f_A f_B, f_B^2$	$f_A^2, f_A f_B, f_B^2$
Number of components for binary	1	6	18	30

EXPERIMENTAL MEASUREMENT OF DIFFUSE SCATTERING

In this section we describe briefly our method for recording diffuse X-ray scattering patterns. Our aim has been to develop a versatile system for measuring large volumes of reciprocal space for a wide range of materials in a reasonably short time using a conventional tube source. The apparatus, shown in Fig. 1a is based on a STOE curved

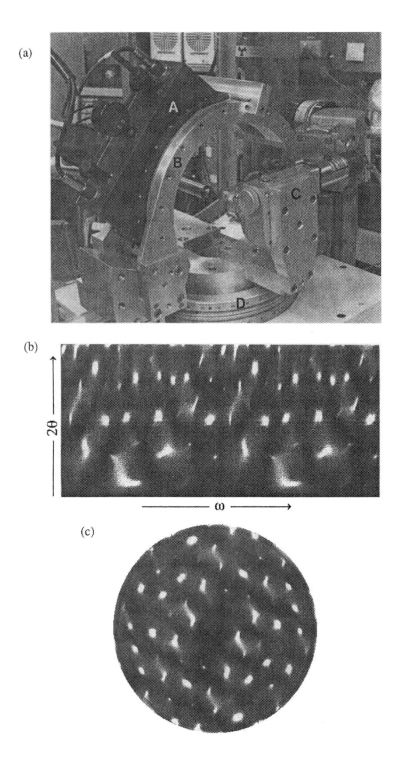

Figure 1. (a) The position-sensitive detector system used to collect diffuse scattering data. (b) Raw data obtained as 1000 stationary 2θ scans at increments of 0.36° in ω. (c) Reciprocal lattice section obtained from (b) by simple geometric transformation.

position-sensitive wire detector (A), and utilises the flat-cone Weissenberg geometry[9]. The detector (PSD) is supported on a semi-circular cradle (B) mounted on one arm of a two-circle goniometer (D). In the zero-position shown the PSD essentially provides a read-out of the diffraction angle 2θ. Mounted on the other arm of D is a goniometer C which provides the rotation (ω) about a pre-aligned crystal rotation axis which is normal to the plane of the PSD. With the PSD in a stationary position a whole reciprocal lattice section can be recorded simply by incrementing ω. To obtain non-zero levels of reciprocal space the two circles of D are rotated together through an angle μ. In Fig. 1b we show a typical section of raw data consisting of 1000 individual PSD scans each of which covers ~50° of 2θ. The counts from 4096 channels of the PSD are binned into 512 data channels each of which corresponds to 0.1° of 2θ. The final raw data image thus consists of 1000×512 pixels. A simple geometric transformation is then applied to these data to yield an undistorted image of the reciprocal section, as shown in Fig. 1c. If higher diffraction angles are required a second set of scans can be made with the PSD moved around the cradle C and the data merged into the same undistorted image. Typical counting times used to record data are ~1-3 days per scan, depending on the size of sample.

Every pixel in the raw data is binned into a corresponding pixel of the transformed image. The size of the latter is usually chosen to be 400×400 pixels, which is close to the maximum that allows complete coverage by the raw data. Each pixel in the transformed image is normalised by the number of raw pixels contributing to it, so that the final image does not require a Lorentz correction. [Note low angle pixels will have several raw data pixels contributing to them while high-angle pixels will generally have only one or two]. The data may be corrected for polarisation and absorption. To facilitate correction of data for absorption the samples used are usually prepared to be either cylindrical, with the cylinder axis coincident with the rotation axis, or spherical. In this way the absorption for any section is a function of 2θ only. Air-scattering is kept to a minimum by having the collimator aperture very close (<2mm) from the crystal and the beam-stop also as close as possible behind the sample.

MODEL SYSTEM SPECIFICATIONS FOR COMPUTER SIMULATION

Our aim when performing computer simulations of disordered systems is to be able to build a computer model from a realisation of which it is possible to compute diffraction patterns of comparable quality (in terms of resolution and smoothness) to the observed X-ray diffraction patterns. Into this model we need to introduce occupational and displacement disorder, but rather than this being specified by the large number of lattice averages which appear in the terms I_0, I_1, *etc.* of Equ. (2), we would like instead for this to be specified by a small number of local interaction parameters which mimic what we imagine is occurring in the real system. In this section, therefore, we first describe the steps necessary to obtain suitable quality diffraction patterns.

If the diffraction pattern of a finite model disordered crystal is computed by direct Fourier transformation of the atomic coordinates, the pattern will in general be very noisy and have a speckly appearance. The typical length-scale of the speckle is of the order of 1/N where N is the number of unit cells of the model in a given direction. By increasing the size of the model crystal the speckle grain becomes finer and the overall appearance of the diffraction pattern becomes smoother. There is a second reason for the smoother appearance, however. If we consider for example a random mixture of two types of atoms, all perfectly positioned at the lattice points then the diffraction pattern is obtained from the I_0 term in Equ. (2) as the sum of a large number of cosines. There is a term for each different inter-site vector ranging from the shortest near-neighbour terms to the longest corresponding to vectors of the maximum dimension of the crystal. For a random distribution all correlations are *expected* to be zero, with the exception of the self-term which produces the uniform scattering well-known as the *Laue monotonic scattering*. For a finite crystal the correlations (*i.e.* the average over the whole crystal of the quantities appearing in Equ. (3)) are not identically equal to zero and the differences from zero will provide contributions which modulate the Laue monotonic scattering. Since the statistical variation of such lattice averages generally goes as $\sim \sqrt{M}$ (where M is the number of lattice sites involved in the averages), as the size of the crystal increases the lattice average for each correlation will be a better approximation to the *expected* zero value. The amount of noise in the diffraction pattern correspondingly decreases. It should be pointed out, however, that the averages for the very long vectors of dimension near to the crystal dimension will still only involve relatively few sites and will approximate to the expected zero rather poorly.

For a model crystal containing say 1000 unit cells the expected accuracy of the lattice averages will be $\sim \sqrt{1000}/1000 = 0.03$ and for 10,000 it will be 0.01. If the model crystal is a realisation of a short-range order model there may be significant low-order correlations but since correlations generally decay towards zero at larger distances many of the higher-order correlations will not be significantly different from zero. Since their presence detracts from the overall appearance of the pattern a means of eliminating these terms from the calculation of the diffraction pattern would be an advantage. This philosophy is taken advantage of in the program DIFFUSE[10]. Rather than calculating the diffraction pattern of a model crystal by Fourier transforming the whole crystal, DIFFUSE instead calculates the average of the diffraction pattern of many small sample 'lots' taken from random positions in the crystal. The lot size is chosen to be large enough to contain all vectors along which significant correlation exists but avoids the many longer-range vectors which contribute to the unwanted noise. If the number of lots is chosen so that in the final average the whole of the crystal has been sampled at least once then the resulting diffraction pattern is the optimum that can be achieved with that size of model crystal.

To demonstrate what these various considerations mean in practice we show in Fig. 2 some example diffraction patterns computed from a model of the Tl cation positions in the disordered structure of $TlSbOGeO_4$, a non-linear optical material[11,12]. Two different model crystals have been used with sizes of 32×32×32 unit cells and 10×10×10 unit cells. Each unit cell contains 8 cation sites. Fig. 2a shows the disposition of those sites occurring in one half of the unit cell. Those in the other half of the cell are related to those shown by an n-glide plane normal to **a**. Each cation site contains a Tl atom in either one or the other of two slightly displaced positions (shown as grey or black circles), and the sites are linked by a honeycomb network of inter-site vectors along which we assume correlations occur. For the example chosen the correlations along the primary vectors A, B, & C shown in the figure were: A = –0.90; B = +0.30; C = +0.20 .

Fig. 2b corresponds to an optimum calculation for a crystal of 32×32×32 unit cells in which 500 lots of 5×5×5 have been averaged. This a reasonably smooth pattern and shows a great deal of detailed correlation structure. A comparable calculation using 500 lots of 5×5×5 from crystal of 10×10×10 unit cells is shown in Fig. 2c. This is considerably noisier and some correlation features are now difficult to discern. [Using 500 lots for this crystal corresponds to sampling each part of the crystal ~125 times and in fact this is mostly a waste of time since the pattern improves very little after the first ~20 lots have been averaged]. In Figs. 2d,e we show for comparison the patterns that are obtained from a single lot of 5×5×5 and 10×10×10 unit cells respectively. It is clear that a diffraction pattern obtained from a crystal with a size as small as either of these single lots would be quite unsatisfactory for comparing observed patterns. It should be noted that Fig. 2b is the average of 500 patterns comparable to that in Fig. 2d. The improvement between Fig. 2e and Fig. 2c largely results from the removal of the high frequency speckle.

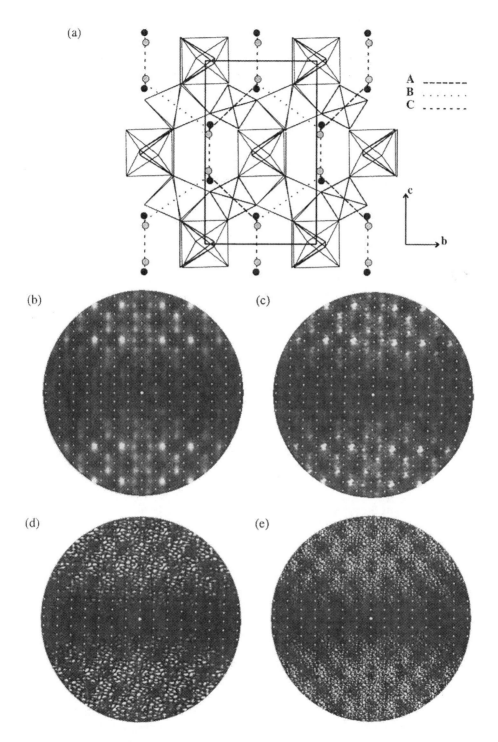

Figure 2. (a) Projection of the structure of TlSbOGeO$_4$, showing inter-cation vectors A, B, C along which correlation was introduced. (b)-(d) *0kl* diffraction patterns calculated from MC simulations of a model for the cation ordering, using different crystal sizes and different 'lot' sizes. (b) 32×32×32 crystal; 500 lots of 5×5×5. (c) 10×10×10 crystal; 500 lots of 5×5×5. (d) 32×32×32 crystal; 1 lot of 5×5×5. (e) 32×32×32 crystal; 1 lot of 10×10×10. The white dots in (b)-(e) indicate the Bragg peak positions.

The general conclusion that can be drawn from this example is that to obtain a calculated diffraction pattern which is comparably smooth and shows a comparable degree of correlation detail to typical observed X-ray patterns, we need to use a system size of at least 32×32×32 unit cells or for a 2D model ~180×180 unit cells. Such system sizes have been close to the maximum that could comfortably be used on the relatively inexpensive work-stations that have been widely available over the last few years. However, more recently, with the reduction in cost of computer memory, computers with much larger memory capacity are becoming commonplace and substantially larger systems can now be contemplated.

SIMULATION OF OCCUPANCY CORRELATIONS

Monte Carlo Simulation

In order to introduce occupancy correlations into a model system like that of the $TlSbOGeO_4$ example described above, we commonly use Monte Carlo (MC) simulation. If the interactions between atoms or molecules could be specified *a priori* with sufficient accuracy then MC could simply be used to determine what correlations would result from these interactions. This level of prescription is rarely achievable in practice and what is required instead is the ability to produce realisations of a model with various different correlation structures so that the importance of different interactions might be assessed. Our general strategy therefore is to try to set up some general Hamiltonian whose individual interaction parameters can be adjusted in order to achieve a particular set of near-neighbour correlation coefficients (or SRO parameters).

To represent the distribution of atomic species we use a set of random variables x_i, where the index i identifies both the particular unit cell and the atomic site within the unit cell. For a binary system x_i may be (0,1) variables where for example $x_i=1$ corresponds to the site i being occupied by atomic species A and $x_i=0$ when it is occupied by B. The site occupancy for species A is then given by $\langle x_i \rangle$ where the average is over all unit cells. It is sometimes more convenient to use Ising spin variables, $s_i = \pm1$. Clearly s_i and x_i are related ($s_i = 2x_i - 1$) and choice between these two types of variable is made for convenience. Using the s_i variables the Hamiltonian takes the form,

$$E = \sum_i \left[a_0\, s_i + \sum_n b_n s_i s_{i-n} + \sum_n \sum_m c_{nm} s_i s_{i-n} s_{i-m} + etc. \right] \tag{4}$$

Here s_{i-n} refers to the spin on the neighbouring site i-n of i. The quantity a_0 is a single-site energy which has the effect of an external field in magnetic Ising systems. The quantities b_n are pair interaction energies corresponding to inter-site vectors defined by i & n. The quantities c_{nm} are three-body interaction energies corresponding to inter-site vectors defined by i, n & m. Four-body terms or higher may be added if required.

As mentioned above the interaction energies a_0, b_n & c_{nm} are generally unknown quantities. In order to achieve the desired realisations therefore an iterative feed-back procedure may be used. Suppose we wish to produce a realisation in which the two-body correlations $P_n=\langle x_i x_{i-n} \rangle$, take particular chosen values, while the occupancy, $\langle x_i \rangle$ is also maintained at a certain chosen value. For the moment we neglect multi-site terms. Initially a_0, b_n are set to zero and Monte Carlo simulation is commenced. After a cycle of iteration (a cycle being that number of individual Monte Carlo steps required to visit each site once on average), the lattice averages $\langle x_i \rangle$, $\langle x_i x_{i-n} \rangle$ are computed and compared to the required values. If the computed averages are too low, then the corresponding a_0, b_n parameter is decreased by an amount proportional to the difference between the calculated and required average. Similarly if the computed averages are too high, then the a_0, b_n parameters are correspondingly increased. Provided that the values required are physically achievable, after a sufficiently large number of cycles the values of a_0, b_n stabilize and the lattice averages $\langle x_i \rangle$, $\langle x_i x_{i-n} \rangle$ converge to the required values. It is frequently the case that the site occupancy $\langle x_i \rangle$ needs to be maintained as a constant. In this case it is often an advantage to initially set the values of the variables to satisfy the required occupancy and then perform all subsequent MC steps by interchanging two variables x_i, x_j on two different sites, i and j, chosen at random.

It should be noted that even the simplest 2D nearest neighbour Ising model, in which only a single b_n term is active, possesses a phase transition. Use of this kind of feed-back technique is fraught with difficulties if the system is close to such a phase transition, and should be avoided. Despite this limitation, we have found the method to have wide application.

Direct Synthesis of Disordered Distributions

An alternative way of treating diffuse scattering is to consider that the intensity that occurs at any point in reciprocal space arises from a periodic modulation of the real-space structure. This idea may be used to *synthesise* a real-space distribution of atomic scatterers which will have a given diffraction pattern. In simple terms the real-space lattice is constructed by applying modulations with wave-vectors corresponding to each elemental volume in the first Brillouin zone of the diffraction pattern. Each modulation is given an amplitude which reflects the intensity at that point, and a phase that is chosen at random. For the case of concentration waves the modulations can be written in the form of a variation from cell to cell of the atomic scattering factors. If the point in reciprocal space is infinitesimally small then the modulation wave must extend over the whole of real space. In practice it is more convenient to consider that a modulation extends over a limited region in real space and that this contributes to a small (but finite) region of the diffuse distribution in reciprocal space. Then the atomic scattering factors in real space may be expressed as a sum of all such modulations:

$$f_\mu(\mathbf{T}) = \langle f_\mu \rangle \left\{ 1 + \sum_{\mathbf{q}} A_{\mathbf{q}} \cos\left(2\pi \mathbf{q}.\mathbf{T} + \phi_{\mathbf{q}}\right) \times \exp\left[-\left(t - t_c\right)^2 \middle/ 2\sigma_c^2\right] \right\} \quad (5)$$

Here $\mathbf{T}$ is a real-space lattice vector. $f_\mu(\mathbf{T})$ is the scattering factor of the atom in the site μ of the unit cell with origin $\mathbf{T}$. The amplitude, $A_{\mathbf{q}}$, of each modulation wave is proportional to the amplitude of the scattering at the point $\mathbf{G}+\mathbf{q}$ in reciprocal space, where $\mathbf{G}$ is a reciprocal space lattice vector. $\phi_{\mathbf{q}}$ is the phase of the modulation of wave-vector $\mathbf{q}$ at the site μ. The summation is over all wave-vectors in the first Brillouin zone. The final term is a Gaussian whose standard deviation σ_c defines the extent of the region in real-space which is modulated (in practice the modulation is truncated at 2.5σ). t is a general vector in real space and t_c defines the randomly chosen centre of the region of modulation. The effect of such a modulation is to contribute to the diffraction pattern a diffuse peak at the reciprocal point $\mathbf{q}$ with a width inversely proportional to σ_c.

If random phases, $\phi_{\mathbf{q}}$, are used in the synthesis (and this is reasonable, since for an incommensurate wave the choice of origin is arbitrary) the value of the atomic scattering factor, $f_\mu(\mathbf{T})$, at a given site that will be obtained from Equ. (5) will be a continuous variable and not just a binary one (representing either atom A or atom B). This is

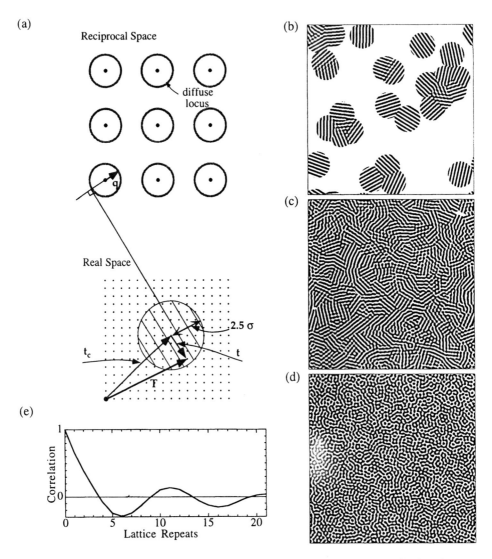

Figure 3. Illustration of the modulation wave method of synthesising a real-space distribution of scatterers which will give a prescribed diffraction pattern. (a) Illustrates the meaning of the different variables in equation (5). (b) Real-space distribution after the addition of 30 modulations. (c) Real-space distribution after the addition of 300 modulations. (d) Real-space distribution after the addition of 30,000 modulations. (e) The correlation structure of the distribution in (d).

overcome in practice by converting the continuous variables representing the scattering factors into binary ones by comparing the value of $f_\mu(\mathbf{T})$ with a threshold, f_T say. For all those sites for which $f_\mu(\mathbf{T}) > f_T$ a scattering factor f_A is assigned and for those at which $f_\mu(\mathbf{T}) < f_T$ a scattering factor f_B is assigned.

The meaning of the different variables in Equ. (5) is illustrated schematically in Fig. 3a. In this figure the required diffraction pattern is seen to consist of a diffuse locus in the form of a circle. The wave-vector $\mathbf{q}$ defines a single point on the locus and this corresponds to a modulation in real-space centred at the point $\mathbf{t}_c$. In Figs. 3b,c,d we show various stages of the synthesis of an example real space distribution. These figures show a region of real space corresponding to 512×512 lattice sites. A given site contains either a black or a white atom. In this case the form of the diffuse diffraction pattern is a diffuse ring centred at the origin with a radius of 0.1 reciprocal lattice units. The value of σ_c was chosen to be 16 lattice repeats so that the area of the region covered by the $2.5\sigma_c$ limit for each modulation was ~5026 unit cells, or ~2% of the total area. Fig. 3b shows the resulting distribution after 30 modulations have been added. Note that for this example the spacing of the modulation planes is the same in each case but the phases are all different. Fig. 3c shows the distribution after 300 modulations, and Fig. 3d after 30,000 modulations. It is clear that Fig. 3c still shows texture relating to the modulation region size, but Fig. 3d shows a completely homogeneous texture. In this latter case each real space site is subject to ~600 modulations on average. In Fig. 3e we show a plot of the correlation coefficients calculated from the distribution shown in Fig. 3d. The correlations are plotted against the number of lattice repeats, since the correlation field is circularly symmetric, i.e. P_{lmn}^{ij} depends only on $|\mathbf{r}_{lmn}|$.

SIMULATION OF DISPLACEMENT CORRELATIONS

Monte Carlo Simulation

The Monte Carlo simulation technique described above for producing distributions containing prescribed occupancy correlations can be carried over quite simply to the case of displacements. In this case we wish to generate a lattice in which predetermined correlations exist between the displacements of neighbouring atoms. For simplicity consider a 2D lattice in which an atom at site i has a displacement, X_i, from its average position. In contrast to the s_i variables in Equ. (4) X_i is a continuous variable and it is convenient to assume that it is Gaussian with zero mean and unit variance. I.e.

$$\langle X_i \rangle = 0; \quad \langle X_i^2 \rangle = 1 \tag{6}$$

In a comparable way to that for the occupancy variables in Equ. (4) a Hamiltonian can be written in terms of interaction parameters:

$$E = \sum_i \sum_n B_n X_i X_{i-n} \tag{7}$$

Here, as before, the second summation is over all n neighbours of site i. In this formulation there is no term dependent solely on X_i, since if it were present it would try induce a shift in the average value of X_i. Similarly there are no higher-order terms such as $X_i X_{i-n} X_{i-m}$ since for Gaussian variables all moments higher than two are zero.

Simulation, using the same kind of iterative feedback described for the case of binary variables, is then carried out as follows. First, random values for X_i, are assigned using a Gaussian distributed pseudo-random number generator. Then MC iteration is carried out using the method in which two sites i and j are chosen at random and the values of X_i and X_j are interchanged. This move is accepted or rejected depending on the change, ΔE, in E and the Boltzmann probability $P=\exp(-\Delta E/kT)$. After a complete cycle of iteration lattice averages for the correlations, $\langle X_i X_{i-n} \rangle$, are computed and values for the interactions B_n are adjusted as described for the case of binary variables in order to move the system toward the target correlations. After a sufficient number of cycles and provided that the values required are physically achievable the values of the B_n stabilise and the lattice averages converge to the required values.

In general an atomic site will have displacements which require a tensorial description (thermal ellipsoid), involving up to six components. Any one of the components for one atom may be correlated with any one for another atom thus making up to 36 possible correlation parameters just for a single atom-atom interaction. Unless sufficient approximations or simplifications can be made this method rapidly becomes unmanageable. One example of a case where the method has been used to good effect is in a description of the distribution of defect clusters in wüstite, $Fe_{1-x}O$. In this case the positions of defect clusters were described in terms of deviations X_i, Y_i, ($\&$, Z_i) away from an underlying regular (incommensurate) lattice in the x, y $\&$ z crystallographic directions respectively. It was possible here to assume that the X_i, Y_i, ($\&$, Z_i) variables were independent, $i.e.$ all cross-correlations such as $\langle X_i Y_{i-n} \rangle$ were zero. Then a suitable description was found in which two types of correlation (and corresponding B_n interaction) were used. These were a transverse nearest-neighbour correlation and a longitudinal nearest-neighbour correlation. Using a nomenclature where i, j, k refer to a site on the primitive cubic lattice, these correlations are defined as:

$$\rho_{Transverse} = \left\langle X_{i,j,k} X_{i,j-1,k} \right\rangle = \left\langle Y_{i,j,k} Y_{i-1,j,k} \right\rangle = \left\langle Z_{i,j,k} Z_{i,j-1,k} \right\rangle$$
$$= \left\langle X_{i,j,k} X_{i,j,k-1} \right\rangle = \left\langle Y_{i,j,k} Y_{i,j,k-1} \right\rangle = \left\langle Z_{i,j,k} Z_{i-1,j,k} \right\rangle \tag{8}$$

$$\rho_{Longitudinal} = \left\langle X_{i,j,k} X_{i-1,j,k} \right\rangle = \left\langle Y_{i,j,k} Y_{i,j-1,k} \right\rangle = \left\langle Z_{i,j,k} Z_{i,j,k-1} \right\rangle$$

Transverse correlation means that X variables which are neighbours in the y $\&$ z directions are correlated, while longitudinal correlation means that X variables which are neighbours in the x direction are correlated. This model was used in a recent study[14] of the distribution of defects in the non-stoichiometric oxide wüstite, $Fe_{1-x}O$. In Fig. 4a we show an example realisation of the model together with the diffraction pattern, Fig. 4b, obtained from it. The values for the transverse and longitudinal correlations for this example were 0.8 and 0.935 respectively and the unit variance Gaussian variables were scaled to give a variance of the defect positions which was 31.5% of the average defect spacing. In addition we show in Fig. 4c a subsequently obtained diffraction pattern using the same distribution after lattice relaxation around the defects had been applied, and in Fig. 4d detail from the corresponding X-ray diffraction pattern for comparison.

Displacements Using Interatomic or Inter-Molecular Forces

A second way in which the method based on Equ. (7) is unsatisfactory is that as it stands it does not allow for the possibility that displacements will be correlated with occupancy. In fact in many disordered systems a large fraction of the displacement disorder results directly from local distortions resulting from relaxation around the different occupational species. Equ. (7) could be combined with Equ. (4) and cross terms involving, $e.g.$ $s_i X_i$, added but this would further add to the unworkability of the method.

In reality, correlations between the displacements of neighbouring atoms result from the forces between the atoms, and a more natural way of simulating such displacements is to create a force-model which mimics the real situation. If there is an interaction between two atoms their displacements away from their average positions along the line of interaction will be strongly correlated, while those normal to it will be weakly correlated. In general the stronger the interaction the stronger the degree of correlation will be. While such displacements $could$ be modelled using Equ. (7), the method would be very cumbersome. The method that we have adopted in numerous studies involving quite a

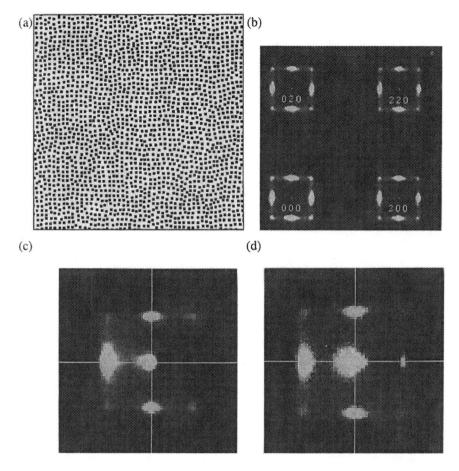

Figure 4. Distribution of defect clusters in wüstite, $Fe_{1-x}O$. The distribution of defects shown in (a) is responsible for the *motif* of scattering that occurs around the Bragg peak positions in the diffraction pattern, as shown in (b). (c) shows a detail from around the (0 2 0) Bragg peak based on the same defect distribution model but with additional effects incorporated to allow for lattice relaxation and inhomogeneity. (d) shows the detail from around the (0 2 0) Bragg peak in the experimental X-ray data. See reference (14) for more details.

diverse range of systems is reviewed briefly in this section.

First we assume that we have prior knowledge of the average crystal structure obtained from Bragg analysis. We then place atoms at these average positions and inter-link them with Hooke's law (harmonic) springs along all interatomic vectors for which we expect significant interaction. Each spring has a force constant, κ, the strength of which can be adjusted as a system variable. At zero temperature the springs should maintain all the atoms in the observed average positions. The force constants, κ, represent the resistance to perturbation away from these equilibrium positions. At an elevated temperature atoms will deviate away from the mean positions, moving most along the softest directions so that even without occupational disorder, thermal diffuse scattering will be generated.

To take account of displacements induced by occupational disorder we adjust the length of the spring depending on the atom types occupying the two sites joined by the

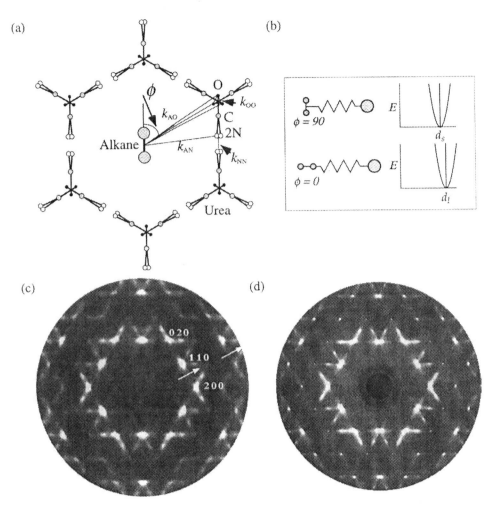

Figure 5. Model of Dibromodecane-Urea inclusion compound. (a) Schematic diagram showing the alkane within the urea channels. Springs of force constants k_{AO} and k_{AN} link the centre of mass of the alkane to the O and N ends of the rigid urea molecules and springs of force constants k_{OO} and k_{NN} mimic the hydrogen-bond network within the urea framework. (b) Schematic diagram showing how the length of the AO and AN springs varies with the angle of rotation ϕ. (c) A diffraction pattern (*hk0* section) calculated using the model. (d) Observed X-ray pattern recorded at 150K for comparison. The arrows in (c) indicate the asymmetry in the scattering across planes of Bragg peaks, due to the size-effect term $\mathbf{I_1}$ in equation (2).

spring. If d_0 is the distance between two sites in the average structure, then we specify the distances for the cases where both sites are occupied by species A, one by A and one by B and both by B, as:

$$d_{AA} = d_0 \left(1 + \varepsilon_{AA}\right); \quad d_{AB} = d_0 \left(1 + \varepsilon_{AB}\right); \quad d_{BB} = d_0 \left(1 + \varepsilon_{BB}\right) \qquad (9)$$

The small increments ε_{AA}, ε_{AB}, ε_{BB} are also quantities which may be adjusted as system variables but are subject to the constraint,

$$1 = P_{AA}\varepsilon_{AA} + 2P_{AB}\varepsilon_{AB} + P_{BB}\varepsilon_{BB} \qquad (10)$$

in order to ensure that the average distance is maintained as d_0.

Typical values for ε_{AA}, ε_{AB}, ε_{BB} may be in the range 0–2%. In a 10Å cell 2% corresponds to 0.2Å or a B-factor of ~3.0, so that such displacements are at least comparable in magnitude to thermal displacements.

As an example of the use of this method we show in Fig. 5a a diagram of the force model used for the intercalated compound urea/dibromodecane[15]. In this 2D simulation the alkane molecule (in projection) is represented as a dumbbell-shaped molecule with atomic scattering factors given by an average of those for the superposed atoms in one unit cell depth of the urea framework. The urea molecules are treated as rigid units which can rotate and translate. Springs of force constants k_{AO} and k_{AN} link the centre of mass of the alkane to the O and N ends of the urea molecules and springs of force constants k_{OO} and k_{NN} mimic the hydrogen-bond network within the urea framework. Fig. 5b shows schematically how the length of the springs used for the interaction with the urea molecules is longer when the plane of the alkane backbone is parallel to the contact vector and shorter when it is normal. Fig. 5c shows a diffraction pattern calculated from a model realisation which corresponds quite closely to the observed X-ray pattern (Fig. 5d) recorded at 150K. Prior to the lattice relaxation an occupational model was used to locally order the orientations of the alkanes in neighbouring channels. The arrows in Fig. 5c indicate the strong asymmetry that is observed in the intensity across the rows of Bragg peaks. This results directly from the lattice relaxation around the alkanes which have assumed different orientations in different channels.

DISORDER IN CUBIC STABILIZED ZIRCONIA

Cubic stabilized zirconias (CSZ's) have extremely simple average structures (the fluorite CaF_2 structure) but they exhibit extremely complex diffuse X-ray diffraction patterns. Despite numerous attempts over the last 30 years to understand the disorder in these materials these efforts have so far failed to yield a completely statisfactory model for the local order, although considerable progress has been made in the last few years[16,17]. In this section we use the case of CSZ's as a pedagogical example to show how various of the simulation methods described above can be used to obtain greater insight into a complex diffraction problem.

The fluorite structure has 8 oxygen sites per cell and 4 cation sites. The oxygen array is primitive but in CSZ's contains vacancies, while the cation array is f.c.c. and in CSZ's, though complete, is disordered. Cubes of oxygens along the three cubic directions are alternately occupied by a cation, or unoccupied, while along $\langle 1\ 1\ 0 \rangle$ there exist chains of cubes all occupied by a cation (see Fig. 6).

In commencing a study of this system it may be noted that in all known fluorite-related superstructure phases (with the one exception of the C-type rare-earth oxide structures) all anion vacancies occur in pairs separated by $\frac{1}{2}\langle 1\ 1\ 1 \rangle$ (in a cube containing a cation). $\frac{1}{2}\langle 1\ 0\ 0 \rangle$ and $\frac{1}{2}\langle 1\ 1\ 0 \rangle$ vacancy pairs are avoided. These $\frac{1}{2}\langle 1\ 1\ 1 \rangle$ cation/vacancy-pair units (see Fig. 6a) may be isolated (as in M_7O_{13}), in linear chains (as in M_7O_{12}), linked into zig-zag chains (as in the pyrochlore structure, see Fig. 6b), helical chains (as in $Ca_6Hf_{19}O_{44}$), or helical clusters (as in $CaZr_4O_9$, see Fig. 6c)[18].

From a chemical point of view it seems reasonable to suppose that the structure of a disordered CSZ of a given composition might consist of small domains of one or other of the known super-lattice phases which occur nearby in the phase diagram. A useful approach to try to understand the complex diffraction patterns that are observed, therefore, is to generate model structures in which such small domains of the various known superstructures occur, and compare the diffraction patterns with the observed patterns. For this purpose we use the example of an yttrium-stabilised cubic zirconia, $Zr_{0.61}Y_{0.39}O_{1.805}$, for which we have recorded numerous sections of X-ray diffraction data. Before proceeding with this survey of possible models for the oxygen vacancy ordering, however, it is necessary to consider how the structure relaxes when an oxygen vacancy is introduced.

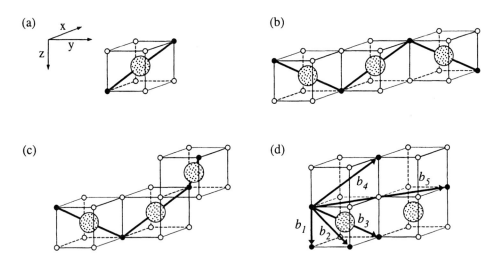

Figure 6. Vacancy pairs in fluorite-related structures. (a) A single $\frac{1}{2}\langle 1\ 1\ 1\rangle$ cation/vacancy-pair unit. (b) A pyrochlore zig-zag chain. (c) A helical cluster, as in Ca_4ZrO_9. (d) The interactions used in the Monte Carlo ordering scheme. Each vector, B_1, B_2, B_3, B_4, B_5 shown is one of a symmetry-related set.

Model for Relaxation

Since Zr and Y differ in atomic number by only one their X-ray scattering factors are practically identical. Consequently to a good approximation the terms that occur in the diffraction Equ. (2) that depend on $(f_Y - f_{Zr})$, *i.e.* I_0 & I_1, are absent. Although mean-square atomic displacements for O are somewhat higher than for the cations, their contibution to the intensity is relatively minor because of their much lower scattering factor. Consequently the observed diffraction patterns are dominated by the displacements of the cations and are described by diffraction terms I_2 and higher in Equ. (2).

We therefore develop a simple relaxation model which describes the way in which cations are displaced away from their average position in response to the removal of an oxygen neighbour. Nearest-neighbour cation pairs are separated by $\frac{1}{2}\langle 1\ 1\ 0\rangle$, with a pair of bridging oxygen sites mid-way between. Our simple relaxation model supposes that if either of these oxygen sites is vacant then the two cations will move further away from each other, while if both are occupied they will tend to be slightly closer than average to compensate. This is shown schematically in Fig. 7a. The vacancy (black circle) near the centre of the figure results in the cations (grey circles) being displaced as indicated by the arrows. The effect is transmitted along the [1 1 0] row of cations. Fig. 7b shows schematically the effect of this on the displacements away from the average positions in such a [1 1 0] row. The top row of vertical lines indicates the actual position of the cations (large open circles) as a result of being displaced by the vacancies (black circles). Small circles indicate the position of bridging oxygens. The lower row of vertical lines indicates the positions of the average cation lattice, *i.e.* a lattice of regular spacing with the same average spacing as the actual lattice. At the bottom of the figure the shift of the actual lattice relative to the average is indicated by the symbols "-" if the displacement is to the left and "+" if it is to the right. This clearly indicates that near-neighbour cations tend to be shifted in the same direction so that the lattice average $\langle X_i X_{i-n} \rangle$ is positive, while for rather longer vectors the shifts tend to be of opposite sign so that $\langle X_i X_{i-n} \rangle$ is negative. Fig. 7c shows how the summation of Fourier terms resulting from these correlations gives a broad diffuse peak resulting from the near-neighbour positive correlations, with a 'dark line' at the centre resulting from the more distant negative correlations.

(a)

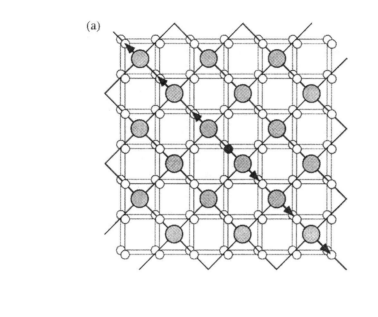

(b)

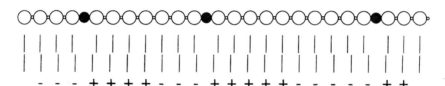

(c)

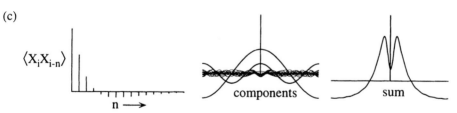

Figure 7. (a) Distortion model for Y-CSZ. Cations (grey circles) tend to move away from vacancy (black circle) resulting in the propagation of displacements along [1 1 0]. (b) Shows how the inclusion of vacancies (black circles) produces displacements away from the average lattice positions (indicated by the lower vertical lines) which are correlated. Near neighbours tend to be displaced in the same direction, giving a positive correlation. More distant neighbours tend to be displaced in the opposite direction giving a negative correlation. (c) Shows how the summation of Fourier terms resulting from these correlations gives a broad diffuse peak resulting from the near-neighbour positive correlations, with a 'dark line' at the centre resulting from the more distant negative correlations.

This same basic distortion model has been used for all subsequent calculations. Examples of calculated diffraction patterns are shown in Fig. 8 and the characteristic 'dark line' is seen to be common to all of them, as well as to the X-ray pattern. It should be stressed that this "dark-line" phenomenon is clearly visible only because of the condition that $f_Y \approx f_{Zr}$ and the $\mathbf{I_0}$ & $\mathbf{I_1}$ diffraction terms are absent. In calcium-stabilised zirconia, although the diffraction pattterns have many similarities, the effect is much less noticeable.

A second feature of all the diffraction patterns shown in Fig. 8 is of note. It may be seen that there is a strong asymmetry in the intensity between the high- and low-angle sides of the dark lines. Since the $\mathbf{I_0}$ & $\mathbf{I_1}$ diffraction terms are absent and $\mathbf{I_2}$ involves only cosine modulations, this asymmetry must originate from odd-order terms $\mathbf{I_3}$ and higher. This asymmetry only occurs when the magnitude of the distortion is sufficiently high for these higher order terms to be important. For the distortion model described above this effect requires a difference in the inter-cation spacing, with and without bridging oxygens, of ~3%. The same patterns computed with a smaller distortion of ~0.5% shows very little asymmetry.

Vacancy Ordering via Monte Carlo Simulation of Pair Correlations

To test the effects of intoducing short-range correlations between oxygen vacancies, a simple pair-interaction model was set up in which a given oxygen site interacts with the five near-neighbour types of site, as shown in Fig. 6d. These different interactions are specified by the parameters b_1–b_5. In addition to the five interactions shown, all symmetry related vectors are included. Note that b_3 & b_4 are both $\frac{1}{2}\langle 1\,1\,1\rangle$ vectors but b_3 is the diagonal of a cube containing a cation whereas b_4 is the diagonal of an empty cube.

Starting with a random distribution of 10% vacancies the effect of adjusting the different b_i can be tested. In Fig. 8 we show the diffraction pattern of the $(h\,k\,0.5)$ section calculated from different realisations. Fig. 8a shows the observed X-ray pattern for comparison. For Fig. 8b the distribution of oxygen vacancies was purely random. For Fig. 8c b_1 & b_2 were set to large positive values in order to induce large negative correlations along nearest-neighbour, $(\frac{1}{2}\langle 1\,0\,0\rangle)$, next-nearest-neighbour, $(\frac{1}{2}\langle 1\,1\,0\rangle)$ vectors. Note that for a concentration of vacancies of 10% the largest negative correlation that can be achieved is −0.11, which corresponds to the total avoidance of vacancy pairs. For Figs. 8d & 8e b_4 was additionally set to a large positive value in order to induce a large negative correlation along third-nearest-neighbour, $(\frac{1}{2}\langle 1\,1\,1\rangle^\dagger)$. Here we use † to denote a vector across a cube of oxygens not occupied by a cation. For Fig. 8d, in order to promote vacancy pairs along $\frac{1}{2}\langle 1\,1\,1\rangle$ in occupied cubes, a target correlation of 0.35 was set and b_3 was adjusted during iteration to achieve this, but b_5 was zero. For Fig. 8e, the same target correlation of 0.35 was set for $\frac{1}{2}\langle 1\,1\,1\rangle$ in occupied cubes and b_3 adjusted accordingly, but in addition a target correlation of 0.0 was set for the $\langle 1\,1\,0\rangle$ vector and b_5 was adjusted to achieve this.

In Fig. 9 we show parts of the two realisations corresponding to the examples of Figs. 8d & 8e. In Fig. 9a, corresponding to the diffaction pattern in Fig. 8e, it is seen that the structure consists of mainly isolated vacancy pairs, while in Fig. 9b, corresponding to the diffaction pattern in Fig. 8d, the vacancy pairs tend to be linked into longer chains. In particular the region marked by the heavy line is a crystallite of the pyrochlore structure.

The diffraction pattern in Fig. 8e qualitatively reproduces all the features in the observed diffraction pattern, although clearly there are quantitative differences. The pattern of Fig. 8d clearly does not and so this model can be eliminated as a suitable model for the structure. Together the calculated patterns in Fig. 8 serve to illustrate how a model can be progressively explored in order to arrive at a structure which gives a calculated diffraction pattern which agrees with the observed pattern. The model for Fig. 8e is still relatively simple and appears capable of further refinement. Comparison with the

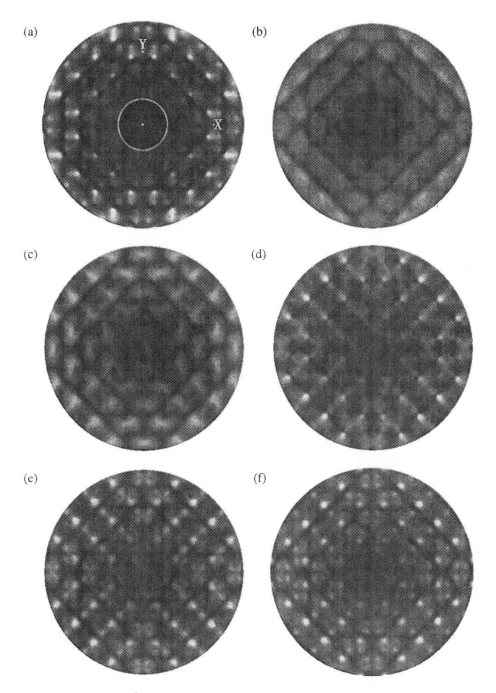

Figure 8. (a) The $(h\ k\ \frac{1}{2})$ section of the X-ray diffraction pattern of an Yttria doped CSZ, (b) Pattern calculated from a MC model with randomly distributed vacancies. (c) A MC model with no nearest- and next-nearest neighbour vacancy pairs. (d) A MC model with pyrochlore-like regions (as in Fig. 9b). (e) A MC model with mainly isolated $\frac{1}{2}\langle 111\rangle$ vacancy pairs (as in Fig. 9a). (f) A model obtained by direct synthesis of the distribution of $\frac{1}{2}\langle 111\rangle$ vacancy pairs. In (a) 'X' and 'Y' mark the points $(4\ 0\ \frac{1}{2})$ & $(0\ 4\ \frac{1}{2})$.

(a)   (b)

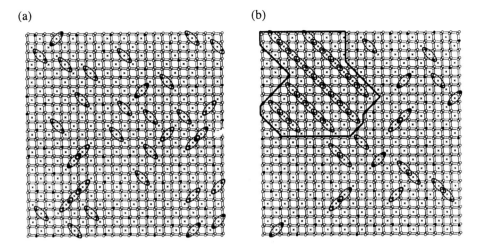

Figure 9. A small portion of two consecutive layers of the 3D distribution of oxygen vacancies (black circles) for the two realisations described in the text. (a) has mainly isolated $\frac{1}{2}\langle 1\ 1\ 1\rangle$ vacancy pairs, indicated by the ellipses. (b) has extended zig-zag chains of $\frac{1}{2}\langle 1\ 1\ 1\rangle$ vacancy pairs and contains microcrystals of a pyroclore-like structure (see outlined example), Note that vacancy pairs are in oxygen cubes that contain a cation (indicated by a cross, $\times$).

observed pattern reveals that the diffuse peaks which are visible either side of the dark lines differ in their detailed shape. In the X-ray pattern the peaks seem sharper in the middle but with rather broader wings whereas in the calculated pattern the peaks are rather less sharp but with most intensity near the peak. Such qualitative differences should provide guidance on how the model could be developed further, but attempts to do this using a more elaborate pair-interaction scheme have been unsuccessful.

Multi-Site Correlations

Since the diffraction pattern of any object is the Fourier transform of the pair-correlation function (see Equ. (1)) it might be argued that a model involving only pair-interactions should be all that is required to satisfactorily reproduce the observed diffraction pattern, and terms in the Hamiltonian such as the 3-site terms shown in Equ. (4) serve no useful purpose. While it is true to say that multi-site correlations do not directly contribute to the diffraction patterns, their effects are felt indirectly, for example in the constraints that are imposed on the 2-site correlations and the way in which these decay with distance. Moreover, multi-site occupational correlations can result in distinctive diffraction effects when additional relaxation displacements are considered.

In the present case, in studying vacancy correlations in CSZ's, it is important to point out that consideration of multi-site correlations is necessary in order to generate some of the known superstructures that occur. For example the $CaZr_4O_9$ structure contains helical clusters, shown in Fig. 6c, in an ordered arrangement in which both left-handed and right-handed helices occur. Since such left- and right-handed arrangements cannot be distinguished by 2-site correlation parameters it is not possible to generate an ordered crystal of this phase using only pair interactions. To define the right-handed helix shown in Fig. 6c, use of the 4-site interaction, $s_i s_{i-n} s_{i-m} s_{i-l}$, where i, i-n, i-m, i-l define the four vacant sites, would be an obvious choice for an interaction term.

Although the inclusion of such multi-site interactions may ultimately prove to be necessary for a full description of the CSZ systems it represents a level of complexity, both conceptually and in terms of implementation in a model, that we would rather avoid if possible. Consequently we consider whether, by redefining the problem sufficiently, it may be possible to remain with the consideration of 2-site interactions only.

Considering again the helix in Fig. 6c, we can see that what defines the handedness of the helix is the dihedral angle between the $\frac{1}{2}\langle 1\ 1\ 1\rangle$ vacancy pairs in the left-most and right-most cubes. Consequently if we consider a vacancy pair as the basic structural unit then each cube (occupied by a cation) can be considered as having a vacancy pair in one of 4 different orientations, together with the possibility of having no vacancy pair. Thus by replacing the simple binary variables representing single vacancies with 5-state variables representing the position and orientation of vacancy-pairs we can again revert to considering only pair interactions, albeit with a considerable increase in complexity. For Monte Carlo simulation it is not clear that such a change of variables offers a distinct advantage. However casting the problem in this new way does allow us to make use of the modulation wave direct synthesis approach to generating disordered distributions that was described earlier.

Modulation Wave Direct Synthesis of Vacancy Distributions

One of the seeming paradoxes of diffuse scattering in CSZ's is that while the X-ray patterns appear complex, electron diffraction patterns appear relatively simple and moreover appear basically the same in many different systems. The electron diffraction pattern shown in Fig. 10a is a typical pattern obtained when the vacancy concentration is ~8-10%. This pattern, corresponding to the $[1\ 1\ \bar{2}]$ zone axis, shows a series of diffuse circles. These are visible both as clearly defined circles but also as pairs of peaks resulting from the intersection of the Ewald sphere with circles which are inclined to the projection axis. All the circles can be shown to occur at positions centred on $\frac{1}{2}\{1\ 1\ 1\}$ and oriented normal to $\langle 1\ 1\ 1\rangle$ as shown in Fig. 10b. The main reason that the electron diffraction patterns are so different from the X-ray patterns is that the scattering, being displacive in origin, has strong azimuthal variation. This azimuthal variation is largely removed in electron diffraction as multiple scattering results in intensity being transferred from one region of the pattern to another translated by a whole reciprocal lattice vector. Thus all reciprocal unit cells in Fig. 10a look essentially the same. On the other hand the pairs of peaks which straddle the dark lines in Fig. 8a may be recognised as being the same features as the pairs of peaks in Fig. 10a.

As a result of these comparisons we might suppose that the circular features observed in the electron diffraction pattern are a reflection of the basic compositional ordering of defects that occurs in CSZ's and that the X-ray patterns give a detailed picture of how the atoms relax about these defects. To test this, therefore, we can use the modulation wave direct synthesis method to generate distributions of defects and the apply the same relaxation procedure as before to obtain the final calculated diffraction pattern.

A particularly simple model is to suppose that each of the four different orientations of diffuse ring shown in Fig. 10b is due to the distribution of defects ($\frac{1}{2}\langle 1\ 1\ 1\rangle$ vacancy pairs) in the particular $\{1\ 1\ 1\}$ plane normal to the $\langle 1\ 1\ 1\rangle$ vector which is bisected by the ring. I.e. we consider the plane consisting of the triangular mesh shown in heavy lines in Fig. 10c, which links all the vacancy pairs drawn as filled circles. Wave-vectors $\mathbf{q}$ corresponding to points on one of the diffuse circles were then used to carry out a direct synthesis of the distibution of vacancy pairs of the corresponding orientation. Since 10% vacancies are required in total only 2.5% are required for one orientation, and for the total synthesis the process is repeated using diffuse circles, planes and vacancy-pair orientations in each of the other three $\{1\ 1\ 1\}$ orientations. One plane from the resulting 3D synthesis is shown in Fig. 10d. Here the black dots indicate the position of the vacancy pairs and the

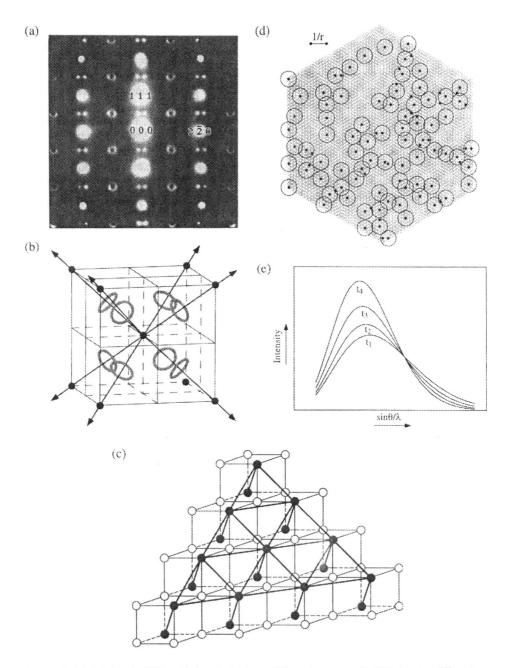

Figure 10. Modulations in CSZ's. (a) A typical electron diffraction pattern of a CSZ showing diffuse rings of scattering together with pairs of satellite peaks. These pairs of peaks are in fact the cross-section of rings in different orientations. (b) Schematic diagram to show that these 'rings' occur in reciprocal space normal to each of the four $\langle 111 \rangle$ directions, mid-way between Bragg positions (black circles). (c) Schematic view to show the pairs of oxygen sites that lie normal to a given (1 1 1) plane. The diffuse 'ring' that occurs around a given [1 1 1] direction corresponds to modulations of these pairs in this (1 1 1) plane. (d) Section from a 3D distribution of oxygen vacancies obtained by the modulation wave synthesis method. The length of the bar indicates the distance that is reciprocal to the radius of the diffuse 'ring' in reciprocal space. (e) Schematic drawing of the time evolution of a diffuse 'ring' peak during spinodal decomposition.

larger circles drawn around most of the vacancies have a diameter corresponding to the reciprocal of the radius of the diffuse circle.

These four separate syntheses are quite independent, but with such a low concentration of defects, the chance of more than one vacancy pair occurring in a given cube of oxygen sites is very low (~0.06%), so that the coordination of the cation site is only very rarely less than a chemically plausible six. A calculated diffraction pattern for the (h k 0.5) section obtain from this kind of synthesised distribution is shown in Fig. 8f. The pattern (and ones calculated for other reciprocal sections, not shown here) is in excellent (qualitative) agreement with the observed pattern. By adjusting the detailed distribution of the modulation amplitudes as a function of the radius r it is likely that the agreement could be further improved, although this has not been attempted. The extreme simplicity of the model, as described, allows new insight into the origins of the disorder to be gained, and it is doubtful whether this would be enhanced at this stage by a more quantitative fit.

Inspection of the distribution of defects in the plane normal to the $\langle 1\ 1\ 1 \rangle$ shown in Fig. 10d provides a clue to a possible mechanism for the disorder. Around each defect (except in a few places where 2 defects occur very close together which we attribute to approximations inherent in the synthesis method) circles have been drawn with a diameter equal to the reciprocal of the radius of the diffuse circle used in the synthesis. It is seen that these circles tend to be predominantly in close contact with each other but with very little overlap. I.e. the figure is very suggestive that around each defect there is a zone of exclusion where it is energetically unfavourable for another defect to occur, but that at a larger distance of, ~1/r, it becomes energetically favourable again. The whole plane appears to consist of two types of region, one defect free and the other consisting of defects closely packed with a mean separation of ~1/r. It should be noted that, in an equilibrium situation, if it were simply that defects tended to repel each other we might expect that the distribution would be more like that of a liquid with a mean inter-defect spacing defined by the concentration. The tendency for defects to avoid each other may be understood in terms of the strain field that will exist around the defect, and which will require a certain distance to dissipate, but the tendency to cluster at a preferred distance of ~1/r is not so easy to understand.

Such phenomena have been explained by Cahn, in the case of alloys, in terms of spinodal decomposition[19]. This occurs when a system is placed in a part of the phase diagram where a homogeneous single phase mixture is unstable and unmixing tries to occur. The description requires the solution of a diffusion equation involving thermodynamic, elastic and interface energy parameters. The outcome however is that compositional modulations develop and, as a function of time, the distribution of the wavelengths of these modulations shows a trend towards quite a narrow spread centred around a dominant wavelength (see Fig. 10e). Although in elastically anisotropic materials the wavelength may be orientation dependent, for cubic materials such as CSZ's the wavelength may be expected to be orientation independent, hence producing the uniform 'ring' of scattering.

CONCLUSION

In this paper we have described the theoretical background, the experimental methods and the computer modelling techniques that we have used to investigate of a wide range of disordered materials. The emphasis in our work is on trying to obtain a physical model which captures the essence of a particular problem so that insight into the mechanism of the disorder can be obtain. Methods to simulate both occupational and displacement disorder have been described, with reference made to particular examples where these methods have been used. The example of Cubic Stablised Zirconias has been described in some detail as it represents an excellent pedagogical example of how several of the different modelling techniques can be used and shows how a simple model can provide real insight into how the complex diffraction patterns occur.

REFERENCES

1. T.R. Welberry, *Rep. Prog. Phys.*, 438:1543 (1985).
2. H. Jagodzinski, *Prog. Cryst. Growth Charact.*, 14:47 (1987).
3. J.B. Cohen and L.H. Schwartz, *Diffraction from Materials*, Springer-Verlag, Berlin (1987)
4. H. Jagodzinski and F. Frey, *International Tables for Crystallography Vol B*, U. Shmueli, ed., A.J.C. Kluwer Academic Publishers: Dordrecht (1993).
5. B.T.M. Willis, *International Tables for Crystallography Vol B*, U. Shmueli, ed., A.J.C. Kluwer Academic Publishers: Dordrecht (1993).
6. J.E. Epperson, J.P. Anderson and H. Chen, *Metall. Mater. Trans. A.*, 25:17 (1994).
7. T.R. Welberry and B.D. Butler, *Chem. Rev.* 95:2369 (1995).
8. F. Frey, *Zeits. f. Krist.*, 212:257 (1997).
9. J.C. Osborn and T.R. Welberry, *J. Appl. Cryst.* 23:476 (1990).
10. B.D. Butler and T.R. Welberry, *J. Appl. Cryst.* 25:391 (1992).
11. T.R. Welberry and S.C. Mayo, *J. Appl. Cryst.* (in press) (1997).
12. E.L. Belokoneva, F.M. Dolgushin, M.Yu. Antipin, B.V. Mill and Yu.T. Struchkov, *Russian J. Inorg. Chem.*, 38:584 (1993).
13. N. Metropolis, A.W. Rosenbluth, M.N. Rosenbluth, A.H. Teller and E. Teller, *J. Chem. Phys.* 21:1087 (1953).
14. T.R. Welberry and A.G. Christy, *Phys. Chem. Minerals.* 24:24 (1997).
15. T.R. Welberry and S.C. Mayo, *J. Appl. Cryst.* 29:353 (1996).
16. T.R. Welberry, B.D. Butler, J.G. Thompson and R.L. Withers, *J. Solid State Chem.* 106:461 (1993).
17. T.R. Welberry, R.L. Withers and S.C. Mayo, *J. Solid State Chem.* 115:43 (1995).
18. H.J. Rossell and H.G. Scott, *J. Phys. Colloq.* 38:C7 (1977).
19. J.W. Cahn, *Trans. Metall. Soc. Amer.* 242:166 (1968).

RECENT ADVANCES IN STRUCTURE REFINEMENT FOR LIQUIDS AND DISORDERED MATERIALS

A. K. Soper

ISIS Facility
Rutherford Appleton Laboratory
Chilton
Didcot, Oxon, OX11 0QX
U.K.

INTRODUCTION

The notion of structure in a material which has no intrinsic long range order is often a difficult concept to grasp. For an army marching in orderly ranks it is easy to see the rows of soldiers with their uniforms and equipment all exactly the same, all moving to a strict pattern, all in the same direction. The order is there and obvious to all. For a crowded shopping mall however, filled with people moving in different directions, with different speeds, dressed in all manner of different clothes, and carrying a whole range of different 'equipment' (bags, strollers, perhaps a walking stick, and so forth), it would be hard to describe this as orderly. Indeed this would correctly be described as a highly *disordered* situation: from a given point in time and place in the mall it would be hard to predict exactly what was going to happen next.

Yet even in the disordered system of a crowded shopping mall there *is* still a degree of order present. Even in dense crowds with people moving in all directions, people normally try to avoid collisions with each other. They do this almost unconciously by walking towards spaces near them which nobody else occupies, but if they do collide then they soon discover that short range, but strong, *repulsive forces* come into play that prevent them occupying the same space! Equally similar kinds of influence determine that people enter shops by going through doors and not by walking through walls or windows. Thus the repulsive forces between the people in the mall, and between the people and the walls of the mall, define a sort of *residual structure*, one that relates to the fact that no two people can occupy the same space at the same time, nor can they pass directly through the walls of the mall.

To carry this analogy one step further, the residual structure of the disordered shopping mall is determined not only by the repulsive forces between people. There are also longer range interactions. A person seeing their child or a friend in the distance might well move towards them and stop and talk to them or hold hands (example of an *attractive force*). If someone came in brandishing a gun, probably people would rush away and towards the exits (*repulsive force*). The **distribution** of people in the mall is therefore also influenced by these *longer range* interactions which may be present.

The atomic-scale structure of a liquid or disordered solid is quite analogous to the example of the crowded shopping mall. Atoms and molecules are in principle free to occupy any spot in the material (and in the case of a liquid they are free to move in all directions with a variety of speeds), but they can only do this subject to the arrangement of neighbouring atoms and molecules, and to the repulsive and attractive forces which act between them. Therefore the problem of defining the structure of a liquid or disordered solid is complicated: on the one hand you really need to know the forces that act between the particles of the material, on the other you need to know how the particles are arranged. The two factors are obviously related, but how?

This is one example out of a huge class of problems which are called *many body problems*: even if you know the interactions between the particles you cannot predict exactly their arrangement. Conversely even if you know the arrangement of the particles you cannot easily determine the forces between them. Perhaps the single most difficult concept to get across for a disordered system is that even if the forces between the particles are known, there still will be *no* single structure (in the sense of a well defined crystal structure) which is consistent with those forces. Instead the structure must be characterised in terms of distribution functions, which measure the extent to which the positions and orientations of the system are *correlated*: a high degree of correlation between atoms and molecules indicates a high degree of structure; conversely weak correlations indicate a weak structure, and of course in practice there will be a whole spectrum of cases from strongly structured to weakly structured. Silica glass is an example of a highly structured disordered system: low density helium gas is an example of a very weakly structured disordered system.

This paper is about ways to measure and characterise the structure of such disordered systems using radiation diffraction experiments. In recent years there have been huge developments in the techniques that can be applied to these systems. Diffraction techniques (which in principle measure directly the radial distribution functions associated with a particular system) are able to map the structure with unprecedented speed and precision. This means that structures over a range of thermodynamic state conditions or through a phase transition are readily obtainable. Data processing techniques, that is the process of reducing the 'raw' data from the diffraction experiment to useful distribution function, have improved enormously. Perhaps the most important development however relates to the fact that computer simulation techniques are now being increasingly applied to the problem of disordered material structure refinement. This technique, which is essentially an approximate way of solving the many-body problem, allows prior information, such as overlap constraints, molecular geometry, dispersion forces (where these are known or can be estimated), inter-atomic bonding constraints (e.g. hydrogen-bonds), possibly even thermodynamic information, to be built automatically into the structure refinement process. The picture of structure which emerges from these new structure refinement tools is therefore correspondingly highly detailed. It would be fair to claim that the ability to refine the structure of a complicated liquid or disordered material has now reached the same level of sophistication as was reached nearly 30 years ago for crystalline structures by Rietveld [1]. Indeed, as can be seen from several of the other papers in this volume, many of these new techniques are now being applied to try to understand the residual disorder in many *crystalline* materials.

In the sections that follow, the notation used to describe the **structure factor** and **distribution functions** of a disordered system is first defined. Then the **Minimum Noise** method for extracting reliable radial distribution functions from diffraction data will be outlined. Note here that there will be no description of the numerous corrections that need to be made to diffraction data before this transformation can be made, since these are already well documented in more specialised texts [2]. The **spherical harmonic expansion** of the structure factor is introduced with the idea that the coefficients of this expansion are to be regarded as the essential "storehouses" of structural information in a system, particularly when molecules or a well defined local geometry is present. Finally the concept of **Empirical Potential Structure Refinement** is described, which introduces the use of computer simulation to refine the structure. The paper is then finished with some examples of applying the full range of techniques to diffraction data from liquid water and amorphous silica.

THE STRUCTURE FACTOR OF A DISORDERED SYSTEM

Single Atomic Component

The scattering amplitude of an array of N atoms each with scattering length b is defined as

$$F(\mathbf{Q}) = \sum_j b_j \exp i\mathbf{Q} \cdot \mathbf{r}_j \tag{1},$$

where $\mathbf{Q} = \mathbf{k}_i - \mathbf{k}_f$ is the change in the wave vector in the scattering event, $|\mathbf{Q}| = 4\pi \sin\theta / \lambda$ with 2θ the scattering angle and λ the radiation wavelength, and $\mathbf{r}_j$ is the position of the jth atom. The scattered intensity, or differential scattering cross-section, per atom of sample is then defined as

$$
\begin{aligned}
I(\mathbf{Q}) &= \frac{1}{N} F(\mathbf{Q}) F(\mathbf{Q})^* = \frac{1}{N} \sum_{jk} b_j b_k \exp i\mathbf{Q} \cdot \left(\mathbf{r}_j - \mathbf{r}_k \right) \\
&= \left\langle b^2 \right\rangle + \frac{1}{N} \sum_{j \neq k} \left\langle b \right\rangle^2 \exp i\mathbf{Q} \cdot \left(\mathbf{r}_j - \mathbf{r}_k \right)
\end{aligned} \tag{2}.
$$

$$\uparrow \qquad\qquad\qquad \uparrow$$
$$\text{"self"} \qquad\qquad \text{"interference"}$$

It will be seen from (2) that in the sum over pairs of atoms, the sum can be separated out into two terms, that for $j = k$, where the atoms correlate with themselves, the so-called "self" term, and that where $j \neq k$, the "interference" term, where the atoms are correlated with other atoms in the system. It is the latter term which provides the useful structural information about the system in question. For the self term, because this corresponds to an atom correlating with itself, the scattering level is proportional to the average of the square of the scattering length, whereas for the interference or distinct terms the scattering level is proportional to the square of the average scattering length. In general for neutrons the scattering length depends on both the isotope and spin state of the nucleus, but these states are not normally correlated with the position of the atom, so the averaging over spin and isotope states can be done outside the sum over atomic sites. It is the ability to vary b, either by isotopes with neutrons, or by anomalous dispersion with X-rays, that gives modern diffraction experiments a significant edge over their earlier counterparts which did not exploit this capability. (For X-rays of course the scattering length is Q dependent, but this dependence can be estimated to a good degree of accuracy [3].)

The contact between the structure factor of the system and the distribution function which represents the arrangement of atoms is made via the density-density autocorrelation function, $G(\mathbf{r})$ [4], which measures the density of atoms a distance $\mathbf{r}$ away from an atom at the origin. This function includes the self and interference terms as in (2):

$$G(\mathbf{r}) = \frac{1}{N} \int d\mathbf{r}' n(\mathbf{r}') n(\mathbf{r}' + \mathbf{r}) = \delta(\mathbf{r}) + \rho g(r) \tag{3}$$

$$\uparrow \qquad \uparrow$$
$$\text{``self''} \quad \text{``interference''}$$

where $n(\mathbf{r})$ is the density of atoms at any point $\mathbf{r}$ in the system, ρ is the average number density of atoms, and $g(r)$ is the well known **radial distribution function** of the system, representing the ratio between the local density of atoms a distance r from an atom at the origin to the average number density of atoms in the system. When written in the form $4\pi\rho r g(r)$, this distribution is also sometimes called the *pair distribution function* (PDF).

The scattering intensity is essentially the 3-dimensional Fourier transform of $G(\mathbf{r})$, bearing mind however the distinction in scattering amplitude between self and distinct terms:

$$I(Q) = \langle b^2 \rangle + \langle b \rangle^2 \left[S(Q) - 1 + N\delta(\mathbf{Q}) \right] \tag{4},$$

where the **structure factor**, $S(Q)$ is defined as:

$$[S(Q) - 1] = 4\pi\rho \int_0^\infty r^2 dr (g(r) - 1) \frac{\sin Qr}{Qr} \tag{5}$$

and the $\delta(\mathbf{Q})$ function in (4) represents the Q=0 Bragg peak that is present in every diffraction experiment, but because the system being measured is normally macroscopic it cannot be observed. Hence this term is normally left out of expressions for the scattering intensity.

Extracting the differential scattering cross-section from a set of diffraction data is typically a non-trivial process and so details of how this is achieved are left to a more specialised account [2]. The primary focus in this paper is the structure factor, $S(Q)$, and the associated radial distribution function, $g(r)$.

Multicomponent Systems

Most systems of interest contain more than one atomic component, and often several of these. If more than one component is present in the system then equations (1)-(5) need to be generalised. For example if there are two atomic components, α, β, with N_α, N_β atoms of each, so that $N = N_\alpha + N_\beta$, and the atomic fractions of each component are defined as $c_\alpha = N_\alpha / N$, $c_\beta = N_\beta / N$, then the differential scattering cross-section is defined exactly as before, but the structure factor can now be split into three terms depending on which pair of atoms are interacting:

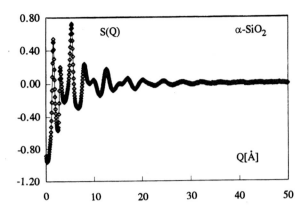

Figure 1. Measured structure factor of silica glass (points), after removal of the self scattering. The line shows the Minimum Noise fit to these data using the procedure described in section 3. The data were obtained on the SANDALS diffractometer at ISIS.

$$
\begin{aligned}
I(\mathbf{Q}) &= \frac{1}{N}\sum_{jk} b_j b_k \exp i\mathbf{Q}\cdot\left(\mathbf{r}_j - \mathbf{r}_k\right) \\
&= c_\alpha\langle b_\alpha^2\rangle + c_\beta\langle b_\beta^2\rangle \qquad \text{"Self"} \\
&\quad + c_\alpha^2\langle b_\alpha\rangle^2\left[S_{\alpha\alpha}(Q)-1\right] \\
&\quad + 2c_\alpha c_\alpha\langle b_\alpha\rangle\langle b_\alpha\rangle\left[S_{\alpha\beta}(Q)-1\right] \quad \text{"Interference"} \\
&\quad + c_\beta^2\langle b_\beta\rangle^2\left[S_{\beta\beta}(Q)-1\right] \\
&\quad + N\langle b\rangle^2\delta(\mathbf{Q})
\end{aligned}
\tag{6}
$$

where, for example, the $\alpha\beta$ *partial structure factor* is defined as

$$
\left[S_{\alpha\beta}(Q)-1\right] = 4\pi\rho\int_0^\infty r^2 dr\left(g_{\alpha\beta}(r)-1\right)\frac{\sin Qr}{Qr}
\tag{7}.
$$

Figure 1 shows the measured structure factor for silica glass, which is a weighted sum of O-O, SiO and SiSi partial structure factors, after removal of the self scattering.

For an M component system there will be $M(M+1)/2$ such partial structure factors needed to characterise the structure of the system. Therefore the data from a single diffraction experiment on the material can be complicated to interpret in terms of the likely atomic structure: it is very likely there will be overlapping interatomic distances involving different pairs of components. Hence a technique such as isotope substitution or anomalous dispersion can be invaluable in sorting out which pairs of component atoms contribute to a particular feature in the diffraction pattern or radial distribution function.

MINIMUM NOISE RECONSTRUCTION OF THE RADIAL DISTRIBUTION FUNCTION

The idea of minimum noise reconstruction (MIN) [5,6] originated from the wide application of Maximum Entropy reconstruction to problems with incomplete and noisy data [7]. The inversion of structure factor data, which are invariably noisy and are available over only a limited range of Q, to radial distribution function is a perfect example where such a procedure might be useful. In fact the experience was that with MaxEnt, residual truncation oscillations always appeared in the reconstructed distributions, and these oscillations were traced to a problem with the functional form of the entropy function itself, which has a quadratic form with respective deviations from the assumed "reference" distribution. This quadratic form meant that large peaks carried proportionally a much larger weighting on the entropy than small ones, so that the larger ones, which might have been real features present in the data tended to get surpressed, while small features due to truncation "ripples" were left unaffected since they made virtually no contribution to the entropy [5].

The trick around this difficulty was to define an alternative form of the entropy functional, this time called the noise functional, in which significant deviations by any particular trial distribution from the reference distribution made only a *linear* contribution to the noise. Thus a peak which was twice as large as another carried a penalty of only *twice* the noise value, and not *four* times the entropy value as in traditional MaxEnt. In addition since the reference distribution is not known beforehand for any particular case, the reference distribution was simply a smoother version of the trial distribution.

To understand how this works, suppose there are a set of data, D_i $(i=1,I)$ with uncertainties σ_i. It is desired to find a model, M_j $(j=1,J)$, or set of models which are consistent with the data. For this there exists a transform $T_{i,j}$ between model and data which produces a

fit $\quad F_i = \sum_j^J T_{ij} M_j \quad$. The quality of fit is determined from $\quad \chi^2 = \sum_i \dfrac{(D_i - F_i)^2}{2\sigma_i^2} \quad$. If the transform

matrix is set up to correspond to the kernal of equations (5) or (7), then simply minimizing χ^2 on its own is effectively the same as performing a direct inverse Fourier transform on the structure factor data, and so there is no gain. However if an additional noise constraint, N, which measures how *noisy* a particular model is, is added to χ^2, and the total **quality factor**, $Q_f = \chi^2 + \lambda N$, is minimised, then the model distribution should not only fit the data within the known uncertainties, but it should also satisfy the specified noise criterion. The factor λ in this definition is simply a variable: too small a value of λ will correspond to good fits to the data but noisy model distributions, while too large a value of λ will correspond to a very

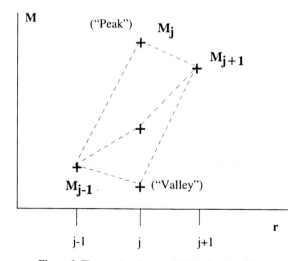

Figure 2. Three points in a model distribution, M

smooth model distribution, but a poor fit to the data. No doubt a more sophisticated Bayesian analysis could be imposed to determine the best value of λ [8], but in practice a simple trial-and-error procedure has worked quite satisfactorily up to now.

The noise function that is used here is based on the idea that the distribution functions of physical systems are expected to be continuous and have continuous derivatives. This is difficult to ensure when by definition the model will be defined only at discrete positions. Nonetheless it is still possible to impose a degree of connectivity between neighbouring values. Fig. 2 shows three points in a model distribution, j-1, j, and j+1. The noise function [6] is defined such that it goes to zero when point j is midway (in the vertical direction) between j-1 and j+1, and initially it grows quadratically as it moves away from this position. However if j is either *above* or *below* either of its neighbours (shown by the horizontal dashed lines in this figure) to form either a "peak" or a "valley" the noise function then grows linearly with the deviation [6]. The result is that each point in the model distribution has an associated noise value, N_j, with the total noise given by $N = \sum_j N_j$.

In fact the original paper [6] which describes this noise function contained a typographical error, so for completeness the full noise function is repeated here again. Initially a reference distribution is defined which is smoother than the trial model distribution:

$$R_j = \frac{1}{4}\left[M_{j-1} + 2M_j + M_{j+1}\right] \tag{8}.$$

The deviation of the jth model value from this reference distribution is given by $X_j = M_j - R_j$. The noise is then defined as

$$N_j = X_j^2/W_j \tag{9}$$

where the weighting function W_j is defined as

$$W_j = \begin{cases} \frac{1}{2}\left(M_{j+1} - M_{j-1}\right) & \text{if } |X_j| \le \frac{1}{2}R_j \quad \text{(a)} \\ |X_j| & \text{if } |X_j| > \frac{1}{2}R_j \quad \text{(b)} \end{cases} \tag{10}$$

The weighting W_j is introduced here so that, in line with the discussion given above, the noise values are down weighted in regions where the model curve is changing rapidly with j, or where a significant peak appears. Cases (a) and (b) above correspond to the situations of when M_j is either (a) between the horizontal dashed lines of Fig. 2, or (b) outside this region. It is this weighting function which makes Minimum Noise reconstruction significantly different from standard MaxEnt, and which allows the MIN reconstruction to remove spurious small truncation peaks while still permitting large real ones. Standard (linearised) MaxEnt would have set $W_j = 1$.

In order to minimize the quality factor, it is necessary to apply a fluctuation δ_j to the model at point j. This causes a change in the fit to the data, $\Delta\chi^2(j)$ and to the noise value, ΔN_j due to the fluctuation at j. The change to the quality factor is then given by

$$\Delta Q_f(j) = \Delta\chi^2(j) + \lambda\Delta N_j = 2A_j\delta_j + B_j\delta_j^2 \tag{11}$$

where

$$A_j = -\sum_i \frac{(D_i - F_i)}{2\sigma_i^2} T_{ij} + \lambda \frac{X_j}{W_j} \tag{12}$$

and

$$B_j = \sum_i \frac{T_{ij}^2}{2\sigma_i^2} + \lambda \frac{1}{W_j} \tag{13}.$$

The form of these equations ensures that discontinuities in the second derivative of ΔQ_f do not occur at $|X_j| = \frac{1}{2} W_j$, and it enables a rapid solution to be found for the problem of minimising equation (11).

Dropping the suffixes in (11) for brevity, it will be noted that because this equation is exactly quadratic in δ, the minimum value of the quality factor will occur at $\delta = -A/B$, with a minimum value of $\Delta Q_f^{(min)} = -A^2/B$. However simply inserting this deviation into the model will not necessarily produce a global minimum to Q_f since neighbouring values of j are coupled by both the data and the noise function. Hence an iterative method is implied, adjusting each j value in turn. In practice the solution converges very rapidly, typically requiring between 5 and 50 iterations to find a global solution.

The form of this equation also prompts the idea of going a step further and attempting to estimate the uncertainty in performing the inversion. Making the substitutions $\Delta Q_f^* = \Delta Q_f + A^2/B$, and $\delta^* = \delta + A/B$ equation (11) becomes

$$\Delta Q_f^* = B\delta^{*2} \tag{14}$$

which is precisely the form for the potential energy of a simple harmonic oscillator with force constant $k = 2B$. Thus instead of choosing simply the minimum value of δ^*, a range of possible values can sampled by choosing randomly (in the Monte Carlo sense) from a probability distribution of the form

$$p(\delta^*) \sim \exp\left(-\frac{B\delta^{*2}}{T_K}\right) \tag{15}$$

where T_K is a global "temperature" applicable to all j values. Experience with using this Monte Carlo sampling method indicates that it actually speeds up the process of convergence quite significantly.

It will now be apparent that the value of B gives a direct indication of the likely uncertainties in the model distribution function: essentially if B is small for a given j value there will be weak constraints on the allowed values of δ, while if B is large, the values of δ will be strongly constrained.

One final point to be noted is that the effective temperature, T_K, used here is not necessarily arbitrary. One of the rules about a harmonic oscillator is that its average potential energy and average kinetic energy are equal. From what was said above the potential energy of an individual oscillator can be written down as $-\Delta Q_f^{(min)}$, so the average potential energy of each of the J oscillators is $T_P = \frac{1}{J}\sum_j A_j^2/B_j$. Thus the likely value of the temperature is

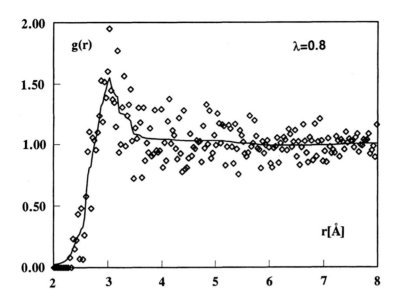

Figure 3. Example of a minimum noise fit to a set of (simulated) very noisy data. Note that in the region where there is apparently a peak the fit has found such a feature, but elsewhere the fit is considerably smoother than the original data and almost featureless. In the region of the peak some small residual "spikyness" is seen because the fit is clearly uncertain precisely where the correct curve should go. Ensemble averaging over a large number of iterations improves this. Increasing λ would also produces a smoother fit, but will also round off the peak more noticeably.

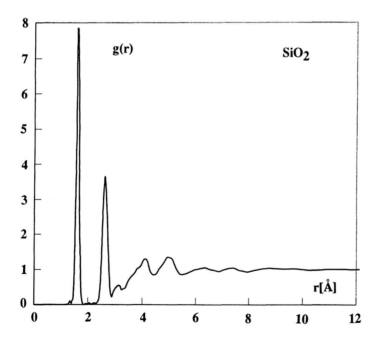

Figure 4. Minimum Noise Fourier transform of the data shown in Fig. 1. Note the sharp SiO and O-O peaks obtained. The area of the first peak corresponds to 3.95±0.21 oxygen atoms around silicon. Note that in obtaining this transform **no** smoothing or truncation functions were emplyed, and the transform matrix, T_{ij} included the resolution function of the diffractometer on which the data were obtained. The small SiSi peak near 3Å is clearly visible in the transform, against the backdrop of much larger peaks.

obtained by setting $T_K = T_P$. Hence if the distribution starts a long way from the global minimum Q_f, the effective temperature has to be re-estimated at each iteration so that the kinetic energy of the system follows the potential energy. Once equilibrium is established then sampling fluctuations from the distribution (15) allows some idea of the uncertainties in the reconstructed distribution function to be generated, and the ensemble average of all the trial model distributions should coincide with the global minimum of Q_f.

A simple example of applying this procedure is shown in Fig. 3, which is the case of trying to generate a reasonable and smoother curve through some very noisy data. In this case the transform matrix is obtained simply by setting $T_{ij} = \begin{cases} 1 & i = j \\ 0 & i \neq j \end{cases}$.

Application of the MIN procedure to evaluating Fourier transforms works in exactly the same way as this, but with a transform matrix corresponding to the appropriate kernal (5) or (7). In this case it is possible to apply additional constraints, such as the fact that $g(r)$ is expected to be precisely zero below a pre-determined value of r: for values of j corresponding to these values of r the model is set initially to zero, and no fluctuations are allowed in this region. The MIN Fourier transform of the data of Fig. 1 is shown in Fig. 4.

THE ORIENTATIONAL PAIR CORRELATION FUNCTION AND ITS SPHERICAL HARMONIC EXPANSION

For systems containing molecules the site-site partial structure factors of section 2 do not provide a complete description of the structure of the material. The theory of molecular fluids [9] indicates that understanding the relative orientation of neighbouring molecules is a vital element in the theory of molecular systems, and that in these cases the site-site

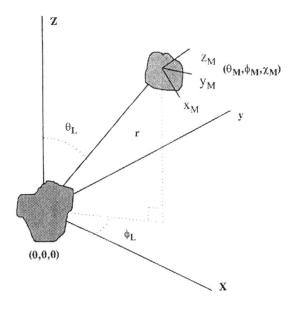

Figure 5. Geometry of the orientational pair correlation function used for reconstruction and plotting. Molecule 1 is at the origin of the coordinate system and the laboratory axes are rotated to coincide with the molecular frame of this molecule. Molecule 2 at position $\mathbf{r}=(r,\theta_L,\phi_L)$ is rotated to an orientation (ϕ_M,θ_M,χ_M) with respect to molecule 1. The orientational pair correlation function represents the density of molecules with this particular orientation as a function of position with respect to the coordinate axes defined by molecule 1.

distributions measureable by diffraction are in fact orientational averages over the orientational pair correlation function, and so contain less information. In the following section a practical way of estimating this orientational correlation function, given a set of diffraction data, will be described. Here the intention is to establish the fundamental ideas.

The orientation of a general molecule relative to a set of coordinate axes is conventionally defined via three Euler angles (ϕ,θ,χ), [9]. For linear molecules this number reduces to 2, but otherwise there is little simplification that can be employed. Therefore if we are to estimate the orientational pair correlation function, $g(\mathbf{r},\omega_1,\omega_2)$, strictly it is a function of 9 variables: 3 Euler angles for the molecule at the origin, $\omega_1=(\phi_1,\theta_1,\chi_1)$, 3 Euler angles for the second molecule, $\omega_2=(\phi_2,\theta_2,\chi_2)$, and 3 coordinates $\mathbf{r}=(r,\theta_L,\phi_L)$ to specify the position of molecule 2 with respect to molecule 1. In practice for an isotropic system such as an homogeneous liquid, glass or a powdered crystal these 9 coordinates are not all irreducible, and for the purposes of describing and plotting the orientational correlation function the number can be reduced to a unique set of 6 coordinates by rotating the laboratory coordinate axes to cooincide with those of molecule 1 at the origin. In this case the *relative* orientation of molecule 2 to molecule 1 is define by the Euler angles ϕ_M,θ_M,χ_M) (see Figure 5).

Attempting to store a function of 6 variables would require an enormous volume of memory even for one particular system. If we imagine digitizing angles into steps of 5° and assume 100 radius values are needed then for the case of an arbitrary shaped molecule with little symmetry the amount of memory required would be $72^3\times36^2\times100 \approx40$ Gbytes, assuming an 8 bit word is sufficient to store the intensities! With molecular symmetry this number will be reduced, but even in the case of water the number is only reduced by a factor of 4. This limitation is no doubt one reason why the orientational pair correlaiton function is rarely displayed in the literature.

However I have pointed out [6,10] that a tremendous memory saving is achieved by emplying the spherical harmonic expansion of the orientational pair correlation function. This was first described by Blum *et. al.* [11], and has been repeated by several authors since then. The notation used here follows closely that used by [9] and [10]. Essentially the orientational pair correlation function is expanded as sum of products of generalised rotational matrices $D_{mn}^l(\omega)$, and the object is to determine the radial dependence of the coefficients of this series, $g(l_1l_2l;n_1n_2;r)$:-

$$g(\mathbf{r},\omega_1,\omega_2) = \sum_{l_1l_2l} \sum_{m_1m_2m} \sum_{n_1n_2} g(l_1l_2l;n_1n_2;r)C(l_1l_2l;m_1m_2m)$$
$$\times D_{m_1n_1}^{l_1}(\omega_1)^* D_{m_2n_2}^{l_2}(\omega_2)^* D_{m0}^{l}(\omega) \tag{16}$$

where $C(l_1l_2l;m_1m_2m)$ are the Clebsch-Gordan coefficients.

The first term of this series, (000;00), corresponds to the radial distribution function of molecular centres around the central molecule, after averaging over all orientations: if there were no orientational correlations this would be the only term needed. Higher order terms introduce the orientational correlations with increasing precision. In practice it is found that the number needed depends on the degree of molecular symmetry and the degree of orientational correlation. Typically the number of coefficients needed to achieve adequate resolution of the orientational correlation function varies between 50 and 1000 for most molecular liquids, and assuming these are specified at 100 radius values, the memory requirement is at most 10^5 words, which is within easy reach of modern computers. Once the coefficients are estimated the orientational pair correlation function can be inspected for particular relative orientations: there is no need to perform a new calculation of coefficients. Thus they are important "storehouses" of structural information on the material.

The actual l,m,n values that need to be stored are determined by the molecular symmetry of molecules 1 and 2: the coefficients for values that are not compatible with the molecular symmetry are identically zero and so do not need to be estimated. Hence the molecular symmetry is built into the expansion by default. This is another reason why the spherical harmonic expansion is a highly economical way of storing structural information.

To see this point, it is instructive to write the site-site structure factor (which *is* measured in the diffraction experiment) in terms of these spherical harmonic coefficients - this expansion serves to illustrate the connection between measured data and the orientational pair correlation function:

$$S(Q) = \sum_{l_1 l_2 l} \sum_{n_1 n_2} \sum_{\alpha\beta} b_\alpha b_\beta f(l_1 l_2 l) C(l_1 l_2 l; 000) j_{l_1}(Q d_\alpha) j_{l_2}(Q d_\beta)$$
$$\times D^{l_1}_{n_1 0}(\omega_{1\alpha}) D^{l_2}_{n_2 0}(\omega_{2\beta}) H(l_1 l_2 l; n_1 n_2; Q) \qquad (17)$$

where

$$H(l_1 l_2 l; n_1 n_2; Q) = 4\pi\rho \int_0^\infty r^2 g(l_1 l_2 l; n_1 n_2; r) j_l(Qr) dr \qquad (18).$$

Here the position of each atom α in molecule 1 is defined by the vector $(d_\alpha, \omega'_{1\alpha})$, where d_α is the distance of the atom from the centre of the molecule, and $\omega'_{1\alpha} = (\theta'_{1\alpha}, \phi'_{1\alpha})$ defines the angular position of this atom *within the molecular coordinate frame*. The product of the rotation matrices and spherical Bessel functions serves to ensure that only coefficients with the correct molecular symmetry are accessible by the experiment. The factor $f(l_1 l_2 l)$ is simply a function of its arguments [10].

This expansion also highlights the distinction between neutron scattering and X-ray scattering from molecular liquids. Hydrogen atoms are largely transparent to X-rays: since hydrogen atoms tend to sit on the outside of organic molecules they will generally be a greater distance from the center of the molecule. Thus X-rays will be more sensitive to the distribution of molecular centres, while neutrons, which are strongly scattered by hydrogen will be more sensitive to orientational correlations. This is a good example of the complementarity between X-rays and neutrons.

It will be noted that despite its apparent complexity, equation (17) is a simple linear transform of the orientational expansion coefficients, and is therefore susceptible to Minimum Noise analysis. Thus the early work with this expansion used the MIN procedure to estimate the coefficients [6,10]. Although this usually gave sensible reconstructions, difficulties were encountered in complex molecular liquids and mixtures in ensuring that the reconstructed distributions were physically reasonable. That is there was little that could be done to ensure they did not go strongly negative in some orientations and positions. Therefore an alternative approach was developed, using computer simulation as the tool, to model distributions of molecules which were consistent with the diffraction data, but which also satisfied specified overlap constraints. The result, which can be regarded as an off-shoot of the Reverse Monte Carlo method [12], is called EmpIrical Potential Structure Refinement, EPSR [13].

ESTIMATING THE SPHERICAL HARMONIC COEFFICIENTS - EMPIRICAL POTENTIAL STRUCTURE REFINEMENT (EPSR)

Computer simulation using Monte Carlo (MC) or Molecular Dynamics (MD) can be regarded as an approximate way of solving the many-body problem alluded to in the introduction [4]. Given a force field for intermolecular interactions molecular assemblies can be generated which in principle are determined by those force fields. In practice there are a couple of approximations involved which may or may not affect the result. Firstly it is realistic to perform the simulation with a few hundred or perhaps a thousand or more molecules. Larger systems *can* be simulated but there is a corresponding increase in computing time. This means periodic boundary conditions must be imposed whereby if a molecule moves out of the simulation box it must immediately reappear at the corresponding point on the opposite side of the box [14]. A more complex correction arises from the fact that the forces between molecules are often long range and go beyond the size of the box being simulated. The precise method of correcting for long-range forces can affect the results of the simulation, particularly if thermodynamic or dynamic information is being calculated. Fortunately for the calculation of structure the result does not seem to be overly dependent on the precise way that the long range corrections are dealt with.

The EPSR technique works by initially performing a standard MC simulation [14] of the system of interest, using an assumed force potential $U_0(r)$, called the reference potential, between the atoms and molecules of the system. This potential will normally have built into it all that is known about the atomic overlap, dispersion, and electrostatic forces between molecules. For molecules it may also have the set of harmonic forces needed to define the molecule, including constraints on bond angles. For molecules with internal rotational degrees of freedom about particular bonds, these are treated as free to rotate at the present time, although there is nothing in principle to stop torsional forces to be included if these are known. This reference potential is therefore an attempt to build into the structure refinement at the outset chemical and physical constraints where these are known.

This simulation is brought to equilibrium, and once this is achieved the radial distribution fucntion(s) of the system can be estimated and compared with those derived from the diffraction measurements. Almost invariably discrepancies are found. These discrepancies will be larger or smaller depending on how good the reference potential is. The EPSR method works by trying to determine a perturbation to the reference potential which when used in the simulation will produce a radial distribution function closer to what has been measured.

It does this by noting that every radial distribution function has a corresponding potential of mean force $\psi_{\alpha\beta}(r) = -kT\ln\big(g_{\alpha\beta}(r)\big)$. Thus there will be a potential of mean force for the simulation $\psi_{\alpha\beta}^S(r) = -kT\ln\big(g_{\alpha\beta}^S(r)\big)$, and one for the data, $\psi_{\alpha\beta}^D(r) = -kT\ln\big(g_{\alpha\beta}^D(r)\big)$. These two potentials can be used to define the perturbation to the reference potential:

$$U_{\alpha\beta}^N(r) = U_{\alpha\beta}^O(r) + \big(\psi_{\alpha\beta}^D(r) - \psi_{\alpha\beta}^S(r)\big) = U_{\alpha\beta}^O(r) + kT\Big[\ln\big\{g_{\alpha\beta}^S(r)/\, g_{\alpha\beta}^D(r)\big\}\Big] \qquad (19)$$

This new potential is now used in the simulation in place of the reference potential, and once again the simulation is brought to equilibrium with the new potential.

This process of refining the potential is repeated many times, and, if all goes well, at the end there are set a radial distribution functions derived from a physical distribution of atoms and molecules which are consistent with a set of diffraction measurements. If, as sometimes happens, the diffraction data (or at least the radial distribution functions derived from the diffraction data) contain artifacts due to a systematic error left over from the original data analysis, then the simulation will show these up by being unable to generate good fits to the data. Thus the simulation is also a check that the data themselves are sensible, to the extent that this is possible.

A few additional features need to be addressed. Perhaps the most immediate problem is the noise in the simulated radial distribution functions, $g_{\alpha\beta}^S(r)$, which arises from the simple fact that the simulation has only a finite number of particles in it. Thus even if the supplied radial distribution functions, $g_{\alpha\beta}^D(r)$ are smooth, repeated direct application of equation (19) as it stands would rapidly lead to a very noisy potential, which certainly would not allow the

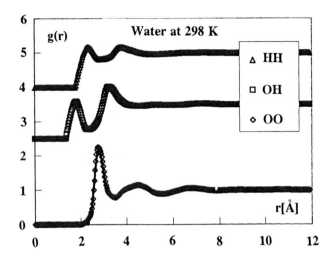

Figure 6. EPSR fits to the HH, OH and OO radial distribution functions for water at ambient conditions as derived from neutron diffraction data from SANDALS.

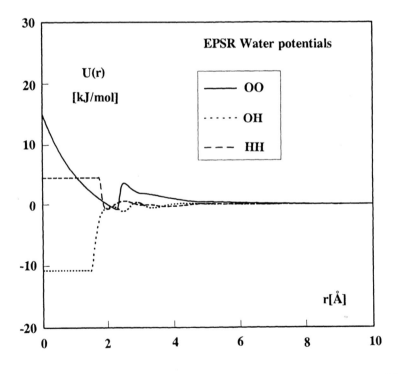

Figure 7. Derived empirical potentials from water under ambient conditions, using the SPC/E potential [15] as a reference potential.

simulated system to proceed on a true random walk through phase space. Therefore in practice it is necessary to first smooth the ratio of radial distribution functions in (19) before applying it to the old potential function. This is conveniently and rapidly achieved using the MIN technique described in Section 3, with the weighting λ being used to control the final degree of smoothness in the empirical potential. If this potential is too noisy the system will fail to proceed on a true random walk through phase space. If it is too smooth then significant features of the data will not be reproduced.

A second aspect concerns the values of thermodynamic quantities such as the configurational energy and pressure. Both these quantities can be estimated in this simulation since the Empirical Potential is a real potential with the units of energy per molecule. In fact it turns out to be straightforward to control the energy: at each iteration of the simulation a factor F_1 is applied to the perturbation potential, and the value of this factor depends on whether the energy of the simulation is above or below the expected configurational energy for this system.

To control the pressure it is necessary to add an extra term to the perturbation potential of the form $U_{\alpha\beta}^{(P)}(r) = F_2 \exp\left(\dfrac{(R_{\alpha\beta} - r)}{\sigma}\right)$, where $R_{\alpha\beta}$ is a characteristic distance for each radial distribution function, and σ is a "hardness" parameter, usually chosen to be wide enough not to have a strong effect on the simulated structure. The factor F_2 controls how strongly this additional term influences the simulation. Because this form has a negative gradient at all r values it can contribute only a positive contribution to the energy and pressure. The combination of factors F_1 and F_2 are capable of controlling both the pressure and energy of the simulation, providing of course the reference potential does not produce values for either

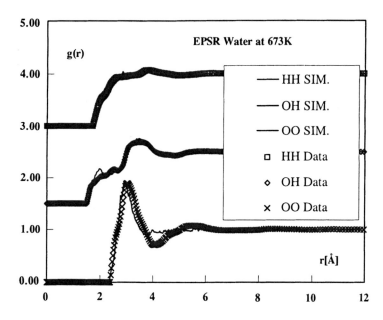

Figure 8. Comparison of the simulated structure of supercritical water using the EPSR potential derived from water data under ambient conditions, with the radial distribution functions of water determined by neutron diffraction from supercritical water. A small OH peak is still seen in the simulation, but it is much weaker than in simulations with one of the standard potentials, such as SPC/E [15]

quantity which are vastly different from what is expected. Thus not only does the EPSR simulation produce a structure consistent with the diffraction data, it will also have a sensible pressure and energy.

Questions about the validity and uniqueness of the EPSR procedure have been raised in several conference discussions recently. This is inevitable whenever a new procedure is introduced. However many of these comments apply to all simulations which use empirical and effective pairwise-additive potentials. In a sense the absence of a *single* structure consistent with the diffraction data is a strength, not a weakness, since it is in accord with all that is known about the disordered states of matter. There **are** no unique structures in a disordered material, and the EPSR techinique is a way of sampling from the ensemble of possible structures which are consistent with the measured data. In that sense the problem is quite the opposite to the solution of a structure in crystallography.

As an example of the application of this approach Fig. 6 shows the EPSR fit to the measured site-site radial distribution functions for water at ambient pressure and temperature. As can be seen the discrepancies between data and fit a very small. It must be borne in mind of course that a nearly perfect fit cannot be taken to imply the data are themselves perfect. It simply means they are consistent with a physical distribution of molecules. Within that constraint it still might be possible for them to contain systematic errors. Obviously the requirement to be fit by a physical distribution of molecules is an important boundary constraint which must be satisfied, but it cannot reveal all the systematic errors in the data.

The reference potential for this particular simulation was the Simple Point Charge (Extended) model, SPC/E of Berendsen *et. al.* [15] The derived perturbations to this potential are shown in Fig. 7. It is found that the required modifications to the SPC/E potential (which has large Coulomb repulsive and attractive terms) are mostly small. Currently a debate exists about the validity of this potential as the temperature and pressure of water are raised. Above the critical point it appears from the experiment that the first peak in the OH radial distribution function disappears and becomes only a smooth shoulder [15], but computer simulations with the SPC/E potential and other potentials have failed to reproduce this trend-they almost always show a distinct peak. It is currently speculated that the discrepency may arise from the lack of a polarizability term in the SPC/E potential (and others like it)

It was instructive therefore to apply the empirical potential derived for water under ambient conditions, and apply it to a simulation of water in the supercritical phase, in this case with *no* further refinement of the potential. The comparison is shown in Fig. 8, and it can be seen that this empirical potential does a remarkably good job at reproducing the supercritical water diffraction data. Thus if polarisability is the clue to understanding water structure at high temperatures, it appears that the empirical potential is able correct for this polarizability in an effective sense.

Once the empirical potential has been found the generated molecular ensembles can be used, by making use of the orthogonality properties of the rotational matrices, to invert equation (16) and so obtain a set of spherical harmonic expansion coefficients. The simulation is then run for many more iterations at equilibrium so that ensemble averages of the coefficients can be built up.

In the next section the methods of this and the preceding sections are combined in the structural refinement of some neutron diffraction data from amorphous silica.

APPLICATIONS: THE STRUCTURE OF AMORPHOUS SILICA

A Reference Potential for Silica

In the present work the original diffraction data (see Figure 1) were obtained from a sample of pure silica on the SANDALS diffractometer at ISIS. After making the standard corrections for attenuation and multiple scattering, they were subject to the MIN procedure to obtain the radial distribution function, Figure 4. It is found that the first two peaks are significantly sharper than what is obtained by conventional Fourier filtering, and the SiO coordination number is very close to 4.

The model chosen as a reference potential for silica was based on the charge model of Tsuneyuki *et. al.* [17]. However instead of the Born-Meyer repulsive potential used by those authors, an equivalent Lennard-Jones potential with the same dispersion forces as [17] was formulated to give the correct Si-O and O-O distances. Of course there have been a number of computer simulations of silica, see for example [18-20], and the RMC method has also been applied to this system [21]. See also the paper by D. Keen in this volume. The decision to use the Tsuneyuki *et. al.* potential was based on its inherent simplicity and the fact that it does not contain any three body potentials. However, as will be seen below, for practical reasons a near-neighbour tetrahedral arrangement atoms was imposed on the EPSR simulation.

In order to form a model about which to measure orientational correlations, the system was assumed to consist of an equimolar mixture of SiO_4 and Si ions, to preserve the stoichiometry. The SiO_4 ions were held to be closely tetrahedral by defining appropriate harmonic forces between Si and O and between O and O. In this model all the silicon atoms have an electronic charge of $+2.4e$, while the oxygen atoms have a charge of $-1.2e$ [17]. Figure 9 shows a representation of this "molecule", and the coordinate axis that was defined for each molecule.

Starting at a very high temperature, 6000K, the simulation was run with the reference potential alone to allow the atomic configurations to reach equilibrium. Then it was cooled to 1000K and all the subsequent simulations were run at this temperature to ensure sufficient atomic mobility for the simulation to proceed on a random walk after a reasonable number of iterations. With the reference potential alone, plus the assumed molecular entities, the result was a surprisingly good representation of the short range part of the measured radial

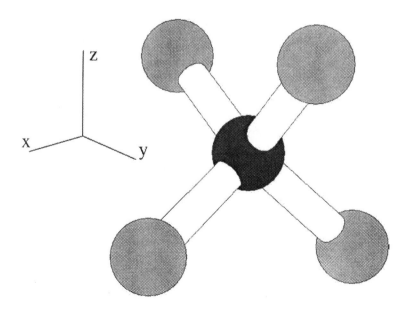

Figure 9. SiO_4 tetrahedron used as the main building block in the EPSR simulation of the amorphous silica diffraction data. The silicon is black and at the centre of the "molecule", and the oxygens are grey. The upper two oxygens lie in the z-y plane while the lower two oxygens lie in the z-x plane. In the simulation this unit is not rigid but is subject to harmonic forces which maintain the tetrahedral geometry on average.

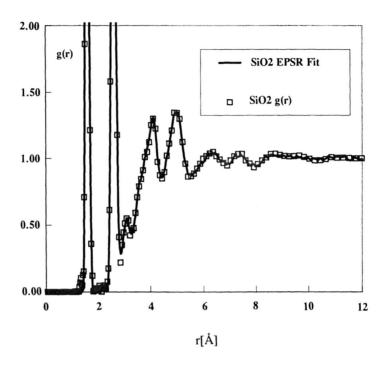

Figure 10. EPSR fit to the data of Figure 4.

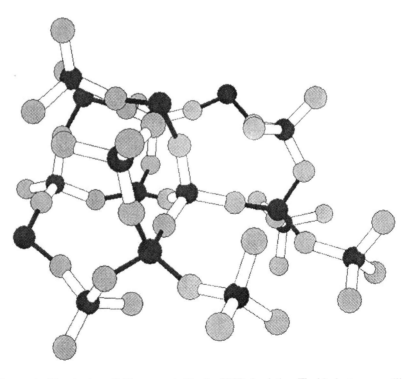

Figure 11 Fragment of the structure of silica generated by the EPSR simulation. The black atoms are silicon atoms and the grey ones are oxygen. The white bonds are those of the SiO_4 molecules used in the simulation, but the black bonds have been generated by the structure refinement. These bonds emphasis the strongly tetrahedral structure of amorphous silica, even though there is no long range order.

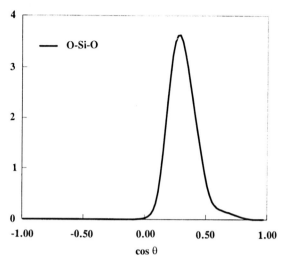

Figure 12 Oxygen-silicon-oxygen angle distribution from the EPSR simulation of amorphous silica. The angle θ is defined such that a linear bond corresponds to θ = 0° (cos θ = 1). Thus the peak near cos θ = 0.333 corresponds to the expected tetrahedral angle. Notice however that it is apparently not precisely symmetric about this value.

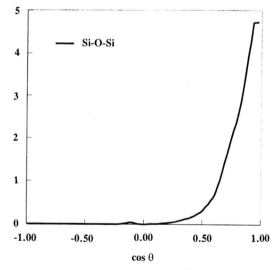

Figure 13 Silicon-oxygen-silicon angle distribution from the EPSR simulation of amorphous silica. The most probable angle is 0°, that is a linear Si-O-Si bond, but the *average* Si-O-Si is about 34°, close to the value that is often quoted in the literature.

distribution function. However it was not quite so accurate at longer distances. Subsequently the empirical potential was switched on and Figure 10 shows the quality of the fit from this procedure.

Comparison of the Fourier transform of this *g(r)* with the original diffraction data showed an excellent fit over most of the Q range. The main discrepancy was that the first diffraction peak was not so sharply reproduced in the EPSR simulation compared to the data. The reason for this is almost certainly that currently EPSR fits exclusively in *r*-space and is affected by the truncation of the empirical potential near half the box length. It would be perfectly feasible to perform the iteration step (19) in the simulation in *Q*-space, and so help to emphasize longer range features in the interatomic potential which need to be simulated.

Figure 11 shows a small fragment of the structure obtained in the EPSR simulation. It can be seen that although the single Si ions were in principle free to move around the simulation box, in fact they mostly ended up in tetrahedral arrangements with the oxygens of neighbouring SiO_4 tetrahedra.

Bond Angle Distributions

Much discussion has occured about the bond angle distributions in amorphous silica. It is generally recognized that the O-Si-O inclusive angle is expected to be close to the tetrahedral value, 109.47°, but the picture is less clear for the Si-O-Si angle, since the *position* of the Si Si peak in the radial distribution function would indicate that this angle is less than 180°. As a result, several computer simulations of amorphous silica [19,20] have introduced somewhat arbitrary angle dependent terms into their potentials. This brings into question the nature of the measured radial distribution function, since it is an *average* over the local structural arrangement. If pronounced tetrahedral units are present in the glass it is not clear that peaks in this distribution function can be used to identify bond angles too precisely

In the present work the bond angle is defined as the *exclusive* angle, θ, between bond directions, so that a value of $\theta = 0°$ corresponds to a linear bond. Thus the bond angle that is frequently displayed in other work is $\Omega=180°-\theta$.

Figure 12 shows the O-Si-O bond angle distribution found in the EPSR simulation, and it can be seen that as expected it peaks near $\cos \theta = 0.333$, ($\Omega=109.47°$) corresponding to the tetrahedral arrangement of O about Si. To some extent this is not surprising since a degree of tetrahedrality was imposed at the outset. Note however that this distribution includes all the free Si ions in the system, so these atoms too are mostly forming into tetrahedra with the neighbouring oxygen atoms.

In direct contrast to this case Figure 13 shows the Si-O-Si angle distribution found in the simulation. Now the distribution peaks strongly at 180°, indicating that the most *probable* arrangement is for there to be *linear* Si-O-Si bonds. Note that this is in spite of the fact that a perfectly acceptable fit to the radial distribution function, including the small Si-Si peak near 3.0Å has been obtained. Therefore there is nothing in the diffraction data to justify the adoption of non-linear Si-O-Si bonds. However as pointed out in the caption of Figure 13 integrating over this distribution indicates that the *average* of this angle corresponds to $\Omega \approx$ 146°, which is close to the value often quoted as being derived from the diffraction data. Hence it is now very clear that simple assignment of peak positions to a particular structural feature is not always valid in a disordered system.

Three Dimensional Structure

The point of the previous section is also emphasized when reconstructing the three-dimensional distributions consistent with the diffraction data. From the simulated distributions of atoms and molecules two sets of spherical harmonic coefficients were calculated. The first set corresponded to the distribution of free Si ions around a SiO_4 tetrahedron at the origin. These were used to make the three dimensional representation of the free ions shown in Figure 14. It can be seen that these Si atoms themselves form, on average, a tetrahedral arrangement around the central SiO_4 tetrahedron, although this is not a rigid unit in the crystallographic sense - quite wide distortions from tetrahedrality will occur. The positions of these Si ions again suggests that the most likely place for one of these ions is opposite the apexes of the central tetrahedron, which also implies a linear Si-O-Si bond.

The second set of coefficients corresponded to the distribution of SiO_4 tetrahedra around a central one at the origin. In this case note that the positional arrangement is important but also the relative orientation of the second molecule with respect to the first. This relative orientation is a much more difficult quantity to visualize, and Figure 15 is an attempt to do this for two particular aspects of the orientation of the second molecule.

The two examples shown correspond to a second SiO_4 tetrahedron lying in the y-z plane of the central tetrahedron, and opposite one of the apexes (oxygen atoms) in this direction, that is $\theta_L = 54.7°$ and $\phi_L = 90°$.

For case (a) the second molecule rotates about its own z axis with the z axes of the central and second molecules parallel, that is by varying ϕ_M. The most pronounced lobes occur at $\phi_M = \pm 90°$: these are also the positions at which the two tetrahedra are at the greatest distance apart, and correspond to one of the lower apexes of molecule 2 pointing towards the upper apex of the central molecule. The lobes which are closer in at $\phi_M = 0°$ correspond to the apex of the central tetrahedron pointing towards a face of the second tetrahedron.

Case (b) corresponds to rotating the z axis of molecule 2 about the laboratory x axis, with averaging taking place over rotations around the molecule's z axis. In this case the tendency for the apex of the central tetrahedron to point towards one of the *edges* of the neighbouring tetrahedron is found.

Discussion

Clearly the orientational maps presented in the previous section represent only a small glimpse into what is a complicated structure. It should be emphasized that the model of the potential used in this simulation, Figure 9, is only one possible representation, chosen for its simplicity and symmetry, and because the SiO_4 tetrahedron is a fundamental building block for this structure. The simplest alternative would have been to treat the system as consisting simply of separate Si and O ions, but in that case it is unlikely that a sensible fit would have been obtained. For example it would likely be found that unphysically short Si-Si distances occur, or the partial charge ordering that is typically observed in other computer simulations of silica [18-20] would not be observed here.

Part of the difficulty about structure refinement in silica is the inability of the diffraction experiment to distinguish the separate Si-Si, Si-O and O-O partial structure factors. To some extent this could be alleviated by employing an X-ray dataset to refine alongside the neutron data. This would provide a different weighting on the three structure factors, but would not of course enable a full separation of the three terms. In any case there will always be systems where the diffraction information is inadequate, however carefully measured it is. Therefore a strong message to be delivered from this paper is the idea that as much additional chemical or physical information as possible should be built into the structure refinement at the outset, and that information should be contested only once it is shown to be incompatible with the data. The example here was that of the Si-O-Si bond angle, which some authorities have maintained should be well defined. In fact as has been seen there is no evidence for this in the diffraction data: the most probable angle is linear, and there is a broad distribution about this value.

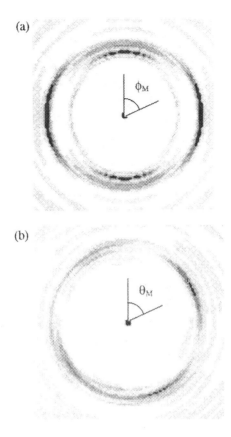

Figure 15. Two reconstructions of the orientational pair correlation function of a second SiO_4 tetrahedron around a central SiO_4 tetrahedron in amorphous silica. Each reconstruction is shown in a box 15Å square. Both correspond to the second molecule being at a position given by $\theta_L = 54.7°$, $\phi_L = 90°$, that is corresponding to the direction of the apex (oxygen atom) of the central tetrahedron which lies in the positive *y-z* plane. In (a) the z-axes of the two tetrahedra are parallel and the map shows the density as a function of rotation ϕ_M of the second tetrahedron about its own z-axis. This case shows that the most probable orientation occurs at $\phi_M = \pm90°$, which corresponds to one of the bottom apexes (oxygen atoms) of the second tetrahedron pointing directly towards that of the central tetrahedron. However values of $\phi_M = 0°$ or $180°$ also occur, indicating that the apex of the central tetrahedron will also point towards one of the *faces* of the neighbouring tetrahedron, presumeably with a Si atom in between to form another tetrahedron. Notice that for the second case the central and second tetrahedra approach each other more closely than the for the first, as seen by the different distances of the dark lobes from the central spot. In (b) the *z*-axis of the second tetrahedron is rotated about its own origin in the *z-y* plane of the central tetrahedron, and the density has been averaged over rotations ϕ_M. Now there are four lobes: those at $\theta_M \approx 60°$ and $-120°$ correspond to the apex of the central tetrahedron pointing towards the *edges* of the second tetrahedron; those at $90°$ to the first two correspond to the apexes of the central tetrahedron pointing roughly towards faces of the second tetrahedron, as in (a).

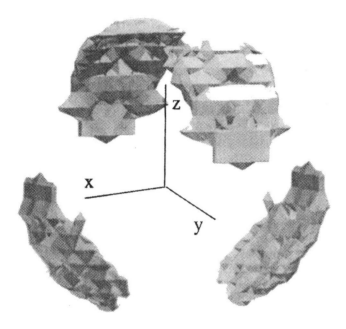

Figure 14. Reconstructed distribution of free silicon atoms about an SiO$_4$ tetrahedron. It is found that the Si atoms congregate near each apex of the tetrahedron. Clearly the Si atoms themselves form a pronounced tetrahedral arrangement about each other. The inset shows the coordinate axes of the central tetrahedron.

One feature of the EPSR technique that has not been exploited so far is the fact that for charge ordered systems, such as the partially charged ions in silica, the contribution to the configurational energy from like-like pairs should be positive, while that from like-unlike pairs should be negative. If it is expected that charges occur in the system then requiring that the site-site interaction energies have the appropriate sign could make an additional constraint on the choice of potential. This idea has yet to be developed.

CONCLUSION

The foregoing material has hopefully given some indication of the enormous amount of information that can be extracted from a diffraction experiment on a disordered material. Conventional computer simulation with assumed forces will go someway to understanding structure and in a sense is almost essential if proper account of the many body problem is to be made. In the end however the simulation data has to be confronted with the diffraction data. The Empirical Potential Structure Refinement technique is a way of testing out the assumed force potential against the data.

Acknowledgements

I am indebted to numerous colleagues for many suggestions regarding the matters discussed in this paper, and particularly to Dave Keen for the suggestion of applying EPSR to the case of amorphous silica, and for supplying some of the relevant literature.

REFERENCES
1. H M Rietveld, *J. Appl. Cryst.*, **2**, 65, (1969)
2. A K Soper, W S Howells and A C Hannon, *ATLAS - Analysis of Time-of-Flight Diffraction Data from Liquid and Amorphous Samples*, Rutherford Appleton Laboratory Report RAL-89-046 (1989)
3. *International Tables for Crystallography*, ed. A. J. C. Wilson, Part C, (Kluwer Academic, 1992)
4. J P Hansen and I R McDonald, *Theory of Simple Liquids*, (Academic Press, London, 1986)
5. A K Soper, *Workshop on Neutron Scattering Data Analysis, March 1990*, ed. M W Johnson, (IOP Conference Series No. 107, 1990)
6. A K Soper, C Andreani, and M Nardone, *Phys. Rev E* **47**, 2598 (1993)
7. J. Skilling, *Workshop on Neutron Scattering Data Analysis, March 1990*, ed. M W Johnson, (IOP Conference Series No. 107, 1990)
8. D.S. Sivia, *Data Analysis: a Bayesian Tutorial*, (Oxford University Press, 1996)
9. C G Gray and K E Gubbins, *Theory of Molecular Liquids Vol I: Fundamentals* (Oxford University Press, New York, 1984)
10. A K Soper, *J Chem Phys*, **101**, 6888-6901 (1994).
11. L Blum and A Toruella, *J Chem Phys*, **56**, 303 (1972)
12. R L McGreevy and L Pusztai, *Mol. Sim.*, **1**, 359 (1988)
13. A K Soper, *Chem. Phys.* **202**, 295-306 (1996); the title of this work was originally called Empirical Potential Monte Carlo, EPMC, but "Monte Carlo" was changed to "Structure Refinement" to emphasise that the primary goal is a structure consistent with the diffraction data, and not a full fledged MC simulation of the material.
14. M P Allen and D J Tildesley, *Computer Simulation of Liquids*, (Oxford, 1987)
15. H J C Berendsen, J R Grigera, T P Straatsma, *J. Phys. Chem.*, **91**, 6269 (1987)
16. A K Soper, F Bruni and M A Ricci, *J Chem Phys.*, **106**, 247 (1997)

17. S Tsuneyuki, M. Tsukuda, H Aoki, and Y. Matsui, *Phys. Rev. Lett.,* **61**, 869 (1988)

18. J D Kubicki and A C Lasaga, *American Mineralogist,* **73**, 941 (1988)

19. P Vashista, R K Kalia, and J P Rino, *Phys. Rev. B*, **41**, 12197 (1990)

20. B Vessal, M Amini, and C R A Catlow, *J Non-Cryst. Solids*, **159**, 184 (1993)

21. D A Keen, *Phase Transitions*, **61**, 109 (1997)

NEUTRON SCATTERING AND MONTE CARLO STUDIES
OF DISORDER IN OXIDES AND HYDRIDES

W. Schweika and M. Pionke *

Institut für Festkörperforschung
Forschungszentrum Jülich
52425 Jülich, Germany

INTRODUCTION

Studies of local order and disorder in materials are essentially dependent on scattering experiments that reveal the more or less structured and typically weak, diffuse intensities that can be observed between the Bragg peaks. In this kind of a tutorial review, we shall discuss in particular the importance and possibilities of neutron scattering experiments. Despite the typical low available flux of neutrons compared, for instance, to the much more intense x-ray radiation from synchrotrons, the selected examples demonstrate the unique and complementary value of thermal neutrons for studying dynamical and structural problems in condensed matter research.

The discussion of the Monte Carlo methods, which enable us to simulate structures and establish realistic Hamiltonians, will be of relevance for the analysis of scattering experiments in general. One important question to answer will be, how well we can really describe structures of real materials from scattering experiments. One has to emphasize that only pair correlations can be obtained from the scattering intensities, at least in the framework of the kinematic theory, which certainly applies to the diffuse scattering of disordered materials. The purpose of the reverse Monte Carlo method is a computer simulation of the structure from experimental information about the pair correlations, while the inverse Monte Carlo method aims at a consistent determination of effective pair interactions. Both methods, their relationship, and their potential will be treated and illustrated for model systems and in applications to interesting materials.

The analysis of the diffuse scattering due to disorder in single crystals is reviewed in another contribution (see Robertson, *ibid.*, and further Refs. [1-3]).

---

*present address: DEBIS, Aachen

SOME METHODOLOGICAL ASPECTS FOR USING NEUTRONS

Scattering methods are the typical tools to investigate the structure of materials. Despite of the increasing share of studies using the intense x-ray radiation from synchrotrons, there are a number of reasons why neutrons remain particularly valuable even for solving structural problems. Here we shall discuss a few examples which are related to the local order in metal hydrides and oxides. Structural problems, where light elements are involved, belong to a typical domain of neutron scattering work because of the favorable scattering properties. In addition, the hydrogen isotopes H and D have very different coherent scattering lengths and also incoherent scattering cross-sections. Often there is a particular interest in the material properties at high temperatures and how these vary under specific conditions, e.g. surrounding atmospheres. The typically low absorption properties of neutrons facilitate such studies which require for a more complicated surface environment and ensures a measure of the bulk properties. Particularly at high temperatures, one can benefit from the ideal energy momentum properties of neutrons, which makes it easy to achieve a sufficient energy resolution to distinguish between elastic and inelastic scattering. Furthermore, the energy analysis provides itself important insight into the dynamic properties of the materials, and also give insight into structural properties. The quasi-elastic broadening due to diffusion or local jumps of atoms reveals the motion itself and also which sites and atoms, regular or interstitial, are involved.

Most of the experiments have been performed at the spectrometer DNS at the Forschungszentrum Jülich, see Fig. 1. With a large set of detectors around the sample, such measurements with energy analysis are done by measuring the neutron's time of flight.

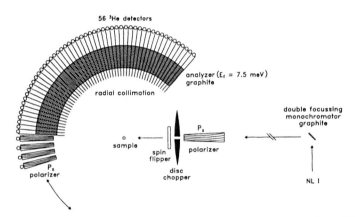

Figure 1. Time-of-flight spectrometer DNS in a schematic representation. The high flux at the expense of modest resolution properties is particular convenient for diffuse scattering problems. Three options can be used (a) the usual time-of-flight mode, (b) alternatively energy analysis by graphite crystals (for $E_f \approx 7.5 \; meV$), (c) polarizers preparing the incident beam and analyzing the scattered neutrons (under development); from Ref. [3].

Compared to triple-axis or four-circle instruments measuring elastic data point by point, the time-of-flight instrument DNS is advantageous particularly when large regions of reciprocal space need to be explored *and* will provide data of higher quality because of the additional energy-dependent information. Various possible influences can be clarified that way, e.g. whether the intensities are related to soft modes and enable one to distinguish and separate multi-phonon contributions which are underneath weak

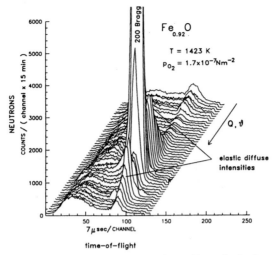

Figure 2. Time-of-flight spectrum as measured for a $Fe_{1-x}O$ single crystal exposed to a reducing atmosphere at high temperatures (raw data). The diffuse elastic intensities, which can be seen at fixed time of flight near channel 90, vary significantly versus scattering vector Q. In addition, there is inelastic (phonon) scattering as well as intense scattering from the 200 Bragg reflection. From Ref. [3].

diffuse elastic signals (like in $Fe_{1-x}O$), see Fig. 2. Vanadium hydride has been measured at the E2 diffractometer at the Hahn Meitner Institut, Berlin. The larger Q-range was more of interest than the energy resolution. In this particularly favorable case, it was not essential to separate the thermal diffuse scattering, which was almost negligibly weak due to the nearly vanishing coherent scattering of the vanadium-host lattice.

The study of water-doped mixed perovskites is a typical neutron application that clarifies the motion of the hydrogen in the bulk material by measuring the quasi-elastic broadening of the diffuse incoherent scattering of the hydrogen.

COMMENTS ON THE REVERSE AND INVERSE MONTE CARLO METHOD

About 25 years ago, attempts have been started to use computer simulations to generate structural models from the measured scattering intensities. One may note the work of Renniger et al.[4] for early simulations of liquids and of Gehlen and Cohen[5] for simulations of short-range order in alloy crystals. Both methods have in common that the models are achieved in an algorithm, which minimizes the (squared) differences between the observed and modeled pair correlations (or scattering intensities). In particular, the method of Gehlen and Cohen has typically been applied in most studies of local order in binary alloys. The interest in such an approach is that it promises a straightforward method instead of proceeding along a way, which might need good intuition when constructing models. One may question, however, whether the structures that are obtained do represent realistic models, and we will discuss in detail such questions further below.

McGreevy and Pusztai[6], for instance, argued that a strict minimization might trap the model and could make the results depended on the chosen initial state of the model. In 1988 they presented their reverse Monte Carlo method for modeling structures based on information from scattering experiments, which made use of a kind of detailed balance condition – as typical for the standard Metropolis Monte Carlo method. The transition probability $w = w(\text{initial} \rightarrow \text{final})$ can be chosen as

$$w = \exp\left[-(\chi_f^2 - \chi_i^2)/2\right] , \qquad (1)$$

with

$$\chi^2 = \sum(\alpha_k^{\text{model}} - \alpha_k^{\text{exp}})^2/\sigma_k^2 ,$$

where σ denotes the standard deviation and α the short-range order parameters, as an example for the pair-correlation function. Then, if $\chi_f^2 < \chi_i^2$ the change is accepted, and if $\chi_f^2 > \chi_i^2$ the change is accepted with probability w. The refinement of the observables α can be made with respect to the correlation functions or to measured intensities. In analogy to the standard MC method this transition probability w replaces the usual Boltzmann weight $w = \exp\left[-(\mathcal{H}_f - \mathcal{H}_i)/k_B T\right]$, where $\mathcal{H}$ is the Hamiltonian describing the configurational energy. This method was originally designed for, and has been frequently applied to, liquids, molecular liquids and glasses, while more recently their program has been applied to disorder in crystals as well. With respect to the above mentioned strict minimization techniques, McGreevy and Pusztai argued that those are comparable to standard Monte Carlo simulations at $T = 0$, where systems typically do not behave ergodic in practice. However, the experimental accuracy not only determines the demands on the accuracy of the correlations in the model but furthermore confines the possible fluctuations. In particular, if a measurement has yielded very precise results for the α_k, the structural simulation by the reverse MC method starts with almost only converging moves being accepted, and the equilibrated structure, which will be found after a long simulation time, could even be frozen in, because any further change causes too large a fluctuation in $\Delta\alpha_k$. In this case the reverse MC method reduces to the criticized algorithm of Renniger et al.[4] or Gehlen and Cohen[5] of only convergent moves. Therefore, different initial states should be tested in the simulations. In case of dense liquids where moves of atoms are correlated for instance, it may become an art of simulation to find appropriate Monte Carlo steps for attempted configurational changes.

Finally, to comment on the analogy of σ_k^2 and T, it is not the temperature but rather the entropy which confines the possible space for fluctuations. Although one should simply recommend the use of the detailed balance principle (and preferred by the author), the suspected minimization approach also yields correct results if it is applied to large models of disordered systems in equilibrium.

Indeed, structural simulations rely on the detailed balance principle and require that the system under study is at thermal equilibrium. If such simulations are based on incomplete knowledge about the correlations in the system, the model will be at least the most probable based on the given information, although it can be systematically wrong. This situation is rather typical, for instance, when partial structure factors of multi-component systems cannot be determined with sufficient contrast from x-ray or neutron scattering experiments. But even for ideal cases, scattering experiments only provide us with information about pair correlations between particles (at least within the first Born approximation) and one may argue that this information may not be sufficient for a unique structure determination. The reader is also referred to the articles of Welberry et al.,[7,8] who have illustrated, at least for two-dimensional structures, with impressive pictures, the caveat that (i) because of the restricted information from scattering experiments alone, the determination of the real structure could be impossible in cases where the structure is governed by many-body correlations, while (ii) even subtleties in the pair-correlation function become apparent in the scattering. This obvious situation in their examples is characterized by completely vanishing pair correlations at all distances, while other many-body terms are present. Before we discuss this aspect further, which seems to invalidate any attempt of structural simulation based on mere pair-correlation functions, it will be useful to discuss the Ising models and atomic interactions which may describe the configurational statistics.

The *inverse Monte Carlo method*,[9] as introduced by Gerold and Kern, is able to determine effective pair interactions which are consistent with the simulated structure. Assuming that the measured and simulated short-range order describes a configuration of thermal equilibrium, one can apply the principle of detailed balance. Essentially one has to establish and to solve numerically a set of l nonlinear equations:

$$\sum_k \Delta p_{kl} w_k = 0 \,, \tag{2}$$

where Δp_{kl} is the change in the number of "bonds" of type l associated with the interaction energy V_l, $w_k(\Delta \mathcal{H}(V))$ is the attendant transition probability, and

$$\Delta \mathcal{H}_k = \sum_l (p_{kl}^{\text{final}} - p_{kl}^{\text{initial}}) V_l \,, \tag{3}$$

is the change of the configurational energy for a particular attempted fluctuation k. Equation (2) only holds for the average over a large number of (virtual) fluctuations. The fluctuations only test the local minimum of the free energy and are not actually performed to keep the system in or near equilibrium.

Alternatively to the method of Gerold and Kern, one may use the fluctuations in the reverse Monte Carlo method (after structural relaxation) to solve the detailed balance for effective pair interactions, or – as proposed by Livet[10] – one may use the standard Monte Carlo method and adjust the interaction model to fit the measured pair-correlation function.

In Eq. (2) we have not specified the Hamiltonian and in principle one could try to solve the system of equations not only for pair interactions but also for many-body interactions. In [11] the question whether this is possible has been analyzed theoretically in high temperature series expansion and has been checked by precise Monte Carlo simulations. One result is that (Ising-like) many-body interactions can be mapped onto effective pair interactions in rapidly converging approximations yielding identical pair correlations as illustrated in Fig. 3 for an example of interactions proposed for the Cu–Au system. Hence the inverse Monte Carlo method will only yield pair interactions in an unambiguous manner. If manybody interactions are included the results for the interactions are strongly correlated. Recently, this ambiguity upon the inclusion of many-body terms has been rediscovered[12] and the usefulness of the inverse Monte Carlo method has been questioned. However, the essential point is that the results for the effective pair interactions – which include not only many-body effects but also other influences like displacements, magnetism etc. – are unique, if the information on the pair correlations is complete. In [11] it was found that the effective pair interactions will depend on composition and temperature, and will increase in range if they have to account for the effects of many-body interactions. When this information is available from experiments, it reversely enables us to determine the possible many-body interactions. Although the effective pair interactions precisely determine the short-range order, they are not appropriate to describe the heat of solution in systems which are dominated by many-body interactions.

The even more striking result of [11] is that the effective pair interactions also describe accurately the specific many-body correlations in the considered example, see

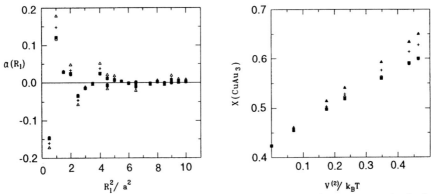

Figure 3. Monte Carlo results for correlations obtained from a many-body interaction model for Cu–Au (*solid squares*) and compared to various approximations of effective pair-interactions (EPI); *Triangles*: bare pair interactions only. *Crosses*: concentration-dependent EPI (exact for $T = \infty$). *Circles*, typically coinciding with *solid squares*: concentration- and temperature-dependent EPI. *Left*: Pair correlations $\alpha(R)$; *right*: Four-body correlations (fraction of Au_3Cu cluster) versus T (from Ref. [11]).

Fig. 3. Because of this we can conclude that even without explicitly determining the effective pair interactions, the true structure is obtained in reverse Monte Carlo simulations if the information about pair correlations is used. There is no proof that this will be valid in general. In the way many-body interactions produce significant pair correlations apart from the related many-body correlations, these will confine also the many-body correlations. Therefore, the examples discussed by Welberry[7,8] of structures with significant many-body correlations but completely vanishing pair correlations are counter examples that demonstrates the impossibility of obtaining the true structure by reverse Monte Carlo methods in these cases. It would be interesting, whether it is possible to generate such examples by a real Hamiltonian and at thermal equilibrium.

To further discuss the possibilities of structural determination from pair correlations, i.e. the essential information available from scattering experiments, we consider another example in which a triangular lattice is randomly decorated by equi-triangular clusters. The Hamiltonian is easiest described by a three-body term that has the character of molecular bonding and is non-Ising-like which is different to the situation which is analyzed in Ref. [11]. The pair correlations are very simple. The first short-range order parameter is equal to 1/3 independent of both temperature and the coverage of the lattice by these clusters. With only this first short-range order parameter, the simulated structures have barely any resemblance to the expected triangular objects. The reason is that, of course, one has to make sure that the simulated structures are also consistent with the demand of vanishing pair correlations at further distances. Inclusion of up to 12 short-range order parameters in the simulation yields a significant increase for clusters of size $3n$ formed by the basic three-atomic cluster units.

The computational effort, at least for Monte Carlo steps based on pair-wise exchanges of atoms, increases dramatically with increasing range considered, and therefore it could become impractical to obtain the perfectly simulated structure, although the information given by the pair correlations is in principle unique and sufficient. With increasing coverage of the lattice by clusters, the formation rules on the structure become less apparent, and not surprisingly, the pair correlations provide a fading significance for the inherent cluster motif. For a further discussion see Ref. [3]. The examples suggest that the reverse Monte Carlo method (RMC) in principle leads to the true structure,

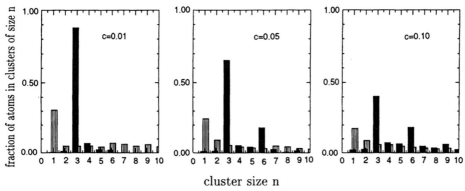

Figure 4. Cluster distributions for RMC simulated structures on a triangular lattice with only nearest neighbor pair-correlations $\alpha_1 = 1/3$. The corresponding true and ideal solution of the structure would be a random distribution of triangular clusters and therefore, only clusters of the sizes 3, 6, 9, etc should appear. From top to bottom the coverage increases from 0.01 to 0.05 and 0.1. The comparison shows that information about multi-site correlations is to be found in the full extent of the pair-correlations (even if these are zero): First (*hatched bars*), only α_1 is simulated, second (*filled bars*), further correlations, $\alpha_l = 0$, *for* $l > 1, ..., 12$, are included. See also Ref. [3].

if the underlying many-body interaction imposes any effect on the pair correlations, although in practice the confinement of a structure relies quantitatively on the error bars of both the experimental results and the simulation.

APPLICATIONS

Defect Structure of Wüstite

The ferrous oxide FeO (wüstite) has a peculiar range of stability. It is supposed to be stable only at high temperatures, and because of the mixed valence of the iron cations (2+,3+) there is an unavoidably large deviation from the ideal stoichiometry. At high temperatures and under appropriate oxygen partial pressures the homogeneous phase, having the NaCl structure, exhibits an iron deficit of $x = 0.05$ up to $x \approx 0.17$. Although wüstite is not the ordinary reddish-brown rust, the investigation of the atomistic micro-structure is of interest for the understanding of the processes of oxidation and corrosion, e.g. morphology changes such as that from metals with smooth surfaces to possibly porous ceramics. Here we meet a situation, where, as already known, cations and cation vacancies constitute the mobile defects. Upon oxidation the iron atoms have to migrate from the metal through the oxide to the surface, where oxygen deposits from the gaseous phase. A more subtle problem is that in those cases where the fraction of defects is not small, as is typical and unavoidable in $Fe_{1-x}O$, correlations among the defects may have an important influence on the ionic transport properties. In fact, there is a very unusual and only minor change of the cation diffusion coefficient with increasing deviation from stoichiometry, i.e. with increasing number of cation vacancies[13]. This example reveals the strong correlations and interactions among the charged defects in this ionic material which are responsible which provides an explanation for its unusual transport properties.

There exists a vast literature proposing various kinds of specific defect arrangements, typically based on measurements of the Bragg intensities[14] but also on theoretical calculations[15-17] of the cluster formation energies. The measurements showed that the oxygen ions are not involved in any disorder of site occupations, however, the number of the iron vacancies is higher than expected from the composition due to additional iron interstitials. The proposed defect clusters have typically one structural element in common, at least in modified forms, i.e. the so-called 4:1 defect cluster, which means that all four regular cation sites around a 3+ tetrahedral cation interstitial are vacant resulting in a net charge of 5− for this unit. Some efforts have been made to investigate the diffuse scattering of this material by x-rays.[18,19] Studies on quenched samples, however, cannot represent equilibrium configurations of the high temperature states because of the high cation mobility; at high temperatures, around 1000°C, the x-ray measurements suffer from a high background of thermal diffuse scattering.

Neutrons are favorable in such kind of studies, because the phonon contributions can be easily separated from the elastic scattering (see the time-of-flight spectra in Fig. 2). In our neutron scattering investigation[20] we used a special furnace, with CO/CO_2 gas mixtures flowing around the sample chamber, to investigate the single crystal specimen at various temperatures and compositions exploring the wüstite phase field. The diffuse elastic scattering intensities measured at 1150°C are shown in Fig. 5.

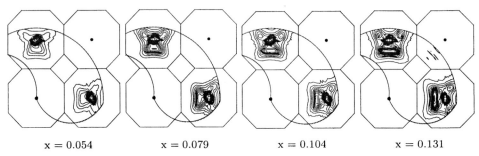

| x = 0.054 | x = 0.079 | x = 0.104 | x = 0.131 |

Figure 5. Observed elastic neutron scattering intensities from an $Fe_{1-x}O$ single crystal at $T = 1150°C$ in the (100) plane (contours are given in steps of 0.1 b/sr). The bcc Brillouin zones of the fcc cation sublattice are indicated in the figure. Vacancy-interstitial correlations (doubling of the reciprocal cell) and displacement fields (anti-symmetries) are required to explain the missing symmetry properties of the Brillouin zones for the diffuse scattering. See also Ref. [20].

The Brillouin zones of the face-centered cubic cation lattice are displayed in the figure to facilitate an understanding of the patterns. The apparent lack of translational symmetry is caused first by additional cation interstitial–vacancy correlations (doubling of the reciprocal cell) and second by strong scattering contributions due to displacements (giving rise to anti-symmetries). Here, we will not discuss the interesting displacement scattering in detail (see Refs. [3,20]), but a few points will be made further below. Of importance for the occupational short-range order is the possibility by Fourier analysis to separate those contributions from those of the displacements. The results, e.g. for $x = 0.079$ shown in Fig. 6 can be distinguished with respect to the different types of correlations. For comparisons the physical boundaries (resulting from the actual composition of vacancies and interstitials) for the short-range order parameters are also displayed as dotted lines in figure. One can see, for instance that there are only negligible correlations among the interstitials, however, the interstitials are very likely surrounded by vacancies, and vacancies tend to form neighbored pairs.

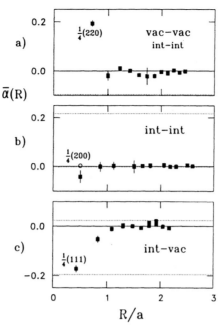

Figure 6. Short range order coefficients $\bar{a}_l$ for $Fe_{0.92}O$ at $T = 1150°C$: result of the Fourier analysis (*filled squares*) and simulation (*open circles*, only one can be seen separately and all others coincide). Dotted lines denote lower and upper bounds. (a) Correlations on the fcc sublattices; these are dominated by those among vacancies, while a minor contribution comes from those among interstitials at the same distances. For example, vacancy neighbors are favored; (b) the part of correlations on the sc sublattice which is purely among interstitials, showing only random correlations; (c) interstitial to vacancy correlations: e.g., a very strong attraction for interstitial vacancy neighbors is found; from [20].

The defect structure has then been simulated in a computer model of of $Fe_{0.921}O$. All measured short-range order parameters were used for this purpose, applying a reverse Monte Carlo algorithm and a large model containing 153 600 lattice sites. In order to analyze the modeled defect structure, we searched for all clusters that are formed by interstitial–vacancy bonds. The probability distribution for these interstitial–vacancy clusters as a function of the sizes is shown in Fig. 7. One of the remarkable features of the cluster distribution is that first, a large fraction of the defects, about 40%, are still unbound vacancies ($m=1$) which will play the dominant role for the transport properties and explains the high cation mobility. Secondly, 4:1 clusters are particularly stable (note the log-scale), consistent with previous theoretical predictions[15–17]. Thirdly, larger vacancy–interstitial aggregates, incorporating the 4:1 motif, are also present. Their existence appears to be, however, mainly a consequence of the high defect content. One may further note details like the obvious decrease in the probability to find the next larger clusters beyond the stable 4:1 cluster. From the negative value of short-range order parameter for next-nearest neighbored interstitial–vacancy pairs alone, one would conclude an opposite trend. There is little or no tendency to form large clusters with well-defined compact shapes. Evidence of this is found from the observation that there is only very low intensity at small scattering vectors.

The picture we obtained, namely that at high temperatures the equilibrium defect structure is characterized by a broad but significantly structured distribution of possible defect arrangements rather than only one specific defect type, is only reasonable and a remarkable achievement that proves the strength of combining diffuse scattering experiments with Monte Carlo simulations of structures.

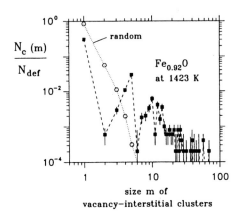

Figure 7. Size distribution of vacancy-interstitial defect clusters (*filled squares*) in $Fe_{0.92}O$ at $T = 1150°C$ in comparison with the random case (*open circles*). The size m equals to the total number of vacancies and interstitials in a single isolated cluster, which is interconnected by nearest interstitial to vacancy bonds. $N_c(m)$ is equal to the number of specific clusters found in the model crystal; N_{def} equals the total number of vacancies and interstitials. E.g., more than 30% of the defects are isolated point defects, – these are essentially Fe-vacancies–, the (isolated) vacancy-interstitial dumbbell is very unlikely. Quite significantly about 15% of the defects are bound in 4:1 vacancy-interstitial clusters, and larger aggregates appear as well. (Lines are only guides for the eye; from [20]).)

Finally, a few remarks should be made on the observed displacement scattering (for detailed discussions see Refs. [3, 20]). The displacements can be analyzed as well and show a displacement field around a cation vacancy that repels the nearest neighbor anions and attracts the second nearest neighbor cations. The displacement fields can be described in Kanzaki force models, and the diffuse intensity can be calculated by using the superposition of the displacement fields for 4:1 defects and the dynamical matrix. This describes nicely the observed pattern, even the change with composition, which manifest particularly in the formation of the diffuse prepeak near to the 200 Bragg-reflexion, *if* one includes in the model a screening of the assumed 4:1 cluster by surrounding this cluster by regular cations for charge compensation in the immediate neighborhood.

Effective Pair Interactions in Vanadium Hydride

Metal hydrides have attracted both scientific and technical interest because of the unusually high hydrogen mobility and the high solubility of hydrogen. Although there are more suitable metal and alloy hydrides, vanadium hydride may be viewed as a model system for a compact solid hydrogen storage medium and technical applications of energy storage. At equilibrium the concentration of hydrogen in vanadium is determined by the temperature and the partial pressure of the surrounding hydrogen gas. With an increase in the H_2 pressure the hydrogen concentration in the metal hydride saturates, all available sites have been filled and almost as much hydrogen has been absorbed as there are metal atoms. As derived from Bragg diffraction data, most of the hydrogen has dissolved more or less randomly on the tetrahedral sublattice. Since this sublattice offers about six times more sites than those used, there are apparently strong correlations and repulsive interactions between the hydrogen which limit the storage capability of the material.

Our studies aimed at a quantitative determination of the effective pair interactions among the hydrogen atoms. At short distances, strong repulsive interactions are expected because of the limited hydrogen solubility, while at larger distances elastic, long-range interactions should dominate the correlations. From an experimental point of view it is a favorable model system, since the vanadium host was almost invisible to the neutrons (its coherent scattering length is negligible small, $b_V = -0.0382 \times 10^{-12}\,cm$). Therefore, the three-dimensional scattering experiments were performed in diffraction mode using the flat-cone diffractometer E2 at the Hahn–Meitner Institute, Berlin. Corrections due to the small inelastic scattering were done carefully. Deuterium, rather than the natural hydrogen, was chosen to reduce the high (spin-)incoherent scattering background and to increase to coherent scattering signal.

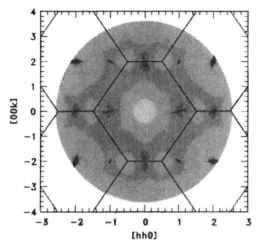

Figure 8. Observed scattering of $VD_{0.781}$ in the (110) plane of the reciprocal lattice at $T = 300$ K. The diffuse scattering in between the Bragg peaks is mainly due to short range order among the deuterium interstitials. The translational symmetry properties for the tetrahedral sublattice are indicated in the figure (from [21]).

Figure 8 shows the diffuse scattering of $VD_{0.781}$ in the (110) plane at room temperature. Diffuse maxima are found near the 200 and 110 reciprocal lattice positions and equivalent positions which coincide with the Bragg peaks of the host lattice. The coherent diffuse scattering due to the deuterium short-range order is very low near the zone centers, which reflects the low compressibility of the deuterium "lattice liquid" at this high deuterium concentration. The translational symmetry of the tetrahedral sublattice, as shown in the figure, agrees well with those of the diffuse scattering. Data were measured in a huge volume of reciprocal space covering in the order of 10^5 data points.

The data were analyzed using a Fourier analysis with respect to short-range order and displacement parameters, based on least-squares methods. Figure 9 shows a strong blocking tendency on the tetrahedral interstitial sublattice for the first three neighboring shells of deuterium. The number of short-range order parameters that have been determined, is as large as 114; the error bars are very small, except for the near neighbors, and do not exceed the symbol size unless it is shown in the figure. These results are also stable and only little modified if the additional contributions from the octahedral sublattice are also taken into account in a more complex model. Actually, about 10% of the deuterium atoms were observed to be distributed on the octahedral sites.

The inverse Monte Carlo simulations were used to determine the effective pair in-

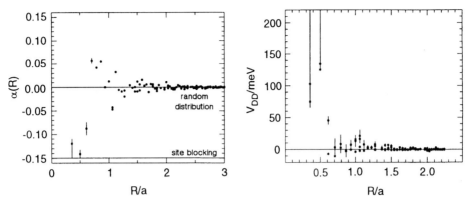

Figure 9. $VD_0.78$ at room temperature. A strong mutual repulsion is found for D on the first three neighbor shells. *Left*: short range order parameters obtained from diffuse neutron scattering. *Right*: effective pair interactions as determined by the inverse Monte Carlo method (*filled squares*) and compared with continuum-mechanical calculations of the elastic interaction (*asteriks*) (from [21]).

teractions from the short-range order parameters. One may note that the mean-field approach using the Krivoglaz–Clapp–Moss formula to determine the D-D interaction energies would be completely insufficient, since for the near-neighbor pairs V_{D-D} is not at all small compared with $k_B T$ but the reverse is true. In addition, there is a breakdown of the usual mean-field prediction that the wave vector of the short-range order peak corresponds to the minimum of the interaction potential.

Our results obtained by the inverse Monte Carlo method from the diffuse scattering data (see Fig. 9) typically show that all the interactions determined from the experiment have a positive sign. This could be attributed to the repulsive part of the interaction due to the (screened) Coulomb interaction between the protons and shows that the Coulomb term is dominant within the whole displayed range of interactions. At short distances the repulsive screened Coulomb interaction between the protons dominates the elastic contributions.

On the other hand, the elastic interactions displayed for comparison in Fig. 9 are typically below the inverse MC result, but they are of comparable magnitude and the variations with R of both data sets reveal much similarity. The elastic interactions were calculated in a continuum-mechanical approach and took account of the dipole–dipole interaction. One has to distinguish the possible relative orientations of the dipoles, for instance in $\langle 110 \rangle$ directions the dipoles are oriented alternately perpendicular and parallel, and further the projections of the dipoles on $\vec{R}$, which give for instance two different results for pairs at distance $\vec{R} = (1, 0, 0)\, a$. However, the surprising agreement of the continuum-mechanical calculations with the experimental result at very close distances seems to be more accidental than really meaningful. The calculations may give a good approximation for the observed values at larger distances.

Hydrogen Motion in Mixed Perovskites

Perovskites of the type ABO_3, such as cerates or zirconates can be ion conducting if they are doped, substituting the tetravalent B ion by a threevalent one, with a concentration of several mol%. The extra charge creates oxygen vacancies which lead to oxygen conduction. Also mixed non-stoichiometric perovskites of the type $A(B'_{(1+x)/3}B''_{(2-x)/3})O_{3-\delta}$ with two different (e.g. divalent and pentavalent) B ions, reveal oxygen vacancies and conduction. Interestingly, some of these perovskites exhibit also significant protonic conductivity, for example if the material is exposed to moist air[22]

The compound $Ba(Ca_{1/3+x} Nb_{2/3-x})O_{3-\delta}$ (BCNO) has attracted special interest particularly in view of possible applications for solid fuel cells. From measurements of the electrical conductivity a low activation energy of $0.54 eV$ was obtained, though these measurements might have some difficulties to separate the true bulk properties. The hydrogen solubility is comparatively good which was determined at lower temperatures by thermogravimetry[23] with values up to 19.4 mol%.

By quasi-elastic neutron scattering experiments it is possible to investigate the hydrogen motion in this compound on an atomistic scale of several Å in a regime of characteristic times between 10^{-8} and 10^{-13} s, namely by measurements of the incoherent neutron scattering intensity as a function of momentum transfer $\hbar\vec{Q}$ and energy transfer $\hbar\omega$; typical energy transfers in the region of 10^{-7} to 10^{-6} eV are measured by backscattering spectroscopy, and 10^{-5} to 10^{-3} eV with the time-of-flight methods. Of course, neutron experiments clearly exhibit the bulk properties and due to the sensitivity to hydrogen, the scattering provides also precise estimates of the hydrogen content itself at pressures not readily accesible to thermogravimetry.

During the neutron scattering experiments over a wide range of temperatures the samples were kept in Al containers and exposed to a controlled water vapor atmosphere with pressures up to 600 mbar. Difference measurements of wet and dried samples (drying time ≥ 5 hours at $500°C$) were performed to subtract the background of the host material and yield essentially only the hydrogen scattering contribution. One may note that 2 mol% H was found to remain tightly bound in the samples as seen also by measurements using the (n,γ)-capture method[24]. From the corrected and calibrated intensity we could determine the hydrogen concentrations $c_H(T)$ (between 10 and 18 mol%) within an accuracy of less than 1%, which agreed with the content extrapolated from thermogravimetry.

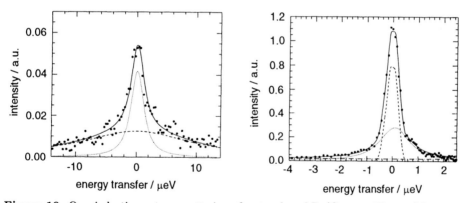

Figure 10. Quasi-elastic neutron scattering of water-doped $Ba(Ca_{1/3+x} Nb_{2/3-x})O_{3-\delta}$ at 664 K shows three characteristic time scales for H-motion. Two slower jump steps are observed (*left*) in high resolution spectroscopy with the backscattering spectrometer BSS1 and (*right*) a fast local motion with the time-of-flight spectrometer DNS; from [27].

Figure 10 shows the measured quasi-elastic broadening due to to the hydrogen motion. Compared to the instrumental resolution of $1 - 2$ μeV for the back scattering spectrometer (BSS1) the left figure shows that two Lorentzians are required to fit the observed intensity. One may suggest a two-state trapping model[25] to describe the diffusion process. This model, however, which has been successfully applied to proton conducting $SrCeO_3$ with dilute Yb doping,[26] failed in the present case completely. Most likely there are two groups of lifetimes related to the randomness of the potential in the lattice of the mixed perovskite.

At higher temperatures the broader component and the faster process dominates. One may note that there are no significant true elastic contributions due to any slower hydrogen motions or any parasitic background. It has been found that at lower Q-values only one Lorentzian was significant and its line width $\Gamma(Q)$ has shown a usual Q-dependence for unbound diffusion $\Gamma(Q) = DQ^2$. The diffusion constants D follow an Arrhenius behavior with $D_0 \approx 0.7 \cdot 10^{-3}$ cm^{2}s^{-1} and an activation energy $E_a = (0.39 \pm 0.05)$ eV. These results agree fairly with the ionic conductivity using the Nernst-Einstein equation (for a detailed discussion see Ref. [27]).

The right side of Fig. 10 displays the much broader quasi-elastic component as measured by time-of-flight spectroscopy. Increasing with temperature the width is about 700 μeV at 400°C and this motion corresponds to a characteristic time of $\tau_r = 10^{-12}$ s, which is a motion much faster than expected from the protonic diffusive step. From the temperature dependence we estimated an activation energy of 0.10 eV. In addition, it has been found that the width of this broad component is independent of Q, which confirms that this motion of the hydrogen is localized. Due to this quasi-elastic contribution, there is a steep decay of the apparent elastic intensity with increasing Q (see Fig. 11) that cannot be explained by the usual Debye-Waller factor due to lattice vibrations.

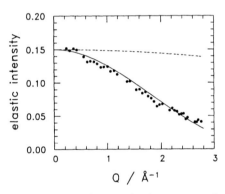

Figure 11. *Filled circles*: measured incoherent elastic structure factor of the time-of-flight spectra versus Q. *Line*: model calculation for a proton motion on a sphere with $R = 0.7$ Å^{-1}. *Dashed line*: Debye-Waller factor with a typical mean square displacement of $\langle u^2 \rangle \approx 0.01$ Å^2; from [27].

The simple model displayed in Fig. 11 results from a rotational motion on a sphere with a radius of 0.7 Å; the remaining small systematic discrepancies indicate that a slightly larger radius is likely for a more refined model, which could account for a preferred OH$^-$ reorientations in elementary steps of 60° and 90° and may be coupling to phonon modes. The results, in summary, are consistent with the Grotthus mechanism (see Ref. [28]). In this picture, the protons are transferred along the OH--O hydrogen bridges with a rate of about $1/\tau \approx 10^{10}$ s^{-1} at 700 K. Combined with this transfer, which is characterized by the observed activation energy of 0.4 eV, there is the rapid rotation of the OH^- ion around its center, orienting the hydroxyl ion towards different neighbors.

FINAL REMARKS

The combination of diffuse neutron scattering experiments and Monte Carlo simulations can provide us with very detailed structural information as illustrated in the study of the defect structure in wüstite, such as many-body correlations that describe the formation of defect clusters that appear to be "spinel embryos" of the neighbored phase Fe_3O_4. By analyzing the modeled structure it was possible to understand the origin of the peculiar transport properties, i.e. the behavior of ionic conductivity as a function of temperature and defect content.

The effective pair interactions in vanadium hydride have been successfully determined, demonstrating the capabilities of the inverse Monte Carlo method. One may add, that the inverse Monte Carlo method has also been applied to liquid systems (see Ref. [29] and Soper, *ibid.*). The evidence for possible many-body interaction terms should be found in the specific temperature and concentration (density) variation of pair correlations and of the effective pair interactions. It will be interesting and challenging to examine materials under this perspective.

The dynamical properties like phonon modes and, as seen in the last example, relaxational, diffusive steps that lead to quasi-elastic broadening, are a typical subject of neutron scattering investigations. Such experiments provide a key to the dynamics *and* the structural properties, which in the present case, for instance, supports the Grotthus mechanism for the hydrogen motion in water-doped perovskites.

One may finally note that besides all of the efforts that are needed, the required beam-time has been comparatively small, each study of the considered examples only took a couple of days.

REFERENCES

1. L. Schwartz and J.B. Cohen, *Diffraction from Materials*, Academic Press, New York (1977).
2. W. Schweika, in: *Statics and Dynamics of Alloy Phase Transformations*, ed. Turchi P.E.A. and Gonis A., Plenum Press, New York (1994).
3. W. Schweika, *Disordered Alloys – Diffuse Scattering and Monte Carlo Simulation*, ed. G. Höhler, Springer Tracts in Modern Physics Vol. 141, Springer, Heidelberg (1997).
4. A.L. Renninger, M.D. Rechtin, and B.L. Averbach, *J. Non-Cryst. Solids* 16:1 (1974).
5. P.C. Gehlen and J.B. Cohen , *Phys. Rev.* 139:844 (1965).
6. R.L. McGreevy and L. Pusztai, *Mol. Sim.* 1:359 (1988).
7. T.R. Welberry and R.L. Withers, *J. Appl. Cryst.* 24:18 (1991).
8. T.R. Welberry and B.D. Butler, *J. Appl. Cryst.* 27:205 (1994).
9. V. Gerold and J. Kern, *Acta Metall.* 35:393 (1987).
10. F. Livet, *Acta Metall.* 35:2915 (1987).
11. W. Schweika and A.E. Carlsson, *Phys. Rev. B* 40:4990 (1989).
12. C. Wolverton, A. Zunger, and B. Schönfeld, *Solid Sate Commun.* 101:519 (1997).
13. W.K. Chen and N. Peterson, *J. Phys. Chem. Solids* 36:1097 (1975).
14. M. Radler, J.B. Cohen, and J. Faber Jr., *J. Phys. Chem. Solids* 51:217 (1990).
15. C.R.A. Catlow and A.M. Stoneham, *J. Am. Ceram. Soc.* 64:234 (1981).
16. R.W. Grimes, A.B. Anderson, and A.H. Heuer, *J. Am. Ceram. Soc.* 69:619 (1986).
17. M.R. Press and D.E. Ellis, *Phys. Rev. B* 35:4438 (1987).
18. F. Koch F. and J.B. Cohen, Acta Cryst. B 25:275 (1969).
19. E. Gartstein, T.O. Mason, and J.B. Cohen, *J. Phys. Chem. Solids* 47:759 (1986).
20. W. Schweika, A. Hoser, M. Martin and A.E. Carlsson, *Phys. Rev. B* 51:15771 (1995).

21. M. Pionke, W. Schweika, T. Springer, R. Sonntag, and D. Hohlwein, *Physica Scripta* T57:107 (1995).

22. T. Norby, in: *Selected Topics in High Temperature Chemistry - Defect Chemistry of Solids*, ed. Øivind Johannesen and Arnfinn G. Andersen, Studies in Inorganic Chemistry, Vol. 9:101, Elsevier, Amsterdam (1989).

23. F. Krug and T. Schober, *Solid State Ionics* 92:297 (1996).

24. F. Krug, T. Schober, R. Paul, and T. Springer, *Solid State Ionics* 77:185 (1995).

25. D. Richter and T. Springer, *Phys. Rev.* B 18:126 (1978).

26. Ch. Karmonik, R. Hempelmann, Th. Matzke, and T. Springer, *Z. Naturforsch.* 50a: 539 (1995).

27. M. Pionke, T. Mono, W. Schweika, T. Springer, and H. Schober, *Solid State Ionics* 97:497 (1997).

28. K.D. Kreuer, *Chem. Mater.* 8:610 (1996).

29. M. Ostheimer and H. Bertagnolli, *Mol. Sim.* 3:227 (1989).

REVERSE MONTE CARLO REFINEMENT OF DISORDERED SILICA PHASES

David A. Keen

ISIS Facility
Rutherford Appleton Laboratory
Chilton, Didcot
Oxon OX11 0QX
U. K.

INTRODUCTION

This chapter will provide a detailed description of the reverse Monte Carlo (RMC) modelling method. RMC has been established over the last ten years as a general method for obtaining models of a wide variety of disordered structures, including liquids, glasses, crystals, polymers and amorphous magnets. Particular emphasis will be given here on how it may be used to refine the disorder in glassy and crystalline materials. In addition to explaining the principles and rationale of the RMC method, results from modelling various polymorphs of silica will be used to demonstrate how the method works in practice. For the crystalline systems, the RMC models will be compared to diffraction patterns from powdered samples, since RMC modelling of single crystal diffraction is covered in the chapter by V. M. Nield in this volume.

The outline of this chapter is as follows. The next section introduces various background formulae which underpin the ideas of total scattering. Then reverse Monte Carlo modelling is introduced and described in detail. This is followed by a section which works through the application of RMC *modelling* to the structure of glassy silica and the results are compared with other existing models. The ideas of RMC *refinement* are then described and applied to glassy silica. Preliminary results from RMC refinement of disordered crystalline phases of silica are then described. Finally, conclusions are made with a discussion of future possibilities for the RMC method.

THEORETICAL FORMALISM

Before describing the modelling method in detail, it is necessary to define various relevant correlation functions. The data used to constrain the RMC models described in this chapter are from 'total scattering' measurements. The ideal total scattering

Local Structure from Diffraction
Edited by S.J.L Billinge and M.F. Thorpe, Plenum Press, New York, 1998

measurement would collect scattering over all momentum transfers $0<|Q|<\infty$, integrating at each $|Q|$ over all possible enegy transfers which may take place within the sample. Most good neutron and X-ray diffraction measurements are a reasonable approximation to this ideal, albeit with a reduced range of $|Q|$. Total scattering therefore contains both Bragg and diffuse components from a crystalline material and, as will be shown later, it is vital to include the diffuse scattering when considering structural disorder. The total structure factor, $F(Q)$, obtained from careful correction of neutron total scattering data[1] from an n-component system, can be defined in terms of Faber-Ziman partial structure factors $A_{\alpha\beta}(Q)$.[2]

$$F(Q) = \sum_{\alpha=1}^{n} \sum_{\beta=1}^{n} c_\alpha c_\beta \bar{b}_\alpha \bar{b}_\beta [A_{\alpha\beta}(Q) - 1] \qquad (1)$$

where $\bar{b}_\alpha$ and c_α is the coherent neutron scattering length and proportion of atom α respectively. The formula still holds for X-ray scattering if the neutron scattering lengths are replaced by X-ray form factors. These partial structure factors are the sine Fourier transforms of the partial radial distribution functions, $g_{\alpha\beta}(r)$

$$g_{\alpha\beta}(r) - 1 = \frac{1}{(2\pi)^3 \rho} \int_0^\infty 4\pi Q^2 [A_{\alpha\beta}(Q) - 1] \frac{\sin(Qr)}{Qr} dQ \qquad (2)$$

where ρ is the number density of atoms. $g_{\alpha\beta}(r)$ are defined explicity as

$$g_{\alpha\beta}(r) = \frac{n_{\alpha\beta}(r)}{4\pi r^2 dr \rho_\beta} \qquad (3)$$

where $n_{\alpha\beta}(r)$ is the number of particles of type β between distances r and $r+dr$ from a particle of type α, averaged over all particles α. ρ_β is the number density of atom type β. It is also useful to define a total radial distribution function $G'(r)$ in terms of $g_{\alpha\beta}(r)$

$$G'(r) = \sum_{\alpha=1}^{n} \sum_{\beta=1}^{n} c_\alpha c_\beta \bar{b}_\alpha \bar{b}_\beta [g_{\alpha\beta}(r) - 1] \qquad (4)$$

such that the total structure factor $F(Q)$ is the sine Fourier transform of $G'(r)$:

$$F(Q) = \rho \int_0^\infty 4\pi r^2 G'(r) \frac{\sin(Qr)}{Qr} dQ \qquad (5)$$

Frequently $G(r)$ is used, which is a normalised total radial distribution function

$$G(r) - 1 = G'(r) / \sum_{\alpha=1}^{n} c_\alpha \bar{b}_\alpha \qquad (6)$$

such that $G(r\rightarrow\infty)=1$ and $G(r<r_0)=0$, where r_0 is the smallest distance two atoms may approach each other. $T(r)$, another popular form of the real space correlation function, is then proportional to $rG(r)$.

$n_{\alpha\beta}(r)$ may be calculated simply from a three-dimensional configuration of atom positions and from these functions all the above functions may be derived. Similarly $F(Q)$, the coherent part of the experimental total scattering can be used either directly, or Fourier transformed to $G(r)$ for comparison with the equivalent functions from the computer generated models.

REVERSE MONTE CARLO MODELLING

Background

The reverse Monte Carlo modelling technique was developed to create three-dimensional structural models of liquid and amorphous (or glassy) materials *without bias*. Liquids and glasses are, in general, macroscopically isotropic and in a diffraction measurement the material will scatter isotropically, varying only as a function of the modulus of the momentum transfer, $|Q|$. Such a one-dimensional scattering function (the structure factor $F(Q)$ - see Equation 1) can be Fourier transformed to provide information about atom-atom distances, but vector information is lost.

Modelling of some form or another was one obvious way to attempt a reconstruction of the three-dimensional local structure of glasses and liquids from the one-dimensional scattering information. The earliest attempts along these lines were by Kaplow *et al* [3] in 1968 (investigating vitreous selenium) and Renninger *et al* [4] in 1974 (on arsenic-selenium glasses). Both these works used small spherical models and further progress was hampered by the limitations in computer power. Subsequently the concepts were developed successfully by McGreevy and Pusztai[5] who modelled the structure of liquid argon in 1988, coined the term 'reverse Monte Carlo modelling' and established the basic method which is used extensively today[6]. McGreevy and Pusztai argued[5] that the RMC method's strength lay in the fact that no interatomic potentials were imposed on the structural model and hence the technique was very different from the more familiar Monte Carlo simulation which uses a set of potentials to constrain the model. In RMC modelling, the model is only required to agree with the structural data. It was believed that, in principle, the resultant RMC models could then be used to determine the interatomic potentials which governed the structure. Instead of working from the potentials to the structure to the structural data, the structural data were used to determine the structure and hence the potentials. The Monte Carlo cycle was operated in 'reverse'. However, because the available structural data do not necessarily describe a unique structure (see later) and three-body terms (which are not contained in the structure factor $F(Q)$) are equally important in determining structure, in practice it is very difficult to use RMC to determine appropriate potential functions.

The RMC approach has many advantages over other modelling techniques, principally because it does not bias the resulting model. If the data do not require a specific structural feature, then the RMC model is unlikely to show such a feature. This has to some extent changed the way that disordered structures, and liquid and amorphous structures in particular, are considered. The local structure of a glass is usually thought of in terms of small structural units which are found in chemically similar crystalline materials. Models are then built by joining these semi-rigid structural units, either explicitly or implicitly from the description of interatomic potential functions. This can result in models which are too ordered and do not have sufficient flexibility to include other possible structural motifs. In contrast, no such features are assumed in RMC models and often exactly the opposite type of structural models are obtained with too much disorder. As will be shown later, this is because the data do not provide sufficient structural information to define uniquely the structure and a disordered structure, although consistent with the data, is more likely to result from random Monte Carlo moves. For many systems neither extreme is entirely satisfactory and a middle ground must be established. As a result, although RMC is a very powerful method, it must be controlled in appropriate ways to yield good, respresentative structural models. Such control can be achieved by using a constrained RMC *refinement* of a carefully constructed structural model which already contains pertinent structural features. Thus the best features of RMC modelling (lack of bias, structural flexibility, consistency with experimental data) are combined with an initial model containing structural elements which RMC modelling may not be good at reproducing, but are nonetheless indisputable. This is clearly common sense when considering disordered crystals, since the long-range periodicity and average crystal structure must be maintained throughout the RMC modelling of the short-range structural disorder. This is analogous to the process of Rietveld refinement[7], where an initial structural model is refined by comparison with a powder diffraction pattern.

The RMC Modelling Method

1. Generate a three-dimensional configuration of N atoms, which is constrained by periodic boundary conditions. The configuration is frequently a cube with lengths L for convenience, but can have different geometries to suit (for example) different crystal symmetries. The practical effect of periodic boundary conditions is that when an atom is moved beyond one side of the configuration box, it moves back into the box at the opposite side. All atoms within the configuration must also satisfy 'closest approach' constraints such that two atom types may only come within a certain distance of each other. The closest approaches may, with ideal data, be determined uniquely from the partial radial distribution functions, but they are more likely to be deduced from a combination of factors, and can be adjusted from examination of how the modelling progresses. It is important at the outset to make them smaller than may seem physical, since they act as an infinite hard-core potential.

2. Calculate the function corresponding to the experimentally determined data, such as the total structure factor $F(Q)$

3. Calculate the difference between the measured structure factor $F_{expt}(Q)$ and that determined from the configuration $F_{calc}(Q)$.

$$\chi^2_{old} = \sum_{i=1}^{n} [F_{calc}(Q_i)_{old} - F_{expt}(Q_i)]^2 / \sigma(Q_i)^2 \qquad (7)$$

where the sum is over all n experimental data points, each with error $\sigma(Q_i)$.

4. One atom is selected at random

5. This atom is moved a random amount in a random direction up to a pre-defined limit. If the atom still satisfies the closest approach constraints, the experimentally determined data is recalculated (e.g. $F_{calc}(Q)_{new}$). Since only one atom is moved at any one time, it is only necessary to calculate the change in $F_{calc}(Q)$ due to the atom's move. This involves a calculation of size ~N compared with ~N^2 for calculating $F_{calc}(Q)$ from scratch at each iteration.

6. The experimentally determined data are compared.

$$\chi^2_{\text{new}} = \sum_{i=1}^{n} [F_{\text{calc}}(Q_i)_{\text{new}} - F_{\text{expt}}(Q_i)]^2 / \sigma(Q_i)^2 \qquad (8)$$

If the new χ^2 is lower than the one determined with the atom in the previous position, $\chi^2_{\text{new}} < \chi^2_{\text{old}}$, the move it accepted and the new configuration becomes the old configuration. If the new χ^2 is higher then the move is accepted with probability $P = \exp(-(\chi^2_{\text{new}} - \chi^2_{\text{old}})/2)$ or else it is rejected. In practice the acceptance is determined by comparing P with a random number R between 0 and 1 generated by the computer. The move is accepted when $P > R$.

7. The algorithm is continued by returning to step 4. Initially χ^2 will decrease until it reaches an equilibrium value and further moves make little change to χ^2. The model is then said to have converged. Moves may then be continued and configurations collected every ~N accepted moves to collect statistically independent configurations.

As can be seen from the description given above, RMC modelling and the well known Metropolis Monte Carlo simulation[8] are very similar. The only difference is that whereas Monte Carlo simulation samples the potential energy, RMC samples the difference between calculated and experimental structural data. The attributes that are particularly important in the Monte Carlo algorithm[9] should be replicated in the RMC algorithm. The most important of which is the use of a Markov chain, so that local minima are avoided and the final configuration is independent of the starting point.

The definition of χ^2 may be generalised to include comparison with extra data sets such as X-ray, neutron and EXAFS structure factors. Equally extra terms may be introduced to constrain the RMC model further such as predefined atom-atom co-ordinations or nearest neighbour distances. χ^2 may then be written as

$$\chi^2 = \sum_{k} \sum_{i=1}^{n} [F_{\text{calc}}(Q_i)_k - F_{\text{expt}}(Q_i)_k]^2 / \sigma_k(Q_i)^2 + \sum_{j=1}^{m} (f_j^{\text{req}} - f_j^{\text{RMC}})^2 / \sigma_j^2 \qquad (9)$$

to include comparison with the k structure factors and the m constraints. f^{req} and f^{RMC} are the required value of the constraint and the value calculated from the RMC generated configuration respectively. σ_j is a weighting term which influences the strength of any particular constraint. In a crude manner co-ordination constraints can be considered as simple three-body terms restraining the model. In crystalline materials it is particularly necessary to constrain the model, for example to maintain the integrity of a molecular fragment or to restrict a molecule to a finite number of possible orientations and these can simply be included in the definition of χ^2 (Equation 9). Equally the model may be constrained by restricting the movements of atoms to specific regions of the configuration, such as disordering an atom along a specific direction, only allowing atoms to swap etc.

The ability to fit different data sets with the same three-dimensional model is particularly important in order to separate the contributions from different partial radial distribution functions. As shown in Equation 1, the total structure factor $F(Q)$ for an n component system is composed of a weighted sum of $\frac{1}{2}n(n+1)$ partial structure factors $A_{\alpha\beta}(Q)$. Therefore $F(Q)$ from a two component system is composed of three partial structure factors or, via Equation 2, three partial radial distribution functions, $g_{\alpha\beta}(r)$. The technique of neutron isotopic substitution[10] can be used to obtain $g_{\alpha\beta}(r)$ experimentally, where, for an n component system, $\frac{1}{2}n(n+1)$ total structure factors $F(Q)$ composed of different weightings (obtained by using different isotopes of the same element with different neutron scattering lengths) of the same partial structure factors, $A_{\alpha\beta}(Q)$, are measured. $g_{\alpha\beta}(r)$ are then obtained by simultaneously solving the set of $F(Q)$ for $A_{\alpha\beta}(Q)$. Even given that suitable isotopes exist for the system of interest, the errors in measurement may mean that the results from such a separation may not provide $g_{\alpha\beta}(r)$ which are self-consistent. The combination of neutron and X-ray $F(Q)$ may provide partial separation of $A_{\alpha\beta}(Q)$, but it can never be unambiguous, even for a binary compound. RMC modelling of the data compensates in part for this loss of information since the $g_{\alpha\beta}(r)$, which come from a single three-dimensional configuration, must be self-consistent. It should however be stressed that the better the data, the more constrained the final model will be and to get the best from RMC modelling, high quality data are required.

MODELLING GLASSY SILICA, SiO_2

Previous Models of Glassy Silica

It was realised very early that glassy materials, although highly disordered and isotropic over large distances, may possess a definite local order. Zachariasen[11] first introduced the idea of a continuous random network (CRN) for a glassy structure where the atoms are bonded locally and form a three-dimensional structure with no periodicity or symmetry. Based on the CRN model, glassy silica SiO_2 is thought of as a continuous network of SiO_4 units joined at the corners such that each Si is surrounded by four O and each O has two Si neighbours. Given that in most silicate crystals the SiO_4 unit forms an approximate tetrahedron, then a single Si-O bond length and O-Si-O bond angle of $109.47°$ would be expected in the glass. This is supported by the experimental neutron $G(r)$ which has one strong peak at 1.617Å (Si-O) and a second one at 2.626Å (O-O) implying a O-Si-O average angle of $108.6°$.[12] Also NMR experiments find little evidence for the Si co-ordination to be different from 4 or the O co-ordination to be different from 2.[13] However the manner in which these SiO_4 units are joined is more difficult to obtain directly from experimental data. Steric constraints prevent face- or edge-sharing tetrahedra so all SiO_4 tetrahedra must be corner-sharing, described by three angles, the Si-O-Si bond angle and two torsional angles which define the orientation of each joined tetrahedra about the Si-O bonds which meet at the common O atom (see Figure 1). The first (Si-O) and third (Si-Si) lowest r peaks in $G(r)$ would give an average value of the Si-O-Si bond angle, but the Si-Si peak is partially overlapped and very weak in the neutron $G(r)$. Synchrotron X-ray data have been used to deduce a broad Si-O-Si bond angle (α) distribution, $V(\alpha)=V_1(\alpha)\sin\alpha$ with V peaking at $143°$ and V_1 at $180°$.[14] Longer-range structure is virtually impossible to extract directly from structural data (apart from the characteristic atom-atom distances). It is therefore not entirely surprising that modelling has been used to investigate the structure of silica further, given that direct experimental information about the structure becomes vague even at distances ~3Å and greater and that it is these distances which play the most important part in glass formation.

One of the earliest three-dimensional models of silica glass was hand-built by Bell and Dean[15]. Such models were in reasonable agreement with the then available diffraction data but were tedious to construct and the density was difficult to control, being critically dependent on the chosen Si-O-Si bond angles. Subsequent models were all computer generated, the most comprehensive of which were by Gladden[16] who followed a complex recipe for joining SiO_4 units to construct 1000 atom clusters. Some of her models had very good agreement with $G(r)$ and could be used to investigate optimal Si-O-Si bond angle distributions and longer-range structure. The other method which has been used extensively is Molecular Dynamics simulation[17]. These simulations vary in configuration size and simulation complexity, but do not, in general, fit the diffraction data as well as the empirical models.

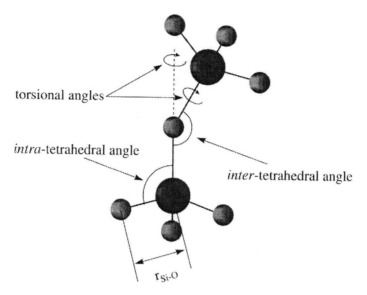

Figure 1. Schematic diagram showing the most likely bonding mechanism in glassy silica where two SiO_4 tetrahedral units are joined at a common oxygen atom.

RMC Models of Glassy Silica

For SiO_2, it is not possible to experimentally separate the Si-Si, Si-O and O-O partial correlation functions, since suitable isotopes for isotopic substitution do not exist and the two available $F(Q)$ (X-ray and neutron) are not sufficient to obtain three $A_{\alpha\beta}(Q)$. The first RMC model of silica[18] used the method in its basic, unconstrained form starting from a configuration of 2596 atoms randomly placed in a cubic configuration box of length 34.017Å. Good agreement was obtained with the X-ray and neutron $F(Q)$ but the partial radial distribution functions $g_{\alpha\beta}(r)$ contained some unsatisfactory features which suggested that the separation of the $g_{\alpha\beta}(r)$ was not correct (for example there was a weak peak in $g_{Si-O}(r)$ at r values where an Si-O correlation was unlikely and which could instead be attributed to the strongest peak in $g_{O-O}(r)$). Also, the average Si-O co-ordination was only 3.7 and the average O-Si co-ordination was 1.8 and some of the SiO_4 tetrahedral unit were somewhat distorted. The O-Si-O bond angle distribution peaked sharply at $109.6°$, but contained a significant tail to higher angles. It is perhaps more significant, given the random starting configuration and lack of any constraints, that any identifiable structural features were found and in fact the majority of the configuration contained joined and structurally correct SiO_4 units. However, some regions of the configuration did not possess the required local connected structure and the RMC modelling was not able to completely connect the structure unaided, since it could find a less-well connected structure with suitable agreement to the data.

Subsequent to this model, a constrained RMC model of 3000 atoms was produced[19], requiring the model to maintain the expected Si-O and O-Si co-ordinations within a defined near-neighbour distance and still fit the data. The second summation term in Equation (9) was used whereby there was a penalty in the χ^2 for all non-perfectly co-ordinated atoms. Again good fits to the $F(Q)$ data were obtained (see Figure 2) and 96.2% of the Si atoms were co-ordinated to four O atoms and 95.4% of the O atoms were co-ordinated to two Si atoms. However, the good connectivity of this model was achieved at the expense of the local order, with, if anything, more distorted SiO_4 tetrahedra than the unconstrained model (including a weak peak on the high-r side of the low-r Si-O peak). Although both these models have considerable merit, they are both flawed in some important respects.

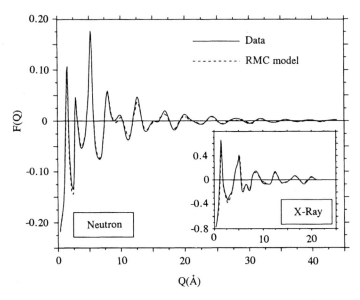

Figure 2. Structure factors for glassy silica calculated from the RMC model of Wicks[19] with co-ordination constraints (dashed lines) compared with those obtained by experiment - neutron[12] and X-ray[20] (inset).

REVERSE MONTE CARLO REFINEMENT

The flaws in the RMC models of glassy silica which have been pointed out in the above section should not be used to discount the RMC modelling method completely. It should be remembered that it was the intention of RMC modelling at the outset to produce valid three-dimensional structural models without bias. The RMC models have demonstrated that the data do indeed suggest that silica glass is composed of ideally connected edge-sharing SiO_4 tetrahedra and they have been used to quantify the Si-O-Si bond angle distribution and further details of the structure. However RMC modelling of the data alone will not be sufficient to produce a model which displays these structural characteristics perfectly. RMC modelling tends to produce the most disordered structure which is consistent with the data and such a model will not produce a glass structure which is composed of very specific structural units. It is therefore necessary to introduce extra constraints into the model in a satisfactory manner without unduly prejudicing the final structure. The constrained model[19] imposed connectivity at the expense of increased SiO_4 tetrahedral distortion.

The two-stage process of RMC *refinement* has therefore been developed[21]. A starting model is created independent of the data which possesses the characteristics which are unambiguously known to be correct and the data are then used to refine the model with RMC. It is clearly essential to constrain the RMC refinement in a suitable way so as not to destroy the essential structural features of the initial model whilst allowing the refinement sufficient flexibility to fit the data. In the case of a glass structure this would usually mean that the initial model had the correct local structure and the RMC refined final model was then used to investigate longer-range structure and amount of distortion. For a disordered crystal, the initial model would normally be the average crystal structure deduced from the analysis of the Bragg peaks and the RMC refined model would provide details about the local deviations from the average structure.

RMC Refinement of Glassy Silica

A form of this method was attempted by Gladden[22] starting from her existing models and Bionducci *et al* [23] attempted to obtain a model of glassy silica starting from the α-quartz structure. However neither of these models started with a completely connected CRN and periodic boundary conditions and during RMC refinement their models were corrupted such that the atoms in Gladden's cluster became too close to each other and the connectivity initially present in α-quartz was partially destroyed.

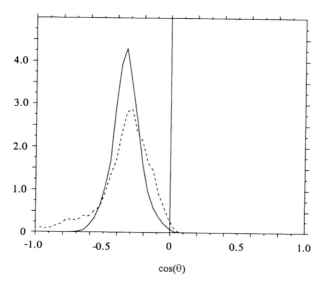

Figure 3. The intra-tetrahedral O-Si-O bond angle distribution from the RMC co-ordination constrained model of Wicks[19] (dashed line) compared with the initial model used for RMC refinement[21] (full line) (arbitrary units).

The most recent RMC refinement of glassy silica starts from a model of 3000 atoms and periodic boundary conditions[21]. The initial model and the method of its construction has already been described[21]. This model has a similar connectivity to the model described by Wicks[19], but has the advantage that the SiO_4 tetrahedra are not distorted (see Figure 3). The $F(Q)$ calculated from this starting model are actually in good agreement with the experimental data except that they do not reproduce the intensity of the peak at lowest Q well. This model was then refined using RMC by slowly increasing the weighting of the

X-ray and neutron $F(Q)$ with respect to the following constraints designed to maintain the integrity of the SiO_4 tetrahedra (compare with Equation 9):

$$\sum (r_{Si-O} - R_{Si-O})^2 / \sigma_{Si-O} + \sum (\theta_{O-Si-O} - \Theta_{O-Si-O})^2 / \sigma_{O-Si-O} \qquad (10)$$

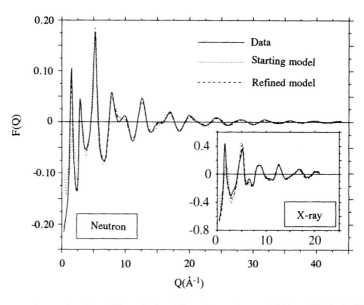

Figure 4. Structure factors for glassy silica calculated from the starting model (dotted line), the RMC refined model[21] (dashed lines) compared with those obtained by experiment - neutron[12] and X-ray[14] (inset).

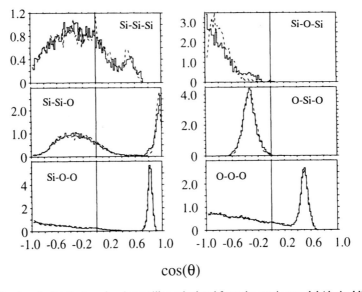

Figure 5. Bond angle distributions for glassy silica calculated from the starting model (dashed line) and the RMC refined model (full line)[21] (arbitrary units).

where R_{Si-O} is the ideal Si-O distance (1.61Å) and Θ_{O-Si-O} is the ideal tetrahedral angle, with σ_{Si-O} and σ_{O-Si-O} chosen to produce qualitatively the appropriate widths of the two lowest-r peaks in $G(r)$. The connectivity was also maintained by not allowing atoms to pass past each other - the CRN could deform but Si-O bonds could not be broken. The model produced in this way gave excellent agreement with the $F(Q)$'s, better agreement than the previous RMC models, with the first structure factor peak fitting well (see Figure 4). There is not much change in the $g_{\alpha\beta}(r)$'s after RMC refinement, with most of the changes in the longer-range details (as reflected by the changes in the first structure factor peak). The most significant difference is found in the Si-O-Si bond angle distribution which becomes broader on RMC refinement (Figure 5).

This RMC refinement of glassy silica demonstrates that it is possible to use RMC modelling, suitably constrained, to improve the fit to data without breaking up an existing structure. The development from the earlier RMC models, which showed that certain structural units are in the glass structure, to the subsequent RMC refinement of a model which set out to incorporate such units, with an improved fit to the data, is a powerful and potentially wide-ranging application of the RMC technique.

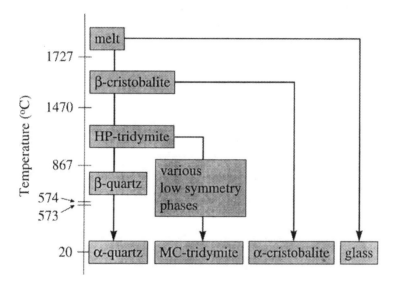

Figure 6. Schematic illustration of the various low-pressure forms of silica after Heaney[24]. The left-hand sequence from the molten phase shows the equilibrium phases whereas the other sequences show phases formed under non-equilibrium quenched conditions.

STRUCTURAL DISORDER IN CRYSTALLINE SILICAS

Introduction

Glassy silica is only one of many structural forms of SiO_2 found at ambient pressure[24]. As shown in Figure 6, at equilibrium and with decreasing temperature, the silica melt transforms in the sequence: melt → β-cristobalite → HP-tridymite → β-quartz → α-quartz. Other phases may be stabilised at room temperature with faster cooling rates: the glass (from the melt), α-cristobalite (from β-cristobalite) and MC-tridymite (from HP-tridymite, via various other low-symmetry tridymite modifications). All of these solid phases are characterised by SiO$_4$ tetrahedra with differing degrees of distortion. The high temperature phases are also believed to be significantly disordered. In the case of quartz, the thermally induced disorder results in the average crystal structure (obtained from Bragg peak intensity analysis) showing a contraction of the Si-O and O-O bond lengths accompanied by an increase in the Si-O-Si bond angle towards 180° with increased temperature (see Table 1). In β-cristobalite, the Si-O bond length is anomalously short and the Si-O-Si bond angle is 180°, with large oxygen displacement parameters normal to the Si-Si bonds[25]. A similar picture is obtained for HP-tridymite (see Figure 7). A 180° Si-O-Si bond angle is known to be unfavourable[26], and various models have been proposed to introduce disorder consistent with a more probable Si-O-Si bond angle of around 145°. One suggestion is that β-cristobalite is composed of domains of the low-temperature α-cristobalite phase[27]. Alternatively, the oxygen atoms do not lie on their average positions but instead they are dynamically distributed around this position to give a longer Si-O bond length and a more physically realistic Si-O-Si bond angle[28]. In this manner the oxygen atoms are disordered around an annulus, which may or may not contain preferred positions[29].

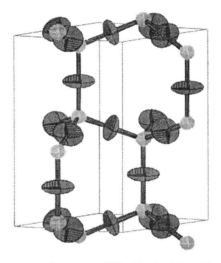

Figure 7. Average crystal structure of HP-tridymite deduced from Rietveld refinement of neutron powder data showing the SiO$_4$ tetrahedral units and the large anisotropic thermal ellipses of the oxygen atoms which elongate perpendicular to the Si-Si directions. [D. A. Keen and M. T. Dove (unpublished)]

In order to distinguish experimentally between possible disorder models the average structure is inappropriate. Bragg intensities arise from elastic scattering and structures deduced from them are time- and space-averaged structures. Structural disorder is then inferred from the variance in the distribution function of instantaneous atom positions that reflects the thermal motion or the partial occupancy of a number of possible sites. Such a structure would not distinguish between an average arising from a superposition of static local domains or from dynamical disorder. In contrast, total scattering directly determines local disorder. Total scattering contains the Bragg and diffuse scattering and integrates over all possible energy transfers between the probe and the sample and gives an instantaneous or 'snap-shot' picture of the structure. The $G(r)$ determined from a total scattering $F(Q)$ will directly determine bond lengths and can be used to distinguish between different models for structural disorder. This is demonstrated in Figure 8, which shows the low-r part of $rG(r)$ from β-cristobalite at 300°C,[30] obtained from a direct Fourier transform of $F(Q)$ measured on the LAD neutron time-of-flight diffractometer at the ISIS spallation neutron source and compared with the positions of the shortest 'bonds' obtained from Rietveld refinement. The three lowest-r peaks in $G(r)$ directly correspond to the Si-O, O-O and Si-Si bond lengths and can be used to deduce the O-Si-O intra- and Si-O-Si inter-tetrahedral angles. These are also shown in Table 1. It should be pointed out that time-of-flight neutron diffraction is unrivalled in this regard. A time-of-flight neutron diffractometer optimised for the study of disordered materials will measure $F(Q)$ up to Q-values of at least 50Å^{-1} (giving good real space resolution in $G(r)$ of order $2\pi/Q_{max}$) and will simultaneously determine the Bragg intensities with good Q-space resolution for reliable Rietveld refinement. The data shown in each column of Table 1 are from a single measurement on LAD.

Table 1. Local structural parameters of various phases of silica. Values which are not in parentheses are from Rietveld refinement of Bragg intensities. Values which are italicised and in round brackets are from the positions of the lowest-r peaks in $G(r)$ obtained from the direct transform of $F(Q)$. ρ is the number of atoms per Å^3. Where the Rietveld refinement gives more than one bond length or angle the range of values are shown [D. A. Keen and M. T. Dove (unpublished)].

	Quartz			Cristobalite		Tridymite	Glass
T/°C(phase)	25(α)	500(α)	620(β)	200(α)	300(β)	550(HP)	25
$\rho(\text{Å}^{-3})$	0.0795	0.0773	0.0761	0.0692	0.0661	0.0655	0.0657
Si-Si (Å)	3.059	3.081	3.093	3.077	3.089	3.068/3.109	
	(3.06)	(3.11)	(3.12)	(3.08)	(3.11)	(3.10)	(3.10)
Si-O (Å)	1.609	1.602	1.588	1.597	1.544	1.534/1.555	
	(1.609)	(1.612)	(1.613)	(1.606)	(1.606)	(1.613)	(1.617)
O-O (Å)	2.616-2.645	2.601-2.628	2.565-2.611	2.590-2.636	2.522	2.532	
	(2.632)	(2.626)	(2.627)	(2.623)	(2.623)	(2.634)	(2.626)
Ô-Si-O (°)	108.7-110.5	108.4-110.3	107.8-110.6	108.3-111.2	109.5	108.8/110.1	
	(109.8)	(109.1)	(109.0)	(109.5)	(109.5)	(109.5)	(108.6)
Ŝi-O-Si (°)	143.7	148.5	153.9	148.9	180.0	180.0	
	(144)	(149)	(151)	(147)	(151)	(148)	(147)

The data from $G(r)$ in Table 1 give a very different picture of the local structure of these silica phases. There is no contraction of the Si-O and O-O bond lengths and no anomalous increase in Si-O-Si bond angles towards 180° with increasing temperature. The local structure is much more physically sensible, and incidentally, much more similar to the glass.

RMC Refinement of Cristobalite

In order to characterise the structural disorder in these systems further, a structural model must be determined which is consistent with the local structure (from $G(r)$) and the average structure (from Rietveld refinement of Bragg intensities). This is an obvious application for RMC refinement since a good starting model is available (the average structure) and RMC modelling will introduce structural disorder into the model in an unbiased manner. The procedure for refinement is as follows:

1. Use the Rietveld method to refine the Bragg peak intensities to obtain the average structure.

2. A configuration based on the ideal average structure is constructed. This may be viewed as a supercell of the crystal unit cell (e.g. 10x10x10 unit cells) and takes no account of the distribution of atoms implied from the thermal parameters in the average structure.

3. The atoms are then moved randomly one at a time so as to satisfy the constraints of Equation 10, without any comparison to $F(Q)$ (i.e. $\sigma_{data}=\infty$). R_{Si-O} in Equation 10 is determined from the lowest-r peak in $G(r)$. σ_{Si-O} and σ_{O-Si-O} are chosen to approximately reproduce the widths of the two lowest-r peaks in $G(r)$. The atom-atom connectivity is maintained throughout.

4. The weighting of fit to $F(Q)$ is slowly increased (σ_{data} is slowly decreased) with respect to the constraints until a good fit to the data is obtained.

The constraints are not too strong to dominate the final structure but are necessary to impose SiO_4 tetrahedra on the structure during the RMC refinement. Indeed the final refined model has broader Si-O and O-O peaks in $G(r)$ and a broader O-Si-O bond angle distribution than would be expected on the basis of the constraints alone.

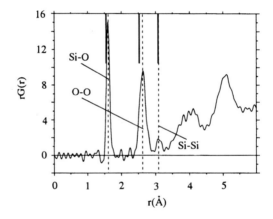

Figure 8. The low-r part of $rG(r)$ from β-cristobalite at 300°C from the Fourier transform of $F(Q)$ compared with the peak positions expected from the average structure deduced from Rietveld refinement (tick marks) [30].

The crystalline long-range order must be effectively accounted for in the model. It is only possible to calculate $G(r)$ from the model out to distances $L_{min}/2$, where L_{min} is the shortest dimension of the configuration box. This is equivalent to the perfect long-range $G(r)$ ($0<r<\infty$) multiplied by a step function $m(r)$ where $m(r)=1$ if $m(r)<L_{min}/2$ and $m(r)=0$ otherwise. The Fourier transform of this section of $G(r)$ is $F(Q)$ convoluted with the transform of $m(r)$, i.e. $M(Q)=\sin(QL_{min}/2)/Q$. This has the effect of broadening the sharp

Bragg peaks and comparison between $F_{expt}(Q)$ and the calculated $F_{calc}(Q)$ from the configuration would be inappropriate. Hence the comparison is made between $F_{calc}(Q)$ and $F_{expt}(Q)$ convoluted with $M(Q)$, i.e. $F(Q)^L$

$$F_{expt}^{L}(Q_j) = \frac{1}{\pi}\int F_{expt}(Q_i)\left(\frac{\sin(|Q_i - Q_j|L/2)}{|Q_i - Q_j|} - \frac{\sin(|Q_i + Q_j|L/2)}{|Q_i + Q_j|}\right)dQ_i \qquad (11)$$

This procedure is not usually necessary for glassy or liquid data where $G(r)$ is flat at $r=L_{min}/2$. The alternative to this is to compare $G_{calc}(r)$ with $G_{expt}(r)$ where $G_{expt}(r)$ has been determined using an inverse method which bypasses the truncation effects of the forward transform $F(Q) \rightarrow G(r)$ when $F(Q)$ is not flat at the maximum Q measured. Such techniques will not be described here, and readers are referred to Soper et al[31].

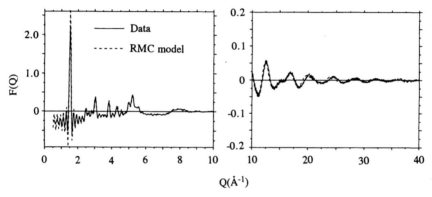

Figure 9. Structure factor of β-cristobalite at 300°C convoluted with the Fourier transform of an appropriate sized step function (full line) (see text for details) and compared with the structure factor from the RMC refined model (dashed line). Note the change of scales between the left-hand (low-Q) and right-hand (high-Q) panels.

The RMC fit to the $F(Q)$ data from β-cristobalite at 300°C is shown in Figure 9. The model consisted of 24,000 atoms (10x10x10 unit cells within a cubic box of sides L=71.351Å). The initial model was created by placing the atoms on the positions determined from Rietveld refinement of the data and then randomly moving atoms to increase the Si-O distance from the Rietveld determined value of 1.544Å to the bond length determined from the lowest-r peak in $G(r)$ (1.606Å) while maintaining the tetrahedral SiO$_4$ arrangement, using constraints described by Equation 10. This starting model was then refined using the $F(Q)$ data and the RMC method described at the start of this section. Figure 10 shows the comparison between $rG(r)$ calculated from the RMC model and $rG(r)$ obtained from the direct transform of $F(Q)$. This shows that very good agreement is also obtained between the real space correlation functions. The bond-angle distribution functions are shown in Figure 11 and the partial radial distribution functions in Figure 12. Three things should be noted. First, as discussed previously[30] the $G(r)$ for β-cristobalite is similar to that of the glass and different from α-cristobalite. This would discount the suggestion that β-cristobalite is an average of α-cristobalite domains. Secondly, the distribution of O around the bond-joining neighbouring Si atoms in [111] directions is isotropic in θ (the torsional angle around the Si-Si bond) and peaks at φ~17° (the angle

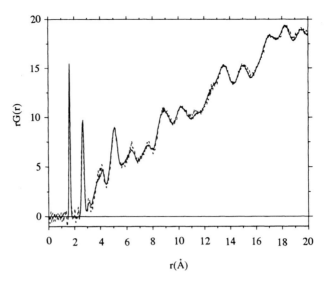

Figure 10. $rG(r)$ from β-cristobalite at 300°C from the Fourier transform of the experimental $F(Q)$ (dashed line) compared with the same function from the model produced from RMC refinement (full line).

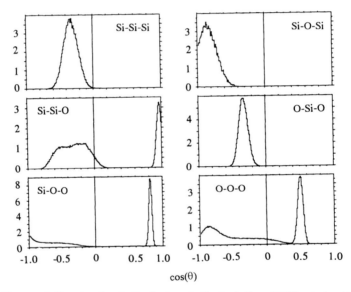

Figure 11. Nearest neighbour bond angle distributions for β-cristobalite at 300°C calculated from the RMC refined model (arbitrary units).

between the Si-O bond and the Si-Si direction). Hence there is no evidence for preferred oxygen sites on this ring of O density in the RMC model. Thirdly, it is possible to use the RMC model to calculate the expected diffuse scattering in planes of reciprocal space. This is equivalent to the scattering which would be measured from a single crystal (if one existed that were large enough) and the calculation is possible because of the three-dimensional nature of the RMC model. The diffuse scattering in the ($hk0$) plane is shown in Figure 13 and compared with electron diffraction results from a very small single crystal grain[32]. There is very good agreement in the positions of the diffuse scattering lines which occur principally in 110 and 100 directions in this plane. This shows that the RMC model is not only able to reproduce the one dimensional $F(Q)$ but also three-dimensional scattering data. Further discussion of the results from these models, and similar models of other phases of crystalline silica will be presented in a later paper[33], although some of the consequences of these models on rigid unit mode theories of silicate minerals are described in another chapter of this book by M. T. Dove *et al*.

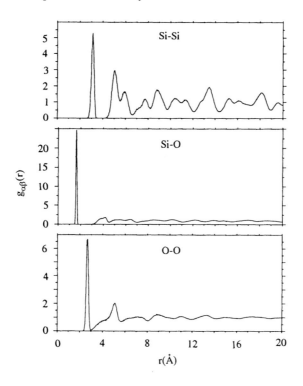

Figure 12. Partial radial distribution functions $g_{\alpha\beta}(r)$ for β-cristobalite at 300°C calculated from the RMC refined model.

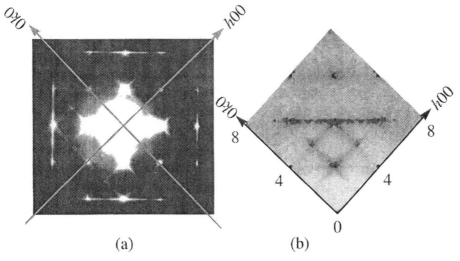

(a)   (b)

Figure 13. Diffuse scattering from β-cristobalite in the $hk0$ reciprocal lattice plane. (a) measured by electron diffraction[32] and (b) from the RMC refined model.

CONCLUSIONS

This chapter has described in detail the techniques of RMC modelling and refinement. Results from glassy silica and the disordered crystalline β-phase of cristobalite have been presented. These results show that with careful application of RMC methods, good representative structures of both locally ordered glasses and locally disordered crystals can be obtained. These methods are completely general and can be applied successfully to a wide range of systems which show structural disorder, with careful consideration to the form of constraints and the construction of appropriate starting models.

Acknowledgements

I am extremely grateful to Martin Dove who has collaborated with me on the study of local disorder in crystalline silica phases and who has permitted me to quote some of our results here prior to full publication. I also thank Jim Wicks for helpful discussions about the structure of glassy silica and the use of some of his unpublished results and Richard Welberry for providing me with the electron diffraction result shown in Figure 13.

REFERENCES

1. M.A. Howe, R.L. McGreevy and W.S. Howells, *J. Phys: Condensed Matter* 1:3433 (1989)
2. T.E. Faber and J.M. Ziman, *Phil. Mag.* 11:153 (1965)
3. R. Kaplow, T.A. Rowe and B.L. Averbach, *Phys. Rev.* 168:1068 (1968)
4. M.D. Rechtin, A.L. Renninger and B.L. Averbach, *J. Non-Cryst. Sol.* 15:74 (1974)
5. R.L. McGreevy and L. Pusztai, *Mol. Sim.* 1:359 (1988)
6. R.L. McGreevy, *Nucl. Instr. and Meth.* A 354:1 (1995)
7. R.A. Young (ed.), *The Rietveld Method* International Union of Crystallography, Oxford University Press, Oxford (1995)
8. N. Metropolis, A.W. Rosenbluth, M.N. Rosenbluth, A.H. Teller and E. Teller, *J. Chem. Phys.* 21:1087 (1953)
9. K. Binder and D.W. Heermann. *Monte Carlo Simulation in Statistical Physics - An Introduction*, (2nd edition) Springer-Verlag, Berlin (1992)
10. J.E. Enderby and G.W. Nielson, *Rep. Prog. Phys.* 44:593 (1981)
11. W.H. Zachariasen, *J. Am. Chem. Soc.* 54:3841 (1932)
12. D.I. Grimley, A.C. Wright and R.N. Sinclair, *J. Non-Cryst. Sol.* 119:49 (1990)
13. E. Dupree and R.F. Pettifer, *Nature* 308:523 (1984)
14. H.F. Poulsen, J. Neuefeind, H.-B. Neumann, J.R. Schneider and M.D. Zeidler, *J. Non-Cryst. Sol.* 188:63 (1995)
15. R.J. Bell and P. Dean, *Nature* 212:1354 (1966)
16. L.F. Gladden, *J. Non-Cryst. Sol.* 119:318 (1990)
17. J.D. Kubicki and A.C. Lasaga, *Am. Min.* 73:941 (1988); B. Vessal, M. Amini and C.R.A. Catlow, *J. Non-Cryst. Sol.* 159:184 (1993); J. Sarnthein, A. Pasquarello and R. Car, *Phys. Rev.* B 52:12690 (1995)
18. D.A. Keen and R.L. McGreevy, *Nature* 344:423 (1990)
19. J.D. Wicks, D.Phil Thesis, Oxford (1993)
20. R.L. Mozzi and B.E. Warren, *J. Appl. Cryst.* 2:164 (1969)
21. D.A. Keen, *Phase Transitions* 61:109 (1997)
22. L.F. Gladden, in: *The Physics of Non-crystalline Solids*, eds. L.D. Pye, V.C. Lacourse and H.J. Stevens, Taylor and Francis, London (1992)

23. M. Bionducci, F. Buffa, G. Licheri, A. Musinu, G. Navarra and G. Piccaluga, *J. Non-Cryst. Sol.* 177:137 (1994)

24. P.J. Heaney, *Reviews in Mineralogy* 29:1 (1994)

25. W.W. Schmahl, I.P. Swainson, M.T. Dove and A. Graeme-Barber, *Z. Kristallogr.* 201:125 (1992)

26. G.V. Gibbs, E.P. Meagher, M.D. Newton and D.K. Swanson, in: *Structure and Bonding in Crystals*, eds. M. O'Keeffe and A. Navrotsky, Academic Press, New York (pages 195-225) (1981)

27. D.M. Hatch and S. Ghose, *Phys. Chem. Minerals* 17:554 (1991)

28. I.P. Swainson and M.T. Dove, *Phys. Rev. Lett.* 71:193 (1993)

29. T.R. Welberry, G.L. Hua and R.L. Withers, *J. Appl. Cryst.* 22:87 (1989)

30. M.T. Dove, D.A. Keen, A.C. Hannon and I.P. Swainson, *Phys. Chem. Minerals* 24:311 (1997)

31. A.K. Soper, C. Andreani and M. Nardone, *Phys. Rev. E* 47:2598 (1993)

32. G.L. Hua, T.R. Welberry, R.L. Withers and J.G. Thompson, *J. Appl. Cryst.* 21:458 (1988)

33. D.A. Keen and M.T.Dove (unpublished)

MODELLING SINGLE CRYSTAL DIFFUSE NEUTRON SCATTERING USING REVERSE MONTE CARLO

V. M. Nield

School of Physical Sciences

University of Kent

Canterbury

Kent CT2 7NR, UK

INTRODUCTION

Ever since the pioneering work of Bragg, the wealth of structural information available from crystalline diffraction data has been recognised. Such diffraction data consists of two components, Bragg and diffuse scattering. The former, which can occur only at integral reciprocal lattice points, is elastic scattering which contains information about average structural features. It is analysed in terms of the average positions in the unit cell which are occupied, fully or partially, by the atoms. The Q-dependence of the Bragg scattering also contains information on thermal displacements. In nearly all cases these are described in terms of thermal ellipsoids, centred on the average positions. The ellipsoids describe the volume inside which the corresponding atom is found with a given probability.

In many crystalline materials there is interest in producing a more accurate description of the static or thermal disorder than is possible using the Bragg scattering alone. This necessitates the analysis of diffuse scattering. Diffuse scattering can be elastic or inelastic, and can occur anywhere in reciprocal space, including beneath the Bragg peaks. It contains information on all deviations from the perfect time-averaged long range crystal structure i.e. it has information on static and thermal disorder, and especially on the correlations between the displacements of different atoms. Interpretation of the diffuse scattering is far less advanced than that of Bragg scattering.

In this chapter the modelling of single crystal diffuse neutron scattering is discussed. Single crystal, rather than powder, diffraction data is used because of the benefits of having three dimensional information, i.e. the information from differently oriented features in real space occurs in different places in reciprocal space. Neutron diffraction data is used because much of the author's interest is in compounds where light and heavy atoms are of equal importance, and which contain hydrogen. The method discussed is, however, equally applicable to X-ray data.

In the past single crystal diffuse scattering was largely used to confirm or reject a static disorder model for a material. Hence a model would be constructed, the scattering it would lead to calculated, and a visual comparison made with the experimental data [1]. More recently methods have been developed which allow refinement of the model to improve the goodness of fit and to allow more information about the disorder to be determined. For example such refinement was employed in studying cubic stabilised zirconia, with the displacements of atoms towards oxygen vacancy sites refined [2]. It was also employed in studying the defect clusters in superionic calcium yttrium fluoride [3,4]. The method to be discussed here, reverse Monte Carlo (RMC), is more ambitious: the positions of several thousand atoms are altered to give agreement with the data. The method was initially pioneered for the study of liquids and amorphous materials [5,6], but it has also been successfully applied to powder diffraction data [7-12]. In this chapter we will look in detail at the RMC single crystal algorithm as it stands at present, with examples from its application to ice, C_{60} and AgBr used to illustrate the problems with the method, and the optimal methods for its use. We will also consider some of the ways in which the algorithm is being improved and methods for future development.

Study of Ice Ih

Ice is by no means simple in structure, even in the usual ice Ih form studied here. The complexity results largely from proton disorder within the perfect hexagonal lattice formed by the oxygens. The oxygens are tetrahedrally coordinated, and the Bernal-Fowler rules [13] require each to be covalently bonded to two protons, with one proton on each O-O bond. The protons thus have a certain freedom as to which sites are occupied and in ice Ih are believed to be randomly disordered whilst obeying these rules [14]. Ice Ih has a hexagonal unit cell, of space group $P6_3/mmc$, but the equivalent 8 molecule orthorhombic unit cell was the one used in the RMC modelling work. The aim of our work was to study the effect of the hydrogen disorder on the local structure, both in terms of the actual bond lengths and angles, and the way it affects the displacements of the oxygen atoms [15-18].

Study of C_{60}

All the C_{60} modelling discussed here was to generated data corresponding to the low temperature phase of C_{60}. In this phase the centres-of-mass of the four C_{60} molecules in the unit cell are on face centred cubic (fcc) sites, but the four molecules have well defined rotations of either 98° or 38° about different threefold ⟨111⟩ directions. The ratio of these two orientations is close to 83%:17% (98°:38°) [19,20]. Attempts were made to model, independently, both orientational and centre-of-mass vibrational disorder [21], with the sole aim of tesing the use of RMC in these cases.

Study of AgBr

AgBr has a rock-salt structure, but as the temperature is increased some of the silver ions move from their usual octahedral sites onto tetrahedral interstitial sites. Powder diffraction data from AgBr was modelled using RMC [8]. The configurations so obtained were used to produce the test single crystal diffuse scattering data modelled here [22].

SINGLE CRYSTAL REVERSE MONTE CARLO (RMCX) [23]

Reverse Monte Carlo is similar to the better known Metropolis Monte Carlo [24] except that one moves atoms to minimize the difference between the measured and calculated structure factors, rather than to minimize the energy of the configuration of particles. Hence no potential function has to be assumed, allowing the technique to be readily applied to a wide variety of materials.

For single crystal data the algorithm is based on the definition of the neutron scattering cross-section in the static approximation, which gives:

$$S(\mathbf{Q}) = \frac{1}{N} \sum_{l,k=1}^{N} < \overline{b_l} \exp(-i\mathbf{Q}.\mathbf{r}_l(0)) \overline{b_k} \exp(-i\mathbf{Q}.\mathbf{r}_k(0)) > \tag{1}$$

where $\overline{b_l}, \overline{b_k}$ are the scattering lengths for atoms l and k, $\mathbf{r}_l(0)$ and $\mathbf{r}_k(0)$ are their positions at $t = 0$, and $< ... >$ denotes an expectation value averaged over all of the initial states of the crystal. N is the number of atoms in the material. Note that the above expression involves an integral over all energies and hence RMCX is used to model total scattering, rather than elastic scattering (Lovesey contains a more detailed discussion of this point [25]). The expression gives a $t = 0$ instantaneous picture, or 'snap-shot', of the whole structure.

To perform any calculations using eqn.(1) a configuration, or super-cell, of atoms is necessary. Periodic boundary conditions are applied, meaning that the configuration is surrounded by images of itself, and so the contribution from any atom and its image must be the same. Hence $S(\mathbf{Q})$ can only be calculated at $\mathbf{Q}$ points which satisfy

$$\mathbf{Q} = 2\pi \left(\frac{h'}{an_a}, \frac{k'}{bn_b}, \frac{l'}{cn_c} \right) \tag{2}$$

for lattice parameters a, b, c and a configuration box with n_a, n_b, n_c unit cells in the three directions. h', k', l' are integers. This means that $S(\mathbf{Q})$ can only be calculated on a grid of points in reciprocal lattice space, with the positions of the points dependent on the number of unit cells. Relating this to a more familiar concept, this means that eqn.(1) can only be used to calculate the scattering at the positions of the super-cell Bragg peaks. The coherent part of $S(\mathbf{Q})$, denoted by $F(\mathbf{Q})$, can be simplified to

$$F(\mathbf{Q}) = \frac{1}{2\pi N'} \left| \sum_i \overline{b_i} \exp(i\mathbf{Q}.\mathbf{R}_i) \right|^2 \tag{3}$$

where the sum is over all N' particles within the configuration box.

The aim of RMCX is to use eqn.(3) to find a configuration of atoms which agrees well with the experimental data. The algorithm is detailed below.

1) Measure and prepare the single crystal diffuse scattering data. As many reciprocal lattice planes of data as have been measured can be used simultaneously in the fitting procedure. (Bragg scattering is only included in analysis in some cases, and Bragg intensities should be extracted following the standard procedure for the instrument in question.)

2) Produce the initial configuration. Routinely these contain about 4000 particles, arranged with the correct crystal symmetry, and using any information which has been

positively determined about their positions in the unit cell. Thus the atoms are normally placed at the average sites as obtained from refinement of Bragg intensities. Periodic boundary conditions require an integral number of unit cells in all three directions. The size of the configuration must be compatible with the spacing of the data points (eqn.(2)).

3) Calculate the coherent scattering of the starting configuration using eqn.(3). The point group is supplied by the user and the algorithm automatically calculates the scattering over the symmetry related directions and averages them.

4) Choose a particle at random and move it a random amount, up to a user specified maximum, along the three axes of the configuration, hence defining a random direction for the move. In molecular systems several types of move can be envisaged. The atoms can be treated individually as described, or the molecule can be kept as a rigid entity, or the molecule can be treated as a flexible body. Molecular motion consists of both centre-of-mass translation and rotation about an arbitrarily chosen axis (using a Euler angle formalism) [16]. In the flexible molecule approach this type of motion is combined with motion of individual atoms [18].

5) The particle-particle distances are checked, and if any particles are unphysically close the move is rejected and step 4 is repeated. This prevents any two atoms from moving too close to one another. The closest approach values are specified by the user, but erroneously large values can be detected by examining the partial radial distribution functions calculated from the super-cell once a fit to the data has been achieved. A sharp spike in the first peak of one of the partial radial distribution functions, sometimes with a corresponding trough elsewhere, indicates that the corresponding closest approach value needs to be reduced.

Additional constraints can readily be incorporated at this stage, for instance in a molecular system the intra-molecular distances can be kept within a specified range [26,17].

6) Eqn.(3) is used to recalculate the coherent scattering for comparison with the data. In practice this step is made faster by calculating the contribution due to the moved particle both before and after its move, and taking the difference.

The goodness-of-fit parameter, χ^2, is defined by:

$$\chi^2 = \sum_{m=1}^{n_E} \frac{(F_E(\mathbf{Q}_m) - F_C(\mathbf{Q}_m))^2}{W(\mathbf{Q}_m)\sigma^2} \tag{4}$$

where the sum is over all n_E data points at positions $\mathbf{Q}_m$. $F_E(\mathbf{Q}_m)$ is the value of the experimental data for point m and $F_C(\mathbf{Q}_m)$ its calculated counterpart (the incoherent scattering is assumed to have a constant flat level and is subtracted from the experimental data before χ^2 is calculated). $W(\mathbf{Q}_m)$ is a weighting factor, and is often taken to be 1 for all $\mathbf{Q}$, or chosen to be $F_E(\mathbf{Q}_m)$ or $F_E^2(\mathbf{Q}_m)$. The standard deviation, σ, is usually taken to be independent of $\mathbf{Q}$ and is treated as a parameter of the modelling. It corresponds to the concept of temperature from simulated annealing [27], but in our case is not normally altered during a modelling run.

The change in χ^2 resulting from the move is $\Delta\chi^2 = \chi_{new}^2 - \chi_{old}^2$. If:

a) $\Delta\chi^2 < 0$ the fit to the data has improved and the move is accepted.

b) $\Delta\chi^2 > 0$ the fit to the data has worsened and the move is accepted with the Boltzmann probability $\exp(-\Delta\chi^2/2)$.

In most of the studies discussed in this chapter σ was taken as 0.01. If extreme values are used this will obviously affect the calculations significantly. With a very large value of σ the data is ignored ($\Delta\chi^2$ is always ~ 0, and so all moves are accepted), with a very small value the configuration is driven to the local minimum closest to the starting point (there is a negligible probability of moves which make the fit worse being accepted).

Soft constraints can be applied if a suitable goodness-of-fit parameter can be defined for them. In this case the move is accepted or rejected depending on a total goodness-of-fit, which is the weighted sum of the goodness-of-fit parameters for all constraints added to the standard χ^2 defined above.

7) The procedure is repeated from step 4 until χ^2 has converged. In practice complete convergence, as indicated by oscillation about a certain χ^2, has rarely been achieved when modelling to single crystal diffuse scattering. In most cases it has been found that the number of moves accepted simply becomes so slow that modelling has to be terminated. No concept of the intrinsic quality of the fit is yet generally applied, except visual inspection. To enable a more qualitative indication it would be useful to define a quantity:

$$QF = \frac{\chi^2}{\chi^2_{bestfit}} \tag{5}$$

where QF is the quality of the fit (and would be close to 1 for a very good fit) and $\chi^2_{bestfit}$ is a measure of the χ^2 that would be achieved if the model showed perfect agreement with the data within the errors.

8) With the same starting point several super-cells are collected, using identical modelling conditions, to improve the statistics on derived quantities. Two types of information are available from the atomic coordinates in the configuration. The model represents an instantaneous structure of the real crystal and many quantities, often of specific interest to the problem under study, can be determined from it. For example in ice the dependence of the displacements of the oxygen atoms on the positions of the neighbouring deuterons was examined. In all cases pair, triplet and higher order correlation functions can be extracted. In fact, it is in principle possible to determine positional probabilities (i.e. the probability that a given atom is in a given position related to a certain configuration of its neighbours) and hence the potentials.

By using the translation vectors of the unit cell all of the unit cells of the independent configurations can be superimposed onto one. All local information is lost, but the average atom density can be examined, and properties of the spatial average structure can be obtained. For example the mean site positions, and mean square displacements in general directions, can be calculated. It needs to be borne in mind that the bond length and angles determined directly from the configuration are not necessarily the same as the distances and angles calculated from the mean site positions, and comparison can give useful information on local, principally static, disorder.

RMC MODELLING

In this section the details of the modelling performed on the three systems of study, ice Ih, C_{60} and AgBr will be reported. Detailed results from these systems can be found in the cited references. In the present chapter, we will solely be comparing the differences in results from the use of different RMCX modelling parameters, to try and elicit the most suitable way to perform single crystal RMC studies.

Modelling of Ice Ih [15-18]

The data used in this modelling work was from a recent investigation of the single crystal diffuse neutron scattering from deuterated ice Ih [28]. The scattering was measured on the SXD time-of-flight Laue diffractometer at ISIS (Rutherford Appleton Laboratory, UK) [29] over a very large volume of reciprocal space. It was corrected and normalised, and then binned to give a data set consistent with the configuration size to be used in modelling. Five experimental planes of data were used in the fitting procedure – the $0\,k\,l$ and its equivalent $h\,h\,l$, the $h\,0\,l$ and its equivalent $h\,3h\,l$ and the $h\,k\,0$ plane – and the data extended out to a $\mathbf{Q}$ of 12Å^{-1}. The hexagonal equivalent planes were used to try and reintroduce the hexagonal symmetry lost by modelling in an orthorhombic unit cell. The starting configuration consisted of either a 6^3 or 10^3 repetition of the unit cell, of dimensions $a = 4.498\text{Å}$, $b = 2a\sin 60°, c = 7.323\text{Å}$, with the deuterium atoms disordered within the Bernal-Fowler rules There were 10000 (25000 in the case of the 10^3 super-cell) data points and so a very low data to parameter ratio (number of parameters per molecule was 9 for free atom modelling, 6 for rigid molecule modelling and 15 for flexible molecule modelling).

A variety of starting configurations was used. The intra-molecular geometry was altered, with O-D lengths and D-O-D angles varied around the values of 0.975Å and $107.0°$, respectively. There were also different inter-molecular geometries, with either all oxygen-oxygen distances equal, or all O-O-O angles equal (these are mutually incompatibly because of the deviation of the c/a ratio in ice from the ideal tetrahedral value). Different closest approach values were examined, and it was found that the most appropriate values were 2.3, 0.5 and 1.0Å for the O-O, O-D and D-D distances respectively.

Moves were made in all of the ways discussed previously, i.e. with independent atoms, rigid or flexible molecules. In general the maximum move size was 0.1Å for translations and $2°$ for rotations. When modelling with rigid molecules relaxation of the atom positions in the molecules was allowed at the end of modelling, with the atoms at this stage moved independently. In nearly all cases σ was 0.01.

Modelling of C$_{60}$ [21]

The C_{60} work discussed here involved RMC modelling with rigid molecules, using computer generated single crystal diffuse scattering data which was based on the low temperature phase of C_{60}. All configurations had 4^3 unit cells, i.e. 256 molecules or 15360 atoms, with the molecule centres in an fcc arrangement. The data used in modelling was a significant section of the hhl plane, containing 6120 diffuse points (the 440 Bragg peaks were only used in fitting in some cases). Hence since in all cases the maximum number of variables per molecule was 3, the data to parameter ratio was close to 8. The generated data was obtained by averaging over the scattering calculated from 8 independent configurations, and no errors were added to it. Different convergence criteria were tested, with all 3 of the suggestions for $W(\mathbf{Q}_m)$ used. In the studies discussed here σ was usually taken as 0.01.

In the first set of tests the data was calculated from configurations of rigid molecules with no centre-of-mass vibrations and with $98°{:}38°$ molecular orientations in the ratio 83:17, within sampling statistics. The RMC modelling was used with fixed molecular centres but with two ways of changing the molecular orientation. Using the standard method molecular rotations were of random orientation and size, up to a user specified maximum, and several orientationally different start points were used. In the second method the RMCX algorithm was adapted so that the orientation of any molecule was altered by randomly picking one set of Euler angles from an array of the 8 possibilities (corresponding to the 98° and 38° orientations of the 4 different molecules in the unit cell). With this method all molecules were initially given random orientations.

For modelling the centre-of-mass vibrations the super-cells had the molecular orientations fixed at 98°:38° in the ratio 83:17 throughout, but the molecule centres were free to move. The generated data corresponded to the C_{60} molecules having the same orientational disorder as before but with centre-of-mass Gaussian vibrations arbitrarily chosen to have a root mean square displacement (rmsd) of 0.1Å. Fitting was to the diffuse scattering only, or to the Bragg and diffuse scattering, and the resulting root mean square displacements were monitored. In general σ was 0.01, the maximum move size was 0.1Å and $W(\mathbf{Q}_m)$ was 1. However all of these parameters were varied and the results monitored.

Modelling of AgBr [22]

The data in this case was calculated by averaging over the scattering from 8 configurations produced by RMC modelling to powder diffraction data [8]. In the study to date, only scattering in the $hk0$ and hhl planes has been used, giving just over 12300 data points. The super-cell used in modelling consisted of 8^3 unit cells, and hence 4096 atoms. Thus the data to parameter ratio is close to 1, although the use of symmetry operations effectively improves this by a factor of 12. In the starting configurations the atoms are arranged in the rock-salt structure. Modelling has been performed with different values of σ, maximum move, and $W(\mathbf{Q}_m)$.

Further work is continuing in which different ranges of data, and powder and/or Bragg scattering are being used in modelling as well as the diffuse scattering.

DETAILED RMCX MODELLING CONSIDERATIONS

In this section we consider in more detail some of the steps mentioned previously. In particular we will look at the failings of single crystal reverse Monte Carlo at present, and some of the means of addressing these problems in the future.

Step 1: Measuring and Preparing the Data

Neutron diffraction measurements can be made in two ways. Conventionally single crystal neutron diffuse scattering studies have been made at reactor sources, where a monochromatic neutron beam is used, and point-wise surveys of reciprocal space are performed using triple-axis instruments to select elastic scattering. To obtain the type of data used in RMC studies these machines are used in two- rather than three-axis mode, so that the scattering is an average over all energies. There is in general a significant limitation on the highest $\mathbf{Q}$ data accessible. For RMCX purposes the grid of the measurement fixes the configuration size except for cases where accurate interpolation of the data is possible. The data is measured at discrete points in reciprocal space, but the resolution of the instrument means that averaging is occurring over a region of reciprocal space near the point of interest.

Time-of-flight (tof) Laue diffraction, as implemented on SXD at ISIS [29], is more flexible and considerably faster. Neutrons of different wavelength (and hence velocity) arrive at the detector at different times. By measuring the number of arrivals as a function of time, the intensity as a function of wavelength, and hence $|\mathbf{Q}|$, is obtained. SXD has a 64×64 pixel position sensitive detector, enabling a large volume of reciprocal space to be surveyed for one sample position, even for samples which are not perfectly aligned. The angular position of the pixel compared to the axes of the instrument, together with the $|\mathbf{Q}|$ assignment just discussed, allow the data to be assigned to a particular $\mathbf{Q}$. Continuous coverage of reciprocal space is achieved and so it is possible to bin the data to allow any reasonable sized configuration to be chosen as the starting point for RMCX. The coarseness of the binning determines how closely the scattering corresponds to that at a discrete point in reciprocal space.

The scattering from either two-axis instruments or time-of-flight Laue diffractometers has to be accurately corrected and normalised. The methods of correction are very similar to those for treating the scattering from liquid and amorphous materials [30]. Corrections are made for background and other non-sample scattering, absorption, multiple scattering and inelasticity. Vanadium scattering is used to normalise for the incident neutron flux profile (in the tof case) and for the detector efficiency. The corrections for tof data are the more difficult, because most corrections are wavelength dependent. On the other hand, with diffraction data from a monochromatic source of wavelength λ, contamination by neutrons of wavelength λ/n, where n is an integer, is difficult to remove completely and can be a considerable nuisance.

The scattering calculated from the model super-cell using eqn.(3) corresponds to energy integrated total scattering, and assumes that the static approximation is valid. For the static approximation to be truly valid the energy of the incident particle needs to be much higher than that of the scattering centres. This is not true for neutron diffraction, and hence the approximation is not completely valid. The correction for this is the Placzek correction. One and multiple phonon scattering can also be a problem, because this is not properly treated. No general evaluation of this problem has been made, and it is very system specific.

As discussed in the previous section, RMC generated single crystal patterns are calculated at discrete points in reciprocal space, determined by the size of the configuration used in the modelling. However experimentally the data is an average over a region of $\mathbf{Q}$ space determined by the resolution of the instrument and any subsequent binning of the data. This difference between calculation and experiment is not believed to be too serious unless diffuse features which are much sharper than the data bin size are involved. In this case the value of the binned data is unlikely to be close to the actual data value for the point at which the calculation is performed.

The range of the data is obviously of crucial importance. High $\mathbf{Q}$ data improves the spatial resolution of models, and ideally data should go out to 30Å^{-1} or more. Data in different regions of reciprocal space contain information on different real space features. Hence in an ideal case the scattering in all unique regions of reciprocal space would be used in modelling with both low and high $\mathbf{Q}$ data included. Traditional reactor data is obtained very slowly and has a restricted high $\mathbf{Q}$ limit. Laue time-of-flight diffraction allows the required volumes of reciprocal space to be measured relatively quickly, but to model these complete volumes would require massive computer memory and CPU. In all published work to date only planes of reciprocal space have been modelled, and especially the principal planes (such as $hk0$ and hhl). However even machines such as SXD do not measure data to sufficiently high $\mathbf{Q}$. To help with these problems an algorithm has been developed for modelling single crystal and powder diffraction data simultaneously. Time-of-flight powder diffraction data not only goes to higher Q, but its use will also constrain the scattering from the model to the correct 3 dimensional average – an advantage when single crystal data over complete volumes of reciprocal space are not being modelled. Testing of this combined approach is now underway for a variety of systems.

Any refinement study also requires a sufficiently high data to parameter ratio. In the present case the number of parameters is generally three times the number of atoms, and in standard Bragg single crystal refinements the preferred data to parameter ratio is greater than 10. This number is large partly because of the errors in the data.

The errors in measured diffuse scattering are generally larger (because the scattering is inherently weaker), which would suggest the need for a greater data to parameter ratio. However, as discussed in the previous paragraph, there are other equally important criteria which affect the quality of the model, and all lead to the requirement of a very large amount of experimental data for accurate RMCX modelling work.

Step 2: Creating the Initial Super-cell

It is important to make the correct choice(s) of initial super-cell, as this can vastly improve the chances of convergence to a sensible minimum in a reasonable time scale. In some cases the choice of starting point is obvious, particularly if any disorder is of relatively small amplitude. For example for AgBr the starting point must be a rock-salt arrangement of atomic centres, because deviations from this are known to be small. However if the percentage of silver ion interstitials was larger, it might be sensible to additionally try a starting point with the silver ions distributed over both octahedral and tetrahedral sites or just on tetrahedral sites. By modelling with a range of starting points it is possible to find a range of models which agree with the data. If agreement is equally good in all cases additional information is needed to decide which of the models is likely to be the most representative of the real structure. A good example of the use of multiple starting points is provided by a reverse Monte Carlo powder diffraction study of AgI [9].

With systems where the amount of structural disorder leads to a poorly defined average crystal structure from Bragg refinement the best starting configuration is harder to define. In work on the structure of ice, Beverley and Nield [16,18] tried a variety of different starting inter- and intra-molecular geometries, spanning the best determined perfect structure [14] to attempt to test the sensitivity of the result to the start point. Table 1 shows some results from using two different initial O-D bond lengths. It can be seen that the final O-D bond length depends strongly on the start point. This means that the model is not converging to a global minimum, which has many possible causes, including the limited Q range of the data, the poor optimisation of the goodness of fit parameter and not modelling the Bragg scattering.

Table 1. Some of the results from modelling ice data with different initial O-D bond lengths and numbers of unit cells. In all cases atoms were moved rather than molecules.

Number of unit cells	Initial O-D lengths (Å)	Final O-D parallel to c (Å)	Final O-D oblique to c (Å)	Oxygen mean square displacement (Å²)
10^3	0.95	0.97	0.97	0.005
10^3	1.00	1.00	1.00	0.005
6^3	1.00	0.98	1.02	0.012

The work on ice also showed a dependence on the configuration size, again as seen in Table 1. It can be seen that on moving from 6^3 to 10^3 unit cells there is a significant change in the resulting bond lengths and mean square displacements. While this might be partly due to modelling the long range correlated static disorder in ice [15], it was also contributed to by the way the moves were made. In the larger configuration a smaller percentage of atoms were moved, so that even when modelling was finished there were a significant number of unmoved atoms. This is reflected in the small oxygen mean square displacement in the final model. This suggests that there is a problem with the way moves were being made, and this is discussed further below.

If the super-cell is too small the statistics of the calculated scattering will be poor [31], although this is alleviated to some extent by averaging over symmetry equivalent directions [23]. The problem is most severe for systems with a great deal of correlated static disorder, such as ice. In this case smaller configurations did not accurately sample the static disorder of the molecules. An evaluation of this problem should be made system by system. Not enough work has yet been done to determine whether this simply imposes a limit on the quality of fit, or can lead to completely fallacious results.

Step 4: Making a Move

As more and more tests are performed, it seems that the method and maximum displacement used in making a move is of crucial importance in obtaining good convergence to a sensible model. There are many ways that moves can be made, some specific to certain systems, such as picking the new orientation of a molecule from an array of possibilities [21] or swapping the species of two atoms [32].

Table 2. RMC results on the effect of different maximum move sizes on the rmsd for test data on C_{60} (root mean square displacement should be 0.1Å) and AgBr (Ag^+ root mean square displacement is given in the table and should be 1.1Å).

System	Maximum Move size (Å)	σ	Root mean square displacement (Å)
C_{60}	0.1	0.01	0.12
C_{60}	0.1	0.0001	0.12
C_{60}	0.01	0.01	0.098
AgBr	1	0.1	1.23
AgBr	1	0.01	0.97
AgBr	0.1	0.01	0.23
AgBr	0.01	0.01	0.03

In the most general case the move is made by choosing a random direction and moving the atom a linearly random amount in that direction, up to a maximum value as specified by the user. Table 2 shows the results obtained on using different maximum move sizes and σ values. In modelling to data corresponding to 0.1Å root mean square centre-of-mass vibrations in C_{60} it was found that when the maximum move size was 0.1Å, with the sizes of the moves chosen randomly up to this maximum, the configuration quickly reached an equilibrium in which the rmsd was 0.12Å [21]. This was independent of the form of $W(\mathbf{Q}_m)$ or of a reduction in σ to 0.001. With the smaller move size of 0.01Å modelling took about ten times longer, but the fit to the data was greatly improved and the final rmsd was 0.098Å, very close to the correct value. However in the modelling of AgBr maximum move sizes of 0.1Å and lower resulted in rmsds far smaller than they should be when modelling was ended – the actual isotropic root mean square displacements of our test data are 1.1Å for the Ag^+ and 0.74Å for the Br^-. Only with the larger maximum move size of 1Å was reasonable agreement to the rmsd obtained. (It should be noted that the goodness-of-fit is different in all cases, and is better for the fits with mean square displacements closer to the true value. Hence by performing modelling with a wide range of maximum move sizes the most appropriate move size can still be found, but this is computationally prohibitively expensive.) This clearly shows that the maximum move size should be chosen using the best available information about the rmsds of the system, i.e. larger for systems with larger rmsds. Proffen and Welberry achieved this in some of their RMC studies by giving atoms initial displacements corresponding to the correct mean square displacement and then swapping the displacements from different atoms of the same species [32]. This method has the disadvantage that only displacements initially input can be present in the final model, which is not ideal for correlated motions.

New work by the author builds on this method, with the atoms in the configuration initially given displacements of the required magnitude and distribution to give the correct probability ellipsoid (as determined from standard Bragg analysis). However moves for atoms which are not believed to exhibit static disorder are made by returning the atom to its average site position and then using a random Gaussian distribution of a shape corresponding to the probability ellipsoids to determine the new move. For atoms with known or suspected static disorder, moves are made either in the standard way, or by combining linear and Gaussian random moves. These methods are still being tested [33]. However, giving the atoms an initial displacement off their sites consistent with the correct mean square displacement avoids the problems noted earlier for ice, in which different sized starting configurations gave very different final mean square displacements, and those for C_{60} and AgBr, where different move sizes and σ values lead to different rmsds. Hence this method is potentially very promising as a means of introducing physically sensible constraints.

Extensive studies on ice Ih have been performed, looking at how the way moves are made affects the dependence of the final RMC model on the starting configuration [16,18]. In separate modelling runs moves were made in three ways, with the molecules initially rigid but later allowed to relax, with flexible molecules and with individual atoms. Table 3 compares the D-O-D intra-molecular water angle and the oxygen mean square displacement from modelling runs which were performed identically except for the method of making moves. In all cases it was found that there was a strong dependence on starting point, with the greatest dependence when the molecules were initially kept rigid and the least when atoms were moved individually throughout, as might be expected. It can also be seen from Table 3 that the constraint of keeping the molecule rigid lead to a larger amount of local distortions and so to larger mean square displacements, and badly incorrect inter-molecular bond lengths and angles. It seems sensible that in cases such as ice, where there is such a large amount of static disorder that the starting bond lengths and angles are intrinsically uncertain from Bragg analysis alone, then these parameters be refined initially. This pre-RMC refinement of, for example, intra- and perhaps inter-molecular bond lengths and angles (including simple static disorder models) could be performed using the same basic algorithm, but with all atoms moved at each step. This would perhaps then define the best starting point for RMC. This presents further convergence problems, but an iterative procedure such as pre-RMC run then perform RMC, then use information from the RMC to make the pre-RMC stage more accurate and so on, may be appropriate. If elastic scattering had been additionally measured for the same system this would enable the pre-RMC step to be more accurately performed without an iterative procedure being necessary.

Table 3. Some of the results from modelling ice data with different ways of making moves – all other factors were identical.

Method of Making Moves	D-O-D angle	Oxygen mean square displacement (Å^2)
Start point	109.5	0.0
Rigid molecule then atoms	109.3	0.020
Flexible molecule	108.8	0.014
Atoms	107.4	0.012

In none of the work so far have orientational and thermal disorder been modelled simultaneously (in the work on ice the moves did not allow large enough rotations of the molecule for re-orientation by more than a few degrees). In all the work of Nield and co-workers, where static and thermal disorder have both been present, they have been modelled simultaneously. In some work of Proffen and Welberry [32] the system had both occupational and thermal disorder, and it was found to be most effective to fit these by modelling first one and then the other and then the first and so on. Hence, initially a number of moves were made in which the species of some atoms were swapped (to deal with the occupational disorder) and then a number of moves were made to model the thermal disorder and then atom swapping was performed again and so on. In this work it was also found to be appropriate to model different regions of the data during the different steps.

In some cases it might be best to make a series of moves before looking at the goodness-of-fit. For examples moves could be made in a way which approximates phonon modes. In some cases the model might need to make a series of moves of nearby atoms for any of those moves to be allowed, for example in moving a section of a molecule that is not being treated as a rigid body. While this is a potentially very powerful method, no work has yet been done in this area to the author's knowledge.

Step 6: Acceptance of Moves

A number of factors are involved in the acceptance of moves. These include the form of χ^2, the Boltzmann criterion used to accept moves that make the fit worse, and the value of σ.

The only RMCX study to look at different forms of χ^2, in this case the different values of $W(\mathbf{Q}_m)$ mentioned previously, was the study on C_{60} [21], some results from which are given in Table 4. Here, when modelling the orientational disorder, the problem immediately encountered was that with $W(\mathbf{Q}_m) = 1$ the modelling would quickly get driven into a local minimum corresponding to an overly disordered configuration of molecular orientations. Hence in some regions of reciprocal space the scattering level was considerably higher than it should be. This occurred even when orientations were picked from an array of the 8 known possibilities, rather than changes being made at random. In this case the final ratio of molecules in the 98°:38° orientations was close to 1. When $W(\mathbf{Q}_m) = F_E(\mathbf{Q}_m)$, the situation was much improved, but still not perfect (see Table 4; if the model was perfect the numbers should be 256:0). The advantage of this weighting seemed to be that it increased the importance of the points where there was very little diffuse scattering, and hence helped to increase the amount of order in the final model. This weighting was found not to cause any significant difference compared to uniform weighting when modelling centre-of-mass vibrations in C_{60}. When the weighting was increased to $W(\mathbf{Q}_m) = F_E^2(\mathbf{Q}_m)$ few moves were accepted and it was found to be impossible to converge the model with any value of σ.

Table 4. RMC results from modelling the orientational disorder of C_{60} with differently weighted χ^2.

Weighting $W(\mathbf{Q}_m)$	σ	Number of molecules in 98°:38° orientations
1	0.01	126:130
1	0.001	123:133
$F_E(\mathbf{Q}_m)$	0.01	155:101
$F_E^2(\mathbf{Q}_m)$	0.01-5	not converged

Welberry and Proffen [31] made the very important comment that the present definition of the goodness-of-fit relates to individual data pixels which correspond to high spatial frequencies, rather than some broader measure of the agreement. Hence perhaps a completely new form of χ^2 needs to be developed.

By using the Boltzmann criterion for accepting moves that decrease the quality of the fit the model can move out of shallow minima, provided σ is greater than zero. In many cases it has been found that adjusting σ by a factor of 10 has little effect on the final model, but this has always involved values of σ of 0.01 or smaller. In most work σ has been kept constant during modelling, but in a few cases σ has been altered [e.g. 32], although the alteration has never been slow enough for the method to have been a true simulated annealing procedure [27]. Simulated annealing would be advantageous in moving the configuration into a global rather than a local minimum, with the main constraint on the use of this technique the large CPU times involved.

THE USE OF CONSTRAINTS

Constraints on the Average Structure

Most RMCX work has modelled the diffuse scattering alone. However diffuse scattering does not contain the average information on the system, and this constraint needs to be fed into the model in some way. One obvious way to do this is to include the Bragg scattering in the modelling. The difficulty with this is in correctly weighting the two contributions (Bragg and diffuse) in the goodness-of-fit. Work on this topic is proceeding [21-23], but at present it seems that there is no ideal way of performing this weighting.

A different way around the problem is to use the average structure (sites plus ellipsoids) as a constraint in the modelling. Hence the deviation from each average atomic site, obtained by averaging over all equivalent atoms in the super-cell, can be used as an additional term in the goodness-of-fit, and ways, such as those discussed above under making a move, found to constrain the atoms to have the correct mean square displacements. This has the disadvantage that, especially in the types of disordered system usually considered in an RMC study, there are often a series of average structures that fit the Bragg scattering equally well. Some of these may involve split sites, or deviations from the thermal ellipsoid model of thermal displacements and this would not be allowed for in the constraints.

It seems that there is no easy solution to allow the inclusion of average information.

CONCLUDING REMARKS

Reverse Monte Carlo modelling to single crystal total (diffuse plus Bragg) scattering in principle has the ability to produce a super-cell of atomic coordinates that is a good representation of a many thousand atom section of a crystal at a certain instant in time. At present the technique is still under development, and very careful evaluation is required of any results obtained. This is in distinction to the application of RMC to powder diffraction data which has been performed successfully in many cases [7-12].

The successful application of the single crystal RMC technique requires the data on which it is to be used to have been carefully measured, with good statistics, and with the accurate performance of all necessary corrections. In conventional refinement studies the data to parameter ratio is the most important concept. In the present case the range of the data is far more important, as even with a high data to parameter ratio only features which give scattering in the planes which are being modelled will be reproduced in the data. High Q data (out to 30Å^{-1}) improves the real space resolution of the models, but is generally not available in single crystal measurements. The use of powder plus single crystal data allows higher Q data to be included and ensures that the model gives the correct 3-dimensionally averaged scattering.

The starting configuration should be a repetition of the time-averaged unit cell obtained by standard refinement. Where this is inaccurately known because of a large amount of static disorder it is suggested that some kind of pre-RMC refinement be performed, ideally on elastic scattering, to obtain the best start point. Where several starting points can be envisaged, modelling should be performed from each, to improve the chances of converging into a global minimum solution.

There are many ways in which the movements of atoms or molecules can be made. These should all lead to the same final configuration, but at present convergence is often into a local minimum, and so they lead to different results. A new method of great promise for many situations is to use the Bragg refinement probability ellipsoids to constrain the thermal displacement component of moves, with standard linear random moves to give the static disorder. This method also allows average information, which is missing from the diffuse scattering, to be included. An alternate method of doing this is to model Bragg and diffuse scattering simultaneously, but optimised relative weightings of the goodness-of-fit parameters for these two types of scattering are proving difficult to establish.

It has been found that a weighted goodness-of-fit parameter is better than the form used initially [21], to give a higher weighting to the regions of reciprocal space where there is little diffuse scattering, and so stop the models from becoming too disordered. The use of a completely different form for this parameter that does not just depend on point by point comparison could improve convergence and merits further work [31]. Similarly a simulated annealing approach would be advantageous in finding the model global minimum, but is computationally expensive.

The use of reverse Monte Carlo on single crystal diffuse scattering is in its relatively early stages, and further development is required along the lines discussed above for it to properly fulfill its potential.

The RMC codes are freeware, and anyone wanting to use the RMCX code is welcome to obtain it from the author or from the reverse Monte Carlo web pages at:

http://www.studsvik.uu.se.

The RMC version of Proffen and Welberry is available as part of the Discus package on:

http://rschp2.anu.edu.au:8080/proffen/discus/discus.html.

ACKNOWLEDGMENTS

The author would like to thank Mark Beverley, Daniel Hesselbarth, Dave Keen and Robert McGreevy for their contributions in the development of the RMCX technique. Useful discussions with Simon Billinge, Bill David, Takeshi Egami, Fredrich Frey, and Chick Wilson are also gratefully acknowledged.

REFERENCES

[1] R.J.Cava, R.M.Fleming and E.A.Rietman, *Solid State Ionics* 9-10:1347 (1983).

[2] T.Proffen, R.B.Neder and F.Frey, *Acta Cryst.* B52:59 (1996).

[3] S.Hull and C.C.Wilson, *J. Solid State Chem.* 100:101 (1992).

[4] M.Hofman, S.Hull, G.J.MacIntyre and C.C.Wilson, *J. Phys.: Condens. Matt.* 9:845 (1997).

[5] R.L.McGreevy R L and L.Pusztai, *Mol. Simul.* 1:359 (1988).

[6] R.L.McGreevy and M.A.Howe, *Ann. Rev. Matter Sci.* 22:217 (1992).

[7] D.A.Keen, R.L.McGreevy, W.Hayes and K.N.Clausen, *Phil. Mag. Lett.* 61:349 (1990).

[8] V.M.Nield, D.A.Keen, W.Hayes and R.L.McGreevy, *J. Phys.: Condens. Matt.* 4:6703 (1992).

[9] V.M.Nield, D.A.Keen, W.Hayes and R.L.McGreevy, *Solid State Ionics* 66:247 (1993).

[10] V.M.Nield, R.L.McGreevy, D.A.Keen and W.Hayes, *Physica B* 202:159 (1994).

[11] L.Karlsson and R.L.McGreevy, (1995). Solid State Ionics, 76, 301-308.

[12] W.Montfrooij, R.L.McGreevy, R.Hadfield and N.H.Andersen *J. Appl. Cryst.* 29:285 (1996).

[13] J.D.Bernal and R.H.Fowler, *J. Chem. Phys.* 1:515 (1933).

[14] W.F.Kuhs and M.S.Lehmann, *Water Sci. Rev.* 2:1 (1986).

[15] V.M.Nield and R.W.Whitworth, *J. Phys: Condens. Matter* 7:8259 (1995).

[16] M.N.Beverley and V.M.Nield, *J. Phys.:Condens. Matt.* 9:5145 (1997a).

[17] M.N.Beverley and V.M.Nield, *J. Phys. Chem. In press* (1997b).

[18] M.N.Beverley, Submitted to Physica Scripta (1997).

[19] W.I.F.David, R.M.Ibberson, T.J.S.Dennis, J.P.Hare and K.Prassides, *Europhys. Lett.*18:219 (1992).

[20] W.I.F.David W I F, R.M.Ibberson R M and T.Matsuo T, *Proc. Roy. Soc. London A* 442:129 (1993).

[21] D.Hesselbarth and V.M.Nield, *Submitted to J. Phys.: Condens. Matt.* (1997).

[22] M.N.Beverley and V.M.Nield, *unpublished work.*

[23] V.M.Nield, D.A.Keen and R.L.McGreevy, *Acta Crystallogr.* A51:763 (1995).

[24] N.Metropolis, A.W.Rosenbluth, M.N.Rosenbluth, A.H.Teller and E.J.Teller, *J. Chem. Phys.* 21:1087 (1953).

[25] S.W.Lovesey, *Theory of Thermal Neutron Scattering from Condensed Matter* Vol. 1. pp. 29-30. Oxford University Press (1984).

[26] V.M.Nield, *Nucl. Instrum. methods A* 354:30 (1995).

[27] S.Kirkpatrick, C.D.Gelatt and M.P.Vecchi, *Science* 220:671 (1983).

[28] J.C.Li, V.M.Nield, D.K.Ross, R.W.Whitworth, C.C.Wilson and D.A.Keen, *Phil. Mag. B* 69:1173 (1994).

[29] C.C.Wilson, *J. Mol. Struct.* 405:207 (1997).

[30] M.A.Howe, R.L.McGreevy and W.S.Howells, *J. Phys: Condens. Matter* 1:3433 (1989).

[31] T.R.Welberry and Th.Proffen, *J. Appl. Cryst. In press.* (1997).

[32] Th.Proffen T and T.R.Welberry, *Acta Cryst.* A53:202 (1997).

[33] V.M.Nield and C.C.Wilson, *unpublished work.*

REAL-SPACE RIETVELD: FULL PROFILE STRUCTURAL REFINEMENT OF THE ATOMIC PAIR DISTRIBUTION FUNCTION

Simon. J. L. Billinge

Department of Physics and Astronomy and
Center for Fundamental Materials Research,
Michigan State University,
East Lansing, MI 48824-1116.

INTRODUCTION

One of the most challenging problems in the study of structure is to characterize atomic short-range order in materials. Long-range order can be determined with a high degree of accuracy by analysing Bragg peak positions and intensities in data from single crystals or powders. However, information about short-range order is contained in the diffuse scattering intensity. This is difficult to analyse because it is low in absolute intensity (though the integrated intensity may be significant) and widely spread in reciprocal space.

The need to persevere and develop reliable techniques for analysing diffuse scattering is becoming necessary. This is because many of the newly emerging materials, including many with potential technological applications, are quite disordered. These include materials such as semiconductor alloys, ferroelectric materials and other transition metal compounds, nanoporous and microporous materials such as zeolites and pyrolitic graphites, and molecular crystals, for example. In order to obtain a full solution of the structure it is necessary to analyze both the long-range crystallographic order and short-range, aperiodic, deviations from the long-range order.

The diffuse scattering can be studied using two broad approaches: analyzing the data from powders or single crystals.[1] The former approach has the advantage that the diffuse scattering is integrated over a solid angle of 4π and so weak diffuse intensities can still be measured accurately. The experiments and analysis are also straightforward (in general) allowing many data points to be obtained to search for local structural dependences on temperature and pressure, for example.[2] Finally, as we will describe, the modelling of the data is quite intuitive because it is in real-space coordinates. The main disadvantage of the powder approach is that directional information, present in the single crystal data, is lost in the powder experiment. This directional information can often be inferred from the data using three dimensional structural modelling. However, a good single crystal experiment will always be the final arbiter of which real-space model is correct.

We will describe an approach for extracting local structural information from powder diffraction data using a full-profile fitting regression technique, where the function which is fit is the atomic pair distribution function (PDF). The PDF technique has been described in detail elsewhere.[3-6] The modelling approach we will describe gives a quantitative solution of the local structure in the form of atomic coordinates, displacement factors (often less accurately called thermal factors) and occupancy factors. It gives motional-correlation factors[7] and can yield correlated short-range ordered displacements where they exist.[2,8-10] Furthermore, because the PDF is fit over a significant range of r (where the r-coordinate measures the distance separating a pair of atoms), the probability of the result being biased by random (or systematic) noise in the function is very small. Structures determined in this way are therefore robust; however, as is true with all structures obtained from powder data, the structural solutions are not unique.

In the following Section we describe the atomic pair distribution function, briefly touching on how it is determined experimentally and how it can be calculated from a model structure. In the subsequent Section we describe the capabilities and implementation of the full-profile fitting program named RESPAR. The next Section contains a number of examples of its use to solve scientific questions. The mathematical Appendix lists the equations used in the RESPAR code to calculate the PDF and determine the least-squares matrices.

PAIR DISTRIBUTION FUNCTION

The atomic pair distribution function, $G(r)$, can be obtained from powder diffraction data through a sine Fourier transformation:[3]

$$G(r) = 4\pi r[\rho(r) - \rho_0] = \frac{2}{\pi} \int_0^\infty Q[S(Q) - 1] \sin Qr \, dQ, \qquad (1)$$

where $\rho(r)$ is the microscopic pair density, ρ_0 is the average number density, $S(Q)$ is total structure function which is the normalised scattering intensity, and Q is the magnitude of the scattering vector, $Q = |\mathbf{k} - \mathbf{k_0}|$. For elastic scattering, $Q = 4\pi \sin \theta / \lambda$, where 2θ is the scattering angle and λ is the wavelength of the scattering radiation.

The PDF is a measure of the probability of finding an atom at a distance r from another atom. It has been used extensively for characterizing the structure of disordered materials such as glasses and liquids. However, the same approach can be applied to study crystalline materials, as discussed elsewhere in this volume.[6] In the past this was rarely the method of choice for studying crystalline materials. However, a need to characterize disorder in new materials, coupled with improved technologies for carrying out the experiments, is now making this approach more interesting. The advent of powerful synchrotron x-ray and pulsed neutron sources are allowing high quality data to be routinely collected over wide ranges of Q, and modern high speed computing is allowing efficient data analysis and modelling to be carried out.

The experimental determination of PDFs has been described in detail elsewhere[3-6] and we will not discuss it here. Of particular importance when partially ordered materials are to be studied is to collect data over a sufficiently wide range of Q. In neutron measurements this is done using pulsed neutron sources which have large fluxes of epithermal neutrons. With x-rays it is necessary to work at x-ray energies above 25 keV and preferrably as high as 40 keV. This is possible at modern synchrotron sources. The data are corrected for experimental effects such as beam polarization, background scattering, sample absorption and multiple scattering, inelasticity effects (this is the incoherent Compton scattering in x-rays and the Plaçek correction in neutron measurements – phonon inelastic scattering is accepted in the usual total scattering experiment), and it is normalized by the incident flux and the number of scatterers in the sample. The data are then divided by the average atomic form factor in an x-ray determination. This analysis procedure is complicated but straightforward and most of the corrections are quite well controlled. The resulting normalised scattering function, $S(Q)$, can then be Fourier transformed according to Eq. 1.

Calculating the PDF from a Structural Model

The radial distribution function (RDF) gives the average atomic density (suitably weighted by atomic scattering factors) in an annulus of thickness dr at a distance r from another atom.[3] This feature gives a calculational scheme for determining the RDF, and from that the PDF, for a given structure.

The structure is specified by a series of delta functions at the positions occupied by atoms in the sample, $\delta(\mathbf{r} - \mathbf{r_i})$, where $\mathbf{r_i}$ is the position of the ith atom with respect to some coordinate system. The PDF is then given by,

$$G(r) + 4\pi r \rho_0 = 4\pi r \rho(r) = \frac{1}{r} \sum_i \sum_j \frac{b_i b_j}{\langle b \rangle^2} \delta(r - r_{ij}), \qquad (2)$$

where the sums go over all the atoms in the sample and r_{ij} is the magnitude of the separation of the ith and jth atoms. The number of atoms in the sample is N, b_i is the scattering length of the ith ion (evaluated at $Q = 0$) and $\langle b \rangle$ is the sample average scattering length. We will refer to the $\delta(r - r_{ij})$ as "atomic pair correlations".

In an experiment it is the ensemble (and temporal) average of a macroscopic sample which is measured. This presents two complications in our calculation of a model $G(r)$ to compare with experiment. First, if the sums were taken over every atom in the sample they would be impractically large, and second, Eq. 2 should be averaged over different possible configurations of the sample. The first problem is generally addressed by specifying a sample of a computationally tractable size and applying some boundary conditions. There are two common approaches to address the second problem. One can specify a model with a large enough number of atoms that a sample-average of the model accounts for all of the static and thermal disorder.[11] Alternatively, the delta functions on each atomic site can be convoluted with a distribution function to account for this disorder as we do here. It is also possible to combine both approaches.[12]

Once $G(r)$ is calculated for the model over some range of r, it can be compared with an experimentally determined $G(r)$ over the same range. An example is shown in Fig. 1. Model parameters such as atom positions can then be varied in such a way as to improve the agreement between the calculated and measured PDFs as we describe below.

REAL-SPACE RIETVELD PROGRAM

In this section we describe our implementation of the above procedure. As we will describe, the approach we have taken is highly analogous to the Rietveld refinement of powder diffraction data. For this reason we have called the method "Real-Space

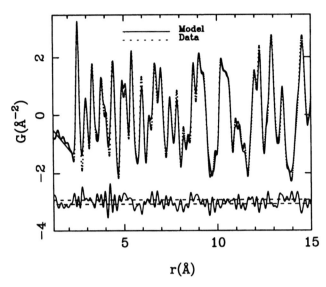

Figure 1. Example of a model PDF (solid line) fit to data (dotted line). Below is the difference curve. The dashed lines indicate the random errors at the level of ± 1 standard deviation. The data are from $YBa_2Cu_3O_{7-\delta}$ collected on HIPD at MLNSC, Los Alamos at 10 K

Rietveld" where real-space refers to the fact that we are fitting $G(r)$ which is a function in real- rather than reciprocal-space coordinates.

The program uses a full-matrix least squares approach for the regression analysis with analytic partial derivatives for all the variables. The mathematical expressions used in the program are laid out in the Appendix at the end of this article. Parameters in the code are varied as desired until the residuals function, which we call the agreement factor, is minimised. The agreement factor, A, is a weighted, full profile, factor similar to the Rietveld weighted agreement factor and defined as,

$$A^2 = \frac{1}{N} \sum_{k=k_{min}}^{k_{max}} \frac{\sigma_k^2}{G_{dk}} \sum_{k=k_{min}}^{k_{max}} \frac{(G_{dk} - G_{mk})^2}{\sigma_k^2}, \tag{3}$$

where the sums go over the N data points from the k_{min}th point at r_{min} to the k_{max}th data point at r_{max}; G_{dk} and G_{mk} are the values of $G(r)$ from the data and the model respectively at the kth point; and σ_k is the estimated random error on the data at the kth point at the level of one standard deviation.

Sample Dependent Parameters

In analogy with crystallography, the model is specified as atom positions in a unit cell: the lattice parameters and cell angles are given and atom positions are then indicated in fractional coordinates. Lattice parameters can be refined; however, in the current implementation of the program the cell angles cannot. These parameters are more reliably determined from a conventional crystallographic analysis anyway. The unit cell could be the crystallographic unit cell. However, in general it is some supercell which allows aperiodic lattice displacements to be incorporated into the model. The unit cell need not be orthogonal.

No space-group symmetry operations are applied to the contents of the unit cell, though periodic boundary conditions are applied to the cell itself. This is an inconvenience when setting up the model because of the number of coordinates which have to be specified; however, it has the advantage that arbitrary distortions (atom displacements) can be introduced into the model without danger of moving off a special position and generating new atomic positions, or having to find the new symmetry subgroup appropriate to the intended distortion.

Each atomic pair correlation is convoluted with a gaussian distribution function to account for thermal and zero-point motion. The width of the gaussian depends on which atoms contribute to the correlation and its direction in space. The gaussian broadening parameters are specified in the model as site-specific anisotropic displacement (thermal) factors, again, in close analogy with Rietveld refinement. The broadening of the atomic pair correlation is calculated by projecting these anisotropic displacement factors along the vector joining the two atoms in the pair, as we describe in detail later.

In the current implementation of the program, only the diagonal elements of the anisotropic displacement-parameter matrix can be determined. This means that the major axes of the thermal ellipsoids are constrained to lie parallel to the lattice vectors. This is a limitation in the code which can be circumvented somewhat by specifying rotated unit cells if a prolate or oblate ellipsoid is suspected whose major axis is not parallel to the original cell axis. This limitation should certainly be taken into account when interpreting anisotropic thermal factors. For example, a cigar-shaped ellipsoid pointing in a [110] direction will appear as a pancake shaped ellipsoid to the program. However, a 45° rotated supercell will recover the correct result.

The width of the gaussian distribution function for the ijth atomic pair correlation (the correlation between the ith and the jth atoms), σ_{0ij}, depends on the pair of atoms involved and on the direction of the vector joining the ith and the jth atom. In detail,

$$\sigma^2_{0\,ij} = x^2_{ij}(\sigma^2_{ix} + \sigma^2_{jx}) + y^2_{ij}(\sigma^2_{iy} + \sigma^2_{jy}) + z^2_{ij}(\sigma^2_{iz} + \sigma^2_{jz}), \tag{4}$$

where σ_{ix} is the x-component of the displacement factor of the ith ion, x_{ij} is the x-component of $\mathbf{r_{ij}}$, and so on.

Again, in analogy with the Rietveld method, site occupancies can be refined. This means that sites with incomplete occupancies can be determined. It also allows uncorrelated atomic displacements to be introduced in some average sense as partially occupied split positions. For example, a symmetry lowering soft phonon introduces correlated displacements which would require a supercell to describe. However, the displacements can be approximated in the undistorted cell by creating two displaced positions in the directions of the phonon distortion, and giving each 50% occupancy. This is a fairly generic way of introducing anharmonicity. It is useful for determining whether significant atomic displacements exist beyond gaussian disorder, before extensive effort is expended on trying to finding complicated correlated short-range ordered local displacements models.

Atomic motion in solids is highly correlated: directly bonded neighbors tend to move in phase with each other. This appears in the PDF as a sharpening of the PDF peaks at low-r.[13] The origin of this effect in the data is thermal diffuse scattering (this can be seen later in Fig. 2(b)). This effect is also discussed in more detail elsewhere in these proceedings[14] where it is being used to extract theoretical potential parameters. It is important to be able to account for this effect in the calculated $G(r)$. We use an empirical approximation for the r-dependence of the PDF peak sharpening:

$$\sigma_{ij} = \sigma_{0\,ij} - \frac{\delta}{r_{ij}^2}, \qquad (5)$$

where σ_{ij} is the corrected broadening for the ijth atomic pair correlation, $\sigma_{0\,ij}$ is the broadening due to the anisotropic displacement factors of the ith and jth atoms defined in Eq. 4 and δ is a parameter.

Experimental Dependent Parameters

Some refinable parameters are included in the program to account for experimental artefacts.

The function $G(r)$ is an absolute function. For example, by integrating the intensity under a peak in the function $4\pi r^2 \rho(r)$, one recovers the number of atomic neighbours contributing to that peak. However, in the process of obtaining $G(r)$ from experimental data it is possible to introduce a scale error. For example, the contribution to the data from multiple scattering events are subtracted before the data are normalized. If this subtraction is not done correctly, an incorrect normalization will result as the additive error is compensated by the multiplicative normalization correction. This will introduce a scale error into the experimentally determined $G(r)$. To allow for this a scale factor is refined in the RESPAR model. The final value of the scale factor should be close to unity and this gives confidence that all of the data corrections were carried out correctly. However, the scale factor can be varied to account for small deviations from unity which may be present.

The finite Q resolution of the measurement introduces a gradual fall-off in the amplitude of $G(r)$ with increasing r. This is probably most easily understood by considering the inverse situation. It is well known that the uncertainty in position of atoms in r-space due to thermal and zero-point fluctuations causes the intensity of elastic scattering to fall off with increasing Q. This is the well known Debye-Waller factor.[3] Thus, a finite resolution in real-space gives rise to a drop off in intensity with Q in Q-space. In a similar way, the resolution function of the measurement in Q is *convoluted* with the real signal, therefore G(r) should be *multiplied* by the Fourier transform of the resolution function. In the RESPAR program a constant Gaussian form is assumed for the Q-space resolution function and $G(r)$ is therefore multiplied by a gaussian:[12]

$$\rho_1(r) = \rho(r)e^{-\frac{(\sigma_Q r)^2}{2}} \qquad (6)$$

where σ_Q is the standard deviation of the Q-space resolution function. This is not a perfect correction since, in general, the instrument resolution function is not Gaussian, nor is it necessarily constant in Q. However, this correction accounts well for the gross features of the effect of the finite Q-resolution, especially when the PDF is being fit over a limited range below ~ 20 Å^{-1}.

The finite Q range of the data also introduces errors into the experimental $G(r)$ in the form of termination errors. These are ripples which appear in $G(r)$ around the base of sharp peaks in the function. The function $G(r)$ is determined from the Fourier transform shown in Equation 1. However in practice the integral does not go from 0 to ∞, but from Q_{min} to Q_{max} which are lower and upper bounds to the available data. The upper bound cutoff can be accounted for by assuming that data to ∞ have been multiplied by a step function, W(Q), which is unity for $Q \le Q_{max}$ and zero for $Q > Q_{max}$. From the convolution theorem we get that the resulting Fourier transform, $G_e(r)$ is a convolution of $G(r)$ with the Fourier transform of the step function. Thus, $G(r)$ should be convoluted with a sinc function according to[17]

$$G_e(r) = \frac{1}{\pi} \int_0^\infty G(r') \left[\frac{\sin Q_{max}(r - r')}{r - r'} - \frac{\sin Q_{max}(r + r')}{r + r'} \right] dr'. \qquad (7)$$

The value of Q_{max} from the data analysis is entered as an input but cannot be refined. If a value $Q_{max} = 0$ is entered, the program changes Q_{max} to 100, effectively eliminating the convolution.

There is currently no account taken in the modelling program for the finite Q_{min}. If significant data is missing in the low-Q region, long wavelength oscillations appear in the difference curve between the calculated and experimental PDFs. This problem is generally minimized if data are collected to below the first large Bragg peak in the data. In this case the long-wavelength oscillations in $G(r)$ are not noticable.

Additive artifacts in the data will also contribute to the systematic errors of $G(r)$. In general, these additive artifacts have a long wavelength in Q since they originate from inadequacies in multiple scattering, or inelasticity (Plaçek) corrections for example. Such long-wavelength contributions to the scattering will give rise to intensity peaks in $G(r)$ close to $r = 0$. Because of the finite Q_{max}, these intensity peaks are convoluted with a sinc function and appear as oscillations in the data which are peaked near $r = 0$ but which die out with increasing r. At present, no account is taken of these errors in the modelling program. It is assumed that they penetrate into the data, but they have some random phase relationship with the real signal. If the data are fitted over a wide enough range of Q so that many correlations are fit at the same time, it is very unlikely that these artificial oscillations will bias the result of the refinement. For example, a noise peak could make a PDF peak have a shoulder which could look like evidence of anharmonic atomic displacements if this single peak is considered alone. However, it is very unlikely that all other peaks elsewhere in $G(r)$ have a perfectly self-consistent noise peak associated with them which would fool the program into introducing a spurious distortion. The presence of these noise peaks will increase the agreement factor and χ^2, but should not affect the refined values for variables. This discussion highlights the danger of attempting to draw conclusions from the appearance of a single peak. It also underscores the importance of fitting data over a reasonable range of r which is one of the advantages this technique has over other local structure techniques such as XAFS.[15]

Difference Modelling

The modelling program has recently been extended in one significant way. It is now possible to model structural *changes* such as might occur at a structural phase transition. In this case, the difference is taken between two data-sets which straddle the structural transition. The resulting difference curve contains features which come from the changes that the structure underwent, as well as random noise.* In this case it can be desirable to fit the difference curve rather than the total PDF, especially if the structural changes are small. A reference structure is found by refining a model to the data in the higher symmetry phase and the PDF, $G_{ref}(r)$, of this model structure is determined. Atomic distortions are then introduced into the model as desired. The difference modelling program calculates the new $G(r)$, subtracts it from $G_{ref}(r)$ and compares the result with the difference curve from the two data-sets. In this way,

---

*Noise in the data from systematice sources can be largely cancelled out when the difference is taken between the two data sets, providing the systematic errors are reproduced between the two measurements

structural *changes* are refined. This approach also works if the reference structure is not known perfectly, since the reference structure cancels out when the difference is taken in both the data and the model. This difference modelling approach will be described in more detail elsewhere[16].

EXAMPLES

Indium Arsenide

The compound semiconductor InAs provides an excellent model system for testing the capabilities of the modelling program. The compound InAs forms in the zinc blende structure. It is a fully ordered compound and so no disorder is expected and the crystallographic structure can be directly compared with that obtained from modelling the PDF. Also, $G(r)$ has been calculated theoretically for this compound using model potentials and the experimentally measured density of states to account for thermal effects.[17,14] This provides a strict test of how well the modelling program accounts for the data and the theoretical calculation of $G(r)$. In particular, we will use this example to test our empirical approach of accounting for correlated thermal motions.

Figure 2(a) shows the experimentally determined $G(r)$ function from InAs. The data were measured at 300 K using x-rays at beamline X7A at the National Synchroton Light Source (NSLS), Brookhaven National Laboratory. The data were collected over a Q-range from 0.074Å^{-1} to 22Å^{-1} in symmetric reflection geometry. The model is a best fit to the data; however, no account has been taken of correlated motion in the material. It is quite clear that the nearest-neighbor peak in the data is significantly sharper than that in the model. This is because in the solid, near neighbour atoms tend to move in phase. Since $G(r)$ is a function of atomic *pairs* it shows the *relative* motion of the atoms in the pair and atoms moving in-phase will give rise to narrow PDF peaks. However, the motion of further neighbors is less correlated and the peaks are relatively broader. The fit has been optimized to account for the uncorrelated motion to highlight the effect of the motional correlations.

The information about the motional correlations in the scattering comes from thermal diffuse scattering (TDS).[3] The high-Q region of the scattering from the same data set, in the form of $i(Q) = Q[S(Q) - 1]$, is shown in Fig. 2(b) and the TDS is clearly evident under the peaks. It should be noted that the TDS is not removed from the data before it is Fourier transformed since it is carrying relevant information about the sample.

In many cases we are not interested in studying the motional correlations and so the correlation-narrowing is accounted for with the empirical expression shown in Eq. 5. We would like to know how well this accounts for the correlation-narrowing. The same data which were shown in Fig. 2(a) are shown in Fig. 3; however, this time the model-PDF has been calculated incorporating Eq. 5. The improvement in fit is clear. The correlation-narrowing can be calculated explicitly in this system using the theoretical model as described above. This has been done assuming a temperature of 300 K. The widths of each of the PDF peaks is then plotted as a function of r. This is shown in Fig. 4. Also plotted are the measured r-dependent PDF peak broadenings from the data. The solid line plotted on top is the empirical relation for the correlation narrowing given in Eq. 5. It is clear that the form used in Eq. 5 accounts quite well for the motional correlations.

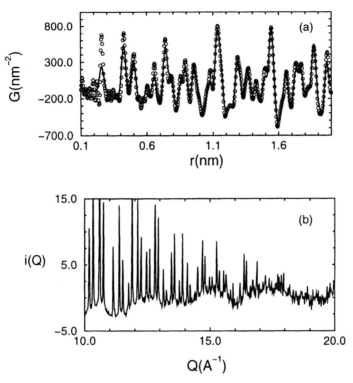

Figure 2. Evidence of correlated atomic motions in $G(r)$. (a) PDF from InAs at 300 K (circles) with calculated PDF (solid line) assuming uncorrelated atomic motions. The low-r peaks appear sharpened in the data. Data collected at beamline X7A, NSLS, Brookhaven. (b) The origin of this information in the scattering data is the thermal diffuse scattering shown here. The high-Q portion of the same scattering data shown as $G(r)$ in (a) is shown as $i(Q) = Q[S(Q) - 1]$.

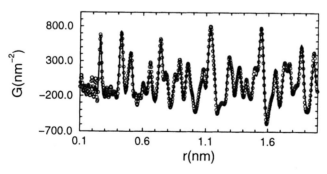

Figure 3. Same data from InAs as shown in Fig. 2 (circles). However, this time the calculated PDF (solid line) has been corrected for correlated atomic motion

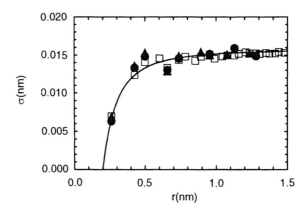

Figure 4. PDF peak width, σ, as a function of pair separation showing the sharpening of peaks at low-r. Solid symbols: values determined from the experimental PDF shown in Figs. 2 and 3. Open circles: values obtained from a theoretical calculation (see Ref. 14 for details). Solid line: obtained using Eq. 5 with $\sigma_0 = 0.0158$ nm and $\delta = 0.00063$ nm^3

Nickel

We have measured nickel at a number of sources to compare the reproducibility of the PDFs we obtain and the ability to model data from different sources. In Fig. 5(a) data collected on the Glass, Liquids and Amorphous Diffractometer (GLAD) at the Intense Pulsed Neutron Source (IPNS) at Argonne National Laboratory at 300 K are shown as a dashed line ($Q_{max} = 30$ Å^{-1}). The data shown in (b) are x-ray data collected at beamline X7A at the NSLS, Brookhaven National Laboratory at room temperature ($Q_{max} = 22$ Å^{-1}), and in (c) from a sealed-tube molybdenum laboratory x-ray source at Michigan State University, again at room temperature ($Q_{max} = 16$ Å^{-1}). The data from the sealed-tube source are interesting because the limited Q-range gives rise to low real-space resolution and large termination ripples. For example, the peak at 3.0 Å evident in Fig. 5(c) is entirely spurious as can be seen by comparing with the data in Figs. 5(a) and (b) where this peak is absent. These well known "termination errors" have been widely discussed.[3,4] Clearly, the best approach to dealing with them is to measure to sufficiently high Q, as is evident in the data in Figs. 5(a) and (b). However, as Fig. 5(c) shows, the termination errors are incorporated into the modelling program (the model PDF is shown as a solid line in the figure) which accounts very well for the spurious peak in the data. It is therefore highly unlikely that the structural refinement will be biased by the presence of these well controlled errors. It should be noted that no damping was applied to S(Q) to diminish termination ripples before Fourier transforming to $G(r)$. It is now our practice to terminate the data at Q_{max} with a sharp step function (no smooth damping) and to apply the simple convolution procedure to the model-$G(r)$.

The values for the thermal factors obtained from refinements carried out with and without the convolution turned on are shown in Table 1. The displacement parameters determined from the x-ray data-sets with the convolution turned on yield very similar values, despite the fact that the data have very different peak widths, as is evident

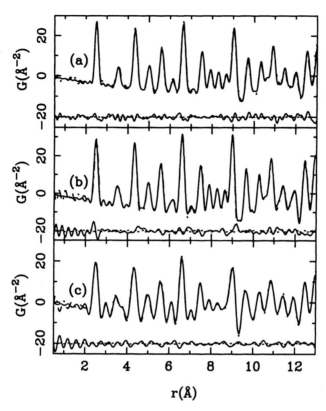

Figure 5. Comparison of data and fits from different diffractometers. All data-sets are from nickel at 300 K. They were collected at (a) GLAD at IPNS, Argonne National Laboratory, (b) beamline X7A at NSLS, Brookhaven National Laboratory and (c) on a laboratory tube source. Difference curves are plotted below the data.

Table 1. Displacement factors determined using the RESPAR program from various nickel data-sets, all collected at room temperature but at different diffractometers (see text for details). Values are shown when the model was convoluted with the measurement resolution and when it was not.

Diffractometer	U_{iso} (Å^2) convolution off	U_{iso} (Å^2) convolution on
GLAD	0.00585	0.00582
NSLS	0.00557	0.00516
in-house	0.00733	0.00480

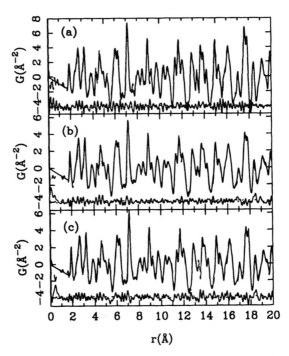

Figure 6. Example of the reproducibility of data and fitting. All data are from La_2CuO_4 at 10 K. Data were collected from the same sample at (a) SEPD at IPNS, Argonne; (b) HIPD at MLNSC, Los Alamos; and (c) from a different sample of the same material at HIPD. Solid lines are the fits, difference curves are plotted below the data.

in Figs. 5(b) and (c). The broadening of the PDF peak widths in the lower resolution (lower Q_{max}) measurements is reflected in the values for displacement parameter determined by the program without the convolution broadening (see Table 1) which are considerably higher for the sealed-tube x-ray data. For some reason, the neutron PDF gives a slightly higher value for U_{iso}, possibly because the temperature of the sample (which was not monitored) was a little higher. It is clear that by the time the data are terminated at $Q_{max} = 30$ $\AA^{-1}$ the effects of the finite data-range are very small: both the termination ripples and the PDF peak broadening due to this effect are insignificant.

La_2CuO_4

We have also used La_2CuO_4 as a model to test the reliability of obtaining data from different neutron sources. This system also should be quite well ordered since it is a stoichiometric compound; however, the structure[18] is considerably more complex than InAs or Ni and therefore provides a more stringent test of the reproducibility of the refinements.

All the data were measured at 10 K at either the High Intensity Powder Diffractometer at the Manuel Lujan, Jr., Neutron Scattering Center (MLNSC) at Los Alamos, or the Special Environment Powder Diffractometer (SEPD) at IPNS. Precisely the same sample was measured at each source, and a second sample of the same material was also measured at MLNSC. The data are shown in Fig. 6. Each data-set has a fully

Table 2. Local octahedral tilt amplitudes from La_2CuO_4 at 10 K determined from the O1 displacements ($|\theta|_{O1}$) and the O2 displacements ($|\theta|_{O1}$) which were refined using the RESPAR program. Values were refined from 2 samples which were measured at two different neutrons sources (see text).

Measurement	Agreement factor	$|\theta|_{O1}$ (degrees)	$|\theta|_{O2}$ (degrees)
sample 1, SEPD (Fig. 6(a))	0.1495	5.226	5.446
sample 1, HIPD (Fig. 6(b))	0.1513	4.852	5.457
sample 2, HIPD (Fig. 6(c))	0.1726	5.110	5.245

converged model-PDF plotted over it and below each one is a difference curve. The dotted lines associated with the difference curve indicate the expected uncertainties due to random counting statistics (random errors) but not systematic errors. In each case there is a satisfactory agreement. There is also good agreement between the values of parameters refined from the three data sets. For example, one parameter of particular interest in these materials is the magnitude of the tilt angle of the CuO_6 octahedra. The 2-dimensional CuO_2 planes in these materials are made up of a network of corner shared octahedra. This network of octahedra collectively buckles at low temperature corrugating the CuO_2 plane. The magnitude of the tilt angle, $|\theta|$ can be extracted from the modelling program independently from the z-displacements of the in-plane (O1) oxygen ions and from the y-displacements of the apical (O2) ions. The values determined from each of the three measurements shown in Fig 6 are given in Table 2. The results show that even subtle parameters such as an octahedral tilt of $< 5°$ can be reproducibly determined.

Close inspection of the low-r region of these curves indicates that there may be systematic differences between the model and the data which are reproduced in the different data-sets.[19] This is currently being investigated.

$La_{2-x}(Sr,Ba)_xCuO_4$

The charge state of copper in the compound La_2CuO_4 can be changed by partially replacing La with a divalent ion such as Sr or Ba. This is known as doping. In the doped phase, the structural transition from a tetragonal phase at high temperature (HTT), in which the average CuO_6 tilts are zero, to an orthorhombic phase at low temperature (LTO) in which the tilts are finite and collective, comes down in temperature so that for $x = 0.15$ it is at $\sim 150K$.[20] This tilting transition has long been understood in terms of the soft-mode, displacive picture,[21] and indeed zone boundary phonon modes are observed to soften at the transition.[21,22] However, our studies using the PDF and analysing the data using the RESPAR program, show that locally, the tilts do not disappear in the HTT phase.[2,9] This is shown in Fig. 7. In this Figure, the behaviour of the average tilts is evident by tracking the integrated intensity of the [032] superlattice peak, shown as circles. This is an order parameter for the phase transition and is proportional to the square of the *average* tilt angle ($\langle\theta\rangle^2$).[21] This goes smoothly to zero at the phase transition temperature of 203 K. However, refinements of the PDF indicate that finite tilts persist to very high temperature, and indeed the magnitude of the local tilt does not vary strongly with temperature.

This is an example where the local structure is different from the average crystal structure and where it is important to study the local structure directly. The full-profile fitting routines described here yielded a quantitative solution of the local octahedral tilt structure as a function of temperature which could not be obtained using conventional Rietveld.

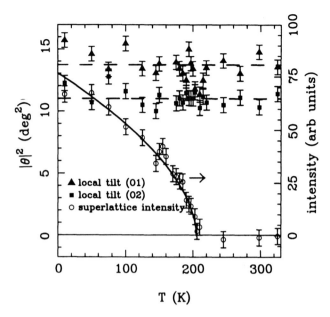

Figure 7. Evolution of octahedral tilt angles with temperature in $La_{2-x}Sr_xCuO_4$. The open circles shows the behavior of the long-range ordered tilt through the [032] superlattice intensity. Solid triangles and solid squares are the local tilt angle refined from displacements of O1 and O2 ions respectively

CONCLUSIONS

We have described a full-profile fitting technique for extracting local atomic structure from the atomic pair distribution function. The implementation described here uses the least-squares regression technique and gives structural information such as atomic coordinates, site specific anisotropic displacement parameters and occupancies. The code is called RESPAR which stands for REal SPAce Rietveld because of its close analogy with the Rietveld method. However, one significant difference should be emphasised. The real-space approach described here yields the *local* short-range order and not the average crystal structure which is obtained from a conventional Rietveld analysis. This approach is straightforward and robust and yields quantitative local structural information. For a well ordered material, it complements conventional crystallography by giving short-range parameters such as displacement (thermal) factors with increased accuracy (because of the wide range of Q which is used in the data). Also, variables in the refinement are differently correlated in real- and reciprocal-space and so a "dual-space" or joint real- and reciprocal-space refinement can yield more reliable values when parameter correlation is a problem. However, the strength of the approach lies in the fact that it goes beyond the approximation of periodicity and yields aperiodic components of the structure when they exist, as was illustrated in the examples presented here.

The program is available for use on a non-commercial basis. Information can be obtained by contacting the author by email (billinge@pa.msu.edu).

ACKNOWLEDGEMENTS

I would like to thank all of those people who contributed to the work described in this paper: E. Božin, J. Chung, D. E. Cox, T. Egami, I-K. Jeong, G. H. Kwei, F. Mohiuddin-Jacobs, M. F. Thorpe, R. Torasso and R. B. Von Dreele. The work at MSU was supported by DOE grant DE FG02 97ER45651 and by the Alfred P. Sloan Foundation. Part of the work described here was carried out whilst the author was a graduate student at U. of Pennsylvania and as a Postdoctoral Fellow at Los Alamos National Laboratory. X-ray data were collected at the National Synchrotron Light Source, Brookhaven National Laboratory which is funded by DOE Division of Materials Sciences and Division of Chemical Sciences through contract DE-AC02-76CH00016. Neutron data were collected at the MLNSC at Los Alamos National Laboratory which is supported by the USDOE offices of Defense Programs and Basic Energy Sciences, and at the Intense Pulsed Neutron Source at Argonne National Laboratory which is funded by the USDOE, BES-Materials Science under contract W-31-109-Eng-38.

APPENDIX: COMPUTATIONAL DETAILS

The equations used in the modelling program to calculate $G(r)$ are laid out below. The program uses the least-squares technique for finding the best-fit of the calculated $G(r)$ to the experimentally determined one. The refinable parameters of the program are listed in Table 3.

The value of the function $G(r)$ is determined at the position $r = r_k$ by G_k:

$$G_k = 4\pi r_k D_k S \left[\rho_k - \rho_0 \right],\tag{8}$$

where D_k is the experimental resolution factor given by $D_k = e^{-\sigma_Q^2 r_k^2 / 2}$, S is a constant scale factor, $\rho_0 = N_u/V_u$ is the average number density where $N_u = \sum_i P_i$ is the number of atoms in the unit cell and V_u is the unit cell volume, $V_u = abc(1 - \cos^2 \alpha - \cos^2 \beta - \cos^2 \gamma + 2 \cos \alpha \cos \beta \cos \gamma)$, where α, β and γ are the unit cell angles. The sum over i is a sum over every atomic site in the unit cell and P_i is the atomic occupancy of that site. Finally, the microscopic density at position $r = r_k$, ρ_k, is the sum of the contributions from all the atomic pair correlations which have significant intensity at this point. Thus,

$$\rho_k = \sum_c \rho_{ck},\tag{9}$$

where the sum over c is a sum over all atomic-pair correlations which contribute intensity at the position $r = r_k$. In practice this means all those correlations which lie

Table 3. List of parameters which can be varied in the RESPAR program.

Parameter	Symbol
scale factor	S
resolution factor	σ_Q
dynamic correlation factor	δ
lattice paramters	a, b, c
anisotropic thermal factors	$\sigma_{ix}, \sigma_{iy}, \sigma_{iz}$
fractional atomic coordinates	f_{ix}, f_{iy}, f_{iz}
site occupancies	P_i

within $5\sigma_c$ of r_k, where σ_c is the gaussian half-width of the cth correlation which is between the ith and jth ions. The contribution to ρ_k from the cth correlation, ρ_{ck}, is given by

$$\rho_{ck} = \frac{1}{4\pi r_k^2} A_c \, g_{ck}. \tag{10}$$

Here A_c is the magnitude of the cth correlation and g_{ck} is the value of the normalised gaussian, centred at the position of the cth correlation, r_c, but computed at the position r_k:

$$g_{ck} = \frac{1}{\sqrt{2\pi}\,\sigma_c} e^{-\frac{(r_k - r_c)^2}{2\sigma_c^2}}. \tag{11}$$

The value of A_c depends on the scattering lengths, b_i, and occupancy factors, P_i, of the two atoms involved in the correlation. In detail,

$$A_c = \frac{P_i P_j b_i b_j}{N_u \langle b \rangle^2}, \tag{12}$$

where $\langle b \rangle$ is the average scattering length of the whole sample. In the case of x-ray data, all scattering lengths are those at $Q = 0$; i.e., the number of electrons in the ion. The position of the cth correlation, r_c, is given by:

$$r_c = \left[(af_{cx})^2 + (bf_{cy})^2 + (cf_{cz})^2 + 2abf_{cx}f_{cy}\cos\gamma + 2bcf_{cy}f_{cz}\cos\alpha + 2acf_{cx}f_{cz}\cos\beta \right]^{\frac{1}{2}}, \tag{13}$$

where f_{cx} is the x-compnent of the separation of the ith and jth atom in the cth correlation, expressed in fractional coordinates, and is given by

$$f_{cx} = f_{jx} - f_{ix} - n_u, \tag{14}$$

where the number of complete unit cells separating these two atoms is n_u and f_{ix} is the x-coordinate of the ith atom within the unit cell. Thus, f_{cx} is the magnitude of the x–component of the vector joining the two atoms, expressed in fractional coordinates.

For completeness, we give the equations for the width of the cth correlation in discrete formalism. These equations can be compared with Equations 4 and 5 given earlier:

$$\sigma_c' = \frac{1}{r_c} \left[f_{cx}^2 a^2 (\sigma_{ix}^2 + \sigma_{jx}^2) + f_{cy}^2 b^2 (\sigma_{iy}^2 + \sigma_{jy}^2) + f_{cz}^2 c^2 (\sigma_{iz}^2 + \sigma_{jz}^2) \right]^{\frac{1}{2}} \tag{15}$$

where $\sigma_c' = \sigma_{0ij}$ defined in Eq. 4. Finally,

$$\sigma_c = \sigma_{ij} = \sigma_c' - \frac{\delta}{r_c^2}. \tag{16}$$

APPENDIX: PARTIAL DERIVATIVES

The least-squares equations use the partial derivatives of the function being fit [in this case, $G(r; \{p\})$] with respect to each of the variable parameters, $\{p\}$, which are given in Table 3. The efficiency of the refinement depends sensitively on these partials being determined accurately and so we have implemented analytic partial derivatives. The partials with respect to each of the parameters are calculated at each cycle of the refinement. The equations for the partial derivative of G_k with respect to each variable parameter are given below.

Scale Factor

$$\frac{\partial G_k}{\partial S} = \frac{G_k}{S}. \tag{17}$$

Resolution Factor

$$\frac{\partial G_k}{\partial \sigma_Q} = -r_k^2 \sigma_Q G_k. \tag{18}$$

Dynamic Correlation Factor

The Dynamic Correlation Factor, δ, appears everywhere that the correlation gaussian broadening, σ_c, appears. Thus,

$$\frac{\partial G_k}{\partial \delta} = 4\pi r_k D_k S \frac{\partial \rho_k}{\partial \delta}, \tag{19}$$

$$\frac{\partial \rho_k}{\partial \delta} = \sum_c \frac{A_c}{4\pi r_k^2} \frac{\partial g_{ck}}{\partial \delta} \tag{20}$$

and

$$\frac{\partial g_{ck}}{\partial \delta} = \frac{g_{ck}}{\sigma_c} \left\{ \frac{(r_k - r_c)^2}{\sigma_c^2} - 1 \right\} \frac{\partial \sigma_c}{\partial \delta} \tag{21}$$

$$= -\frac{g_{ck}}{\sigma_c r_c^2} \left\{ \frac{(r_k - r_c)^2}{\sigma_c^2} - 1 \right\}. \tag{22}$$

Finally,

$$\frac{\partial G_k}{\partial \delta} = \frac{-D_k S}{r_k r_c^2} \sum_c \frac{g_{ck}}{\sigma_c} \left[\frac{(r_k - r_c)^2}{\sigma_c^2} - 1 \right]. \tag{23}$$

Lattice Parameters

Lattice parameters appear in r_c, σ_c and also in the cell volume, V. Consider the a parameter first.

$$\frac{\partial G_k}{\partial a} = 4\pi r_k D_k S \left[\frac{\partial \rho_k}{\partial a} - \frac{\partial \rho_0}{\partial a} \right], \tag{24}$$

where,

$$\frac{\partial \rho_k}{\partial a} = \sum_c \frac{\rho_{ck}}{\sigma_c} \left\{ \left[\frac{(r_k - r_c)^2}{\sigma_c^2} - 1 \right] \frac{\partial \sigma_c}{\partial a} + \left[\frac{(r_k - r_c)}{\sigma_c} \right] \frac{\partial r_c}{\partial a} \right\}, \tag{25}$$

and

$$\frac{\partial \rho_0}{\partial a} = -\frac{\rho_0}{a}. \tag{26}$$

Now,

$$\frac{\partial \sigma_c}{\partial a} = \frac{\partial \sigma_c'}{\partial a} + \frac{2\delta}{r_c^3} \frac{\partial r_c}{\partial a}. \tag{27}$$

Substituting this back into Eq. 25 we get

$$\frac{\partial \rho_k}{\partial a} = \sum_c \frac{\rho_{ck}}{\sigma_c} \left\{ \left[\frac{(r_k - r_c)^2}{\sigma_c^2} - 1 \right] \frac{\partial \sigma_c'}{\partial a} + \left[\frac{(r_k - r_c)}{\sigma_c} + \frac{2\delta}{r_c^3} \left(\frac{(r_k - r_c)^2}{\sigma_c^2} - 1 \right) \right] \frac{\partial r_c}{\partial a} \right\}, \tag{28}$$

with,

$$\frac{\partial \sigma_c'}{\partial a} = \frac{a f_{cx}^2}{\sigma_c' r_c^2} (\sigma_{ix}^2 + \sigma_{jx}^2) - \frac{\sigma_c'}{r_c} \frac{\partial r_c}{\partial a} \tag{29}$$

and

$$\frac{\partial r_c}{\partial a} = \frac{1}{r_c} (a f_{cx}^2 + b f_{cx} f_{cy} cos\gamma + c f_{cx} f_{cz} cos\beta). \tag{30}$$

The relationships for b and c are obtained by permuting a, b, c and x, y, z

Anisotropic Thermal Factors

These appear only where σ_c appears. Thus,

$$\frac{\partial G_k}{\partial \sigma_{ix}} = 4\pi r_k D_k S \frac{\partial \rho_k}{\partial \sigma_{ix}}, \tag{31}$$

where,

$$\frac{\partial \rho_k}{\partial \sigma_{ix}} = \sum_c \frac{\rho_{ck}}{\sigma_c} \left\{ \left[\frac{(r_k - r_c)^2}{\sigma_c^2} - 1 \right] \frac{\partial \sigma_c}{\partial \sigma_{ix}} \right\}, \tag{32}$$

and

$$\frac{\partial \sigma_c}{\partial \sigma_{ix}} = \frac{a^2 f_{ix}^2 \sigma_{ix}}{\sigma_c' r_c^2}. \tag{33}$$

Thus,

$$\frac{\partial G_k}{\partial \sigma_{ix}} = 4\pi r_k D_k S \left\{ \sum_c \frac{\rho_{ck} \sigma_{ix} a f_{ix}^2}{\sigma_c \sigma_c' r_c^2} \left[\frac{(r_k - r_c)^2}{\sigma_c^2} - 1 \right] \right\}. \tag{34}$$

The expression will be the same for $\frac{\partial G_k}{\partial \sigma_{jx}}$. The expressions in the y and z directions are recovered by cycling a, b, and c and x, y, and z.

Fractional Atomic Coordinates

The fractional coordinates, f_{ix} appear in r_c and in σ_c.

$$\frac{\partial G_k}{\partial f_{ix}} = 4\pi D_k S \frac{\partial \rho_k}{\partial f_{ix}}, \tag{35}$$

where

154

$$\frac{\partial \rho_k}{\partial f_{ix}} = \sum_c \frac{\rho_{ck}}{\sigma_c} \left\{ \left[\frac{(r_k - r_c)^2}{\sigma_c^2} - 1 \right] \frac{\partial \sigma_c}{\partial f_{ix}} + \left[\frac{(r_k - r_c)}{\sigma_c} \right] \frac{\partial r_c}{\partial f_{ix}} \right\}, \tag{36}$$

$$\frac{\partial \sigma_c}{\partial f_{ix}} = \frac{\partial \sigma_c'}{\partial f_{ix}} + \frac{2\delta}{r_c^3} \frac{\partial r_c}{\partial f_{ix}}, \tag{37}$$

$$\frac{\partial \sigma_c'}{\partial f_{ix}} = -\frac{a^2 f_{cx}}{\sigma_c' r_c^2} (\sigma_{ix}^2 + \sigma_{jx}^2) - \frac{\sigma_c'}{r_c} \frac{\partial r_c}{\partial f_{ix}}, \tag{38}$$

and

$$\frac{\partial r_c}{\partial f_{ix}} = -\frac{1}{r_c} (a^2 f_{cx} + ab f_{cy} \cos \gamma + ac f_{cz} \cos \beta). \tag{39}$$

The expressions for $\partial G_k / \partial f_{jx}$ are the same except that

$$\frac{\partial \rho_k}{\partial f_{jx}} = -\frac{\partial \rho_k}{\partial f_{ix}}. \tag{40}$$

Site Occupancies

Site occupancy factors appear in ρ_0 and in the weights of the correlations, A_c. Thus,

$$\frac{\partial G_k}{\partial P_i} = 4\pi r_k D_k S \left[\frac{\partial \rho_k}{\partial P_i} - \frac{\partial \rho_0}{\partial P_i} \right], \tag{41}$$

where,

$$\frac{\partial \rho_0}{\partial P_i} = \frac{1}{V} \tag{42}$$

and

$$\frac{\partial \rho_k}{\partial P_i} = \sum_c \rho_{ck} \left(\frac{1}{P_i} - \frac{1}{N_u} \right). \tag{43}$$

Thus,

$$\frac{\partial G_k}{\partial P_i} = 4\pi r_k D_k S \left\{ \left[\sum_c \rho_{ck} \left(\frac{1}{P_i} - \frac{1}{N_u} \right) \right] - \frac{1}{V} \right\}. \tag{44}$$

In the case where the parameter P_i is not involved in the correlation, it still contributes to the derivative. In this case, the derivative is given as,

$$\frac{\partial G_k}{\partial P_i} = -4\pi r_k D_k S \left[\left(\sum_c \frac{\rho_{ck}}{N_u} \right) + \frac{1}{V} \right]. \tag{45}$$

REFERENCES

[1] e.g., see other papers in this volume.
[2] E. S. Božin, S. J. L. Billinge, and G. H. Kwei, to appear in *Physica B*.
[3] B. E. Warren, *X-ray Diffraction*, Dover, New York, (1990).

[4] H. P. Klug and L. E. Alexander, *X-ray Diffraction Proceedures for Polycrystalline Materials*, Wiley, New York, 2nd edition, (1974).

[5] D. Grimley et al., *J. Non-Cryst. Solids* 119:49 (1990).

[6] T. Egami, This volume.

[7] I.-K. Jeong and S. J. L. Billinge, unpublished.

[8] B. H. Toby, T. Egami, J. D. Jorgensen, and M. A. Subramanian, *Phys. Rev. Lett.* 64:2414 (1990).

[9] S. J. L. Billinge, G. H. Kwei, and H. Takagi, *Phys. Rev. Lett.* 72:2282 (1994).

[10] S. J. L. Billinge, R. G. DiFrancesco, G. H. Kwei, J. J. Neumeier, and J. D. Thompson, *Phys. Rev. Lett.* 77:715 (1996).

[11] D. Keen, This volume.

[12] B. H. Toby and T. Egami, *Acta Cryst. A* 48:163 (1992).

[13] R. Lagneborg and R. Kaplow, *Acta. Metall.* 15:13 (1967).

[14] M. F. Thorpe, J. Chung, S.J.L.Billinge, and F. Mohiuddin-Jacobs, This volume.

[15] F. W. Lytle, D. E. Sayers, and E. A. Stern, *Phys. Rev. B* 11:4825 (1975).

[16] R. DiFrancesco and S. J. L. Billinge, unpublished.

[17] J. S. Chung and M. F. Thorpe, *Phys. Rev. B* 55:1545 (1997).

[18] K. Yvon and M. Francois, *Z. Phys. B* 76:413 (1989).

[19] S. J. L. Billinge, G. H. Kwei, and H. Takagi, *Physica C* 235-240:1281 (1994).

[20] T. Egami and S. J. L. Billinge, in: *Physical Properties of High-Temperature Superconductors V*, ed. by D. M. Ginsberg, World-Scientific, Singapore, (1996).

[21] R. J. Birgenau, C. Y. Chen, D. R. Gabbe, H. P. Jenssen, M. A. Kastner, C. J. Peters, P. J. Picone, T. Thio, T. R. Thurston, and H. L. Tuller, *Phys. Rev. Lett.* 59:1329 (1987).

[22] T. R. Thurston, R. J. Birgeneau, D. R. Gabbe, H. P. Jenssen, M. A. Kastner, P. J. Picone, N. W. Preyer, J. D. Axe, P. Böni, G. Shirane, M. Sato, K. Fukuda, and S. Shamoto, *Phys. Rev. B* 39:4327 (1989).

ADVANCES IN PAIR DISTRIBUTION PROFILE FITTING IN ALLOYS

M.F. Thorpe,[1] J. S. Chung,[2] S.J.L. Billinge,[1] and F. Mohiuddin-Jacobs[1]

[1]Department of Physics and Astronomy,
and Center for Fundamental Materials Research,
Michigan State University, East Lansing, Michigan 48824

[2]Department of Physics, Chungbuk National University,
Cheongju, 360-763, Republic of Korea

INTRODUCTION

Semiconductor alloys have been studied extensively due to their importance in applications. These materials have received considerable attention because physical properties, such as the band gap, mobility and lattice parameter, can be continuously controlled.[1] Having such continuous controls is of importance in applications such as electronic devices or optical devices. For example, the energy gap of the pseudobinary compound $Ga_{1-x}Al_xAs$ can be varied between 1.4 and 2.2 eV by varying the composition x, and the wavelength of the solid-state laser made from this material can thus be tuned accordingly.

Unlike pure crystals, the difference in the bond lengths, associated with different chemical species in alloys induces internal strain. The structural characterization of alloys dates back to the work of Vegard[2] who found that the lattice constants of some alloys change linearly with the concentration of the constituents. A simplistic explanation of this phenomenon is the virtual-crystal approximation,[3] in which all the atoms are located on an ideal lattice with the lattice constant given by the compositional average of the constituents. This approximation completely neglects the local deformations which would be expected to occur. In tetrahedrally coordinated semiconductor compounds, deformation also occurs in the bond angles. A better understanding of Vegard's law in random alloys was achieved recently, and the conditions under which Vegard's law is expected to hold was given.[4] These studies used a harmonic potential, which accounts for the bond-stretching and the bond-bending forces. It has been shown that Vegard's law is strictly obeyed when the force constants for bond-stretching and the bond-bending forces are the same for the end members from which the alloy is made, and are independent of the composition. In semiconductor alloys, these simplifications were exploited in a series of recent papers.[5-8]

Most semiconductor alloys from III-V and II-VI elements follow Vegard's law very closely.[6]

The experimental structural characterization of alloys has been accomplished mainly using Bragg x-ray diffraction, and also using Bragg neutron diffraction. These experiments measure the structural quantities which are correlated over *long* distances, such as the lattice constant. More recently, extended x-ray absorption fine structure (XAFS) experiments have been used to study semiconductor alloys.[9] Such experiments investigate the *short* range order, such as the mean near-neighbor, and occasionally next nearest neighbor spacings.

However, the diffuse background in diffraction experiments has not drawn much attention because it is more difficult to obtain and hard to analyze. Nonetheless, the diffuse background exists in all experimental data on alloys due to the local displacements. This information can be analyzed using the pair distribution function (PDF). PDF analysis has been used mainly in the characterization of atomic arrangements in amorphous materials such as non-crystalline alloys or liquids.[10-12] Although it has long been known that the PDF method is well suited for analyzing crystalline as well as amorphous materials, it has only recently been applied to study the local structure of disordered crystalline materials.[13] Because the real-space resolution is inversely proportional to the highest momentum data, it is essential to have *high momentum-transfer* scattering data to study local structures. High momentum-transfer scattering data have become available with the advent of sources such as synchrotron x-ray sources and spallation neutron sources. These data not only give information about local structures from the diffusive background but also allow accurate data normalization and thus reduce systematic errors in the experimentally determined PDF. Since it has now become rather routine to access high momentum-transfer scattering data, PDF analysis is becoming a good candidate for characterizing semiconductor alloys.

There are several advantages to using PDF analysis. On the one hand, it covers a wide range of pair distances. Therefore, it can be used to study local structural characteristics such as nearest neighbor distances or intermediate-range structures such as clustering. On the other hand, it gives a complete description of the structure in that not only the average distance between a pair but the *width* of the length distribution can also be obtained. Furthermore, this method is not subject to any arbitrary fitting parameters and the result of a theoretical calculation can be *directly* compared with the experimental data.

In this paper, we present a method of calculating the PDF of binary semiconductor crystals in the zinc-blende structure, AC, and the associated pseudobinary alloys $A_{1-x}B_xC$. Calculated PDFs are then compared with experimentally determined PDFs from the same compounds. We also discuss the advantages and limitations of using the PDF analysis[14] in investigating local structures. To account for the local strain, we use a simple valence bond model,[6] which has been successful in describing the local strain in semiconductor alloys.

Since experimental data is always subject to thermal broadening, thermal averaging should be taken into account also. This is done in this paper by using the proper Bose factors and also by employing the Debye-Waller theorem. We limit ourselves in our discussion of the PDF of semiconductor alloys to the zinc-blende structure in the form of *random solid solutions*. In particular, we focus on $Ga_{1-x}In_xAs$ as an important example. But our method can be easily modified to any crystals and crystalline alloys, with or without local clustering present.

The reason for choosing $Ga_{1-x}In_xAs$ is that it is one of the largest bond-length mismatched alloys among III-V and II-VI compounds. Therefore, the effect of bond-length disorder will be most pronounced. Also it is one of the standard systems and has been studied quite extensively.[6,9] One of the experimental advantages is that the two end members, GaAs and InAs, are completely miscible and form a random solid solution at *all* concentrations.

158

MODEL

To account for the forces between atoms tetrahedrally coupled by the valence bonding in zinc-blende structures, we adopt the Kirkwood model.[15] The potential energy in this model is given by

$$V = \sum_{\langle i,j \rangle} \frac{\alpha_{ij}}{2} (L_{ij} - L_{ij}^0)^2 + L_e^2 \sum_{\langle ij,il \rangle} \frac{\beta_{ijl}}{8} (\cos\theta_{ijl} - \cos\theta^0)^2. \tag{1}$$

Here, the first term describes the energy due to the bond-stretching force with the force constant α_{ij} between atoms i and j. The lengths L_{ij} and L_{ij}^0 refer to the actual and natural (unstrained) bond lengths between atoms i and j, respectively. The second term in (1) is due to the bond-bending force with the force constant β_{ijl} between the bonds ij and il. θ_{ijl} and θ^0 are the actual and natural (109.5°) angle between the bonds ij and il. L_e is the nearest neighbor distance as given by the virtual crystal approximation. L_e is inserted in the second term to make β_{ijl} have the same dimension as α_{ij}. The angular brackets under the summations denote counting each configuration only once to exclude double counting. The potential (1) has been used extensively in discussing the elastic strain in semiconductor alloys.[5-8]

The harmonic approximation can be applied because we expect *small* positional changes in the alloy from the virtual crystal, which is used as a reference. There are two causes for the distortions, the *static* one from the bond-length mismatch and the *dynamic* one from the thermal motion. The change from bond-length mismatch is small since it is less than 10% of the unstrained bond lengths although $Ga_{1-x}In_xAs$ is one of the largest length-mismatched semiconductors. The thermal broadening is also quite small since we are interested in the low- to room-temperature range. It is therefore reasonable to use a harmonic approximation for the potential due to the displacements.

Let $\mathbf{u}_i$ be the displacement vector of atom i from its perfect crystalline position. Expanding up to linear terms in $\mathbf{u}_i$, we have

$$L_{ij} = L_e + \hat{r}_{ij} \cdot \mathbf{u}_{ij}, \tag{2}$$

where $\hat{r}_{ij}$ is a unit vector in the perfect crystal pointing from atom i to its nearest neighbor j, and $\mathbf{u}_{ij} = \mathbf{u}_j - \mathbf{u}_i$. Then the potential energy (1) can be expanded to have the form,[6]

$$V = \sum_{\langle i,j \rangle} \frac{\alpha_{ij}}{2} (L_e - L_{ij}^0 + \hat{r}_{ij} \cdot \mathbf{u}_{ij})^2 + L_e^2 \sum_{\langle ij,il \rangle} \frac{\beta_{ijl}}{8} \left[\hat{r}_{ij} \cdot \mathbf{u}_{il} + \hat{r}_{il} \cdot \mathbf{u}_{ij} + \frac{1}{3} (\hat{r}_{ij} \cdot \mathbf{u}_{ij} \hat{r}_{il} \cdot \mathbf{u}_{il}) \right]^2. \tag{3}$$

We[5-8] have preferred to use the Kirkwood model (3) rather than the Keating model[16] because of the cleaner separation of length and angular displacements. Since the nearest-neighbor central force alone is not enough to stabilize the zinc-blende structure, this model is one of the simplest force models for the zinc-blende structure. This model (3) is not good enough to produce very exact phonon dispersion relations. However, it has been proved to be accurate enough to describe the local structure quantitatively.[6] It also provides a clear picture for the important microscopic forces. Therefore, the model is a good starting point for our purpose and may be refined later as needed. It appears that this simple model can capture all the essential features in the PDF.

Equation (3) can be recast into a concise matrix form,

$$V = \tfrac{1}{2}\mathbf{u}^{+}\mathbf{M}\mathbf{u} + \mathbf{u}^{+}\mathbf{F} + E_{0} \qquad (4)$$

where $\mathbf{u} = (\mathbf{u}_1, \mathbf{u}_2, \cdots)$ is the displacement field vector and $\mathbf{M}$ is a matrix derivable from Eq. (3). The components of the force vector field $\mathbf{F} = (\mathbf{F}_1, \mathbf{F}_2, \cdots)$ are defined by

$$F_i = -\sum_j \alpha_{ij}\left(L_e - L_{ij}^0\right)\hat{r}_{ij} \qquad (5)$$

which expresses the internal strain due to the disorder. The length disorder only appears through this vector. This form of the potential (4) is useful in that it gives a simple form from which to find the relaxed equilibrium positions of strained systems, namely,

$$\mathbf{M}\mathbf{u} = -\mathbf{F}, \qquad (6)$$

and also in that the dynamical matrix is defined through the matrix $\mathbf{M}$ as discussed in the next section.

DEFINITIONS

Since different definitions are used in literature,[17] we give the definitions we use in this study. To define the dynamical matrix $\mathbf{D}(\mathbf{k})$, we need to distinguish the Bravais lattice and the basis to which the atom i belongs to. Let us divide N atoms into $\mathcal{N}$ Bravais lattice points, each containing p basis atoms $[N = \mathcal{N}p]$. Let $l(l')$ and $\mu(\mu')$ be the Bravais lattice and the basis labels of the atom $i(i')$, respectively. Denoting the position of the atom i as $\mathbf{r}_i$, we use the following definition of the dynamical matrix;

$$\mathbf{D}_{\mu\alpha,\mu'\alpha'}(\mathbf{k}) = \left(M_\mu M_{\mu'}\right)^{-1/2} \sum_{l'} M_{i\alpha,i'\alpha'} e^{-i\mathbf{k}\cdot[\mathbf{r}_i - \mathbf{r}_i]}. \qquad (7)$$

Here, a and a' denote three Cartesian coordinates, and hence the dynamical matrix is a $3p \times 3p$ matrix.

Since different nomenclature has been used for the PDF, we hereby give the definition.[14] For the sake of simplicity, we begin with an arrangement of N identical atoms.[10-13] Then, the atomic density function $p(\mathbf{r})$ is given by

$$p(\mathbf{r}) = \sum_{\mathbf{r}_i} \delta(\mathbf{r} - \mathbf{r}_i), \qquad (8)$$

and the density-density correlation function $C(r)$ can be written as

$$C(\mathbf{r}) = \frac{1}{N}\left\langle \int p(\mathbf{r}'+\mathbf{r})p(\mathbf{r}')d\mathbf{r}' \right\rangle = \frac{1}{N}\sum_{\mathbf{r}_i}\sum_{\mathbf{r}_i}\left\langle \delta(\mathbf{r} - \mathbf{r}_{ij}) \right\rangle = \rho(\mathbf{r}) + \delta(\mathbf{r}), \qquad (9)$$

where $\mathbf{r}_{ij} = \mathbf{r}_j - \mathbf{r}_i$ is a vector from atom i to j and $\langle \cdots \rangle$ denotes the *statistical* average which implies *both* configurational and thermal averages. The correlation function $C(\mathbf{r})$ describes the probability of finding an atom at position $\mathbf{r}$ from a chosen atom at the origin. This probability density is further averaged by taking each atom in turn as the origin. The delta function $\delta(\mathbf{r})$ is the probability of an atom itself and appears as a constant background in the momentum space of experiments. The function $\rho(\mathbf{r})$ defines the PDF. Defining $\rho_{ij}(\mathbf{r}) \equiv \langle\, \delta(\mathbf{r}\text{-}\mathbf{r}_{ij})\, \rangle$, the PDF can be rewritten as

$$\rho(\mathbf{r}) = \frac{1}{N} \sum_{i,j}{}' \rho_{ij}(\mathbf{r}), \tag{10}$$

where the prime in the summation means that $i = j$ is excluded.

In this study, the major interest lies in macroscopically *isotropic* materials, such as pure randomly-oriented crystallites or random solid solutions. In such materials, $\rho(\mathbf{r})$ depends only on the magnitude r. It is convenient to define the radial distribution function (RDF) as $J(r) = 4\pi\, r^2\, \rho(r)$. Then the probability should be interpreted as per unit length rather than per unit volume. The average number of atoms in a shell with radius r and thickness dr is given by $J(r)\, dr$.

However, the RDF tends to obscure the correlations between atoms as r gets larger because it grows rapidly. Hence, it is customary to define the PDF as

$$G(r) = 4\pi r[\rho(r) - \rho_0] = \frac{1}{r}[J(r) - 4\pi r^2 \rho_0], \tag{11}$$

where ρ_0 is the average *number* density of the material. Since the average density is subtracted, $G(r)$ oscillates around zero and shows the correlations more clearly than does the RDF. Usually, it is this function to which the experimental data are transformed through the relation

$$G(r) = \frac{2}{\pi} \int_0^\infty F(q) \sin qr\, dq . \tag{12}$$

Here q is the magnitude of the scattering vector and $F(q)$ is the reduced scattering intensity defined by

$$F(r) = q\left[\frac{I(q)}{Nf^2} - 1\right], \tag{13}$$

where f is the atomic form factor and $I(q)$ is the experimentally measured scattering intensity given by the square of the scattering amplitude;

$$I(q) = \left|\sum_i^N f(q) e^{i\mathbf{q}\cdot\mathbf{r}_i}\right|^2 . \tag{14}$$

In the next section, the Kirkwood model of the previous section is utilized to calculate the PDF and results are given mainly in the form of the PDF which can be directly compared with experiments.

In case of multicomponent systems,[10-13] the definition of the PDF is generalized to

$$\rho(\mathbf{r}) = \frac{1}{N} \sum_{\mathbf{r}_i} \sum_{\mathbf{r}_j} w_{ij} \rho_{ij}(\mathbf{r}). \tag{15}$$

Here, w_{ij} is given by $f_i f_j / \bar{f}$, where f_i is the scattering strength of the atom i and $\bar{f}$ denotes the arithmetic mean of f_i's in the sample. Here f_i is the scattering factor in x-ray scattering, and the scattering length in neutron scattering. Eq. (15) is exact for neutron diffraction, where the scattering is from the nucleus which may be considered as a point. For x-ray scattering (15) is only an approximation as the f_i are due to the electron density associated with each atom, which in reality have different $\mathbf{q}$ dependence. Nevertheless we will use this approximation here so, that the f_i are proportional to the atomic charges Z_i.

EXPERIMENTAL DETERMINATION OF THE PDF

The experimental determination of PDFs has been described extensively elsewhere[10,13,18]. Data were collected from powders of Ni, InAs and $Ga_{0.5}In_{0.5}As$. The nickel and InAs samples were measured at room temperature using synchrotron x-rays at beamline X-7A at the National Synchrotron Light Source. The nickel was also measured at room temperature using a molybdenum sealed laboratory x-ray source. In each case the data were collected in symmetric reflection geometry. Data were corrected for polarization, absorption, multiple scattering, background and Compton scattering, divided by the average form factor and normalized for flux and sample volume to obtain S(Q), the total structure function which was Fourier transformed to obtain G(r). No correction was made for thermal diffuse scattering since this contains important information about the correlations of the atomic dynamics.

The $Ga_{0.5}In_{0.5}As$ sample was prepared by quenching a mixture of InAs and GaAs from the melt in an evacuated quartz ampoule. The sample was then annealed for a week under vacuum at 950°C, just below the solidus temperature. This annealing procedure was repeated twice with an intermediate grinding to obtain complete homogenization of the sample. Sample homogeneity was checked using x-ray diffraction. The $Ga_{0.5}In_{0.5}As$ sample was measured at 10K with neutrons using the High Intensity Powder Diffractometer (HIPD) at the Manuel Lujan, Jr., Neutron Scattering Center at Los Alamos National Laboratory. Approximately 10g of sample were sealed in a vanadium tube in the presence of He transfer gas. This was cooled using a closed cycle helium refrigerator to 10K. The data were corrected for absorption, multiple scattering, backgrounds and inelasticity effects, and normalized to recover S(Q), which was then Fourier transformed to obtain G(r).

CALCULATION OF THE PDF

In this section we evaluate the PDF of crystalline systems in the zinc-blende structure. The system may be either a pure binary semiconductor such as GaAs and InAs, or a pseudobinary alloy such as $Ga_{1-x}In_xAs$. We first rewrite $\rho_{ij}(r)$ as

$$\rho_{ij}(r) = \frac{1}{2\pi} \int dq \, e^{-iqr} \left\langle e^{iqr_{ij}} \right\rangle. \tag{16}$$

This function would be a δ-function located at r_{ij} if all the atoms were stationary in a perfect crystal. However, this δ-function is broadened by the *thermal motions* since the atoms move about the equilibrium positions even at zero temperature. Moreover, it is further broadened by the *internal strains* due to the bond-length mismatch in the case of alloys.

For the thermal motions, the Debye-Waller theorem can be utilized within the harmonic approximation. As shown in Appendix A, this leads to a *Gaussian* peak for $\rho_{ij}(r)$ centered at r_{ij} with a width σ_{ij} given by

$$\sigma_{ij} = \left\langle \left[\mathbf{u}_{ij} \cdot \hat{r}_{ij} \right]^2 \right\rangle^{1/2}. \tag{17}$$

Accordingly, the total PDF consists of a series of Gaussians from each pair in the system with an appropriate weight w_{ij}.

To proceed further, we make use of the quantum mechanical representation of the displacement. Let us divide N atoms into $\mathcal{N}$ Bravais lattice points, each containing p basis atoms as before. Let $\mu(\nu)$ be the basis label of atoms $i(j)$ in a unit cell. Rewriting $\mathbf{u}_{ij}$ in terms of phonon operators, it can be shown that

$$\sigma_{ij}^2 = \frac{2\hbar}{N} \sum_{k,s} \frac{1}{\omega_s(\mathbf{k})} \left(\langle n_{k,s} \rangle + \frac{1}{2} \right)$$

$$\times \left[\frac{1}{2} \left\{ \frac{\left| e_\mu(\mathbf{k},s) \cdot \hat{r}_{ij} \right|^2}{M_\mu} + \frac{\left| e_\nu(\mathbf{k},s) \cdot \hat{r}_{ij} \right|^2}{M_\nu} \right\} - \frac{\left| e_\mu(\mathbf{k},s) \cdot \hat{r}_{ij} \right| \left| e_\nu(-\mathbf{k},s) \cdot \hat{r}_{ij} \right| e^{i\mathbf{k} \cdot \mathbf{r}_{ij}}}{\sqrt{M_\mu M_\nu}} \right]. \tag{18}$$

where $\omega_s(\mathbf{k})$ is the eigenvalue of the dynamical matrix (7) with the wavevector $\mathbf{k}$ in branch s; $n_{k,s}$ is the number operator in that mode, and $e_\mu(\mathbf{k},s)$ is the corresponding eigenvector associated with the basis μ and mass M_μ. In the summation, $\mathbf{k}$ runs from 1 to $\mathcal{N}$ and s runs from 1 to $3p$. Now, the problem of finding the effect of *thermal* broadening is reduced to solving the eigenvalue problem of the dynamical matrix. Solving the eigenvalue problem analytically for general $\mathbf{k}$ with a large supercell has to rely on numerical methods.

Below, we distinguish between the pure and disordered systems. In the former, the only reason for line broadening is the thermal motion. In the latter, there are two reasons, the thermal motion and the static internal strains.

Example: Ni

As a example, we first consider powdered crystalline Ni. Crystalline Ni has the fcc structure and the phonons are well described by a nearest neighbor central force model. The experimental phonon density of states was found by measuring the phonon dispersion curves[19] in a single crystal of in the (100), (110) and (111) directions using inelastic neutron scattering. These dispersion relations wee fit with a model with many parameters, but the density of states is very close to that obtained with a single nearest neighbor central force,[20] and we will use this simple model here.

We compute the eigenvalues and eigenvectors by diagonalising the 3 x 3 dynamical matrix for Ni, and use the value of the central force constant $\alpha = 38$N/m.[19,20] The various σ_{ij} for each neighbor set is obtained by doing the complete Brillouin Zone integration as given in Eq. (18) with $p=1$. There are no adjustable parameters and the result is shown in Figure 1. We have convoluted this result with the appropriate q_{max} using Eq. (A6) and the results are shown for two different values of q_{max} in Figures 2 and 3. It can be seen that the agreement with experiment is very good, particularly considering that there are no adjustable parameters.

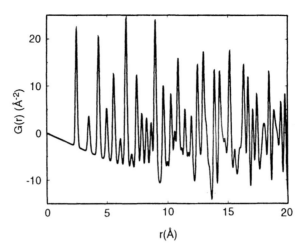

Figure 1. Theoretical calculation of the reduced PDF of Ni at room temperature.

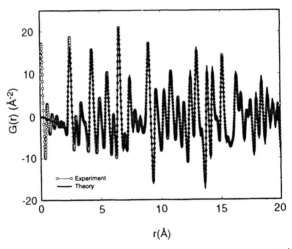

Figure 2. Comparison of theoretical result with experiment in Ni using $q_{max}=16Å^{-1}$ at room temperature (with an X-ray source).

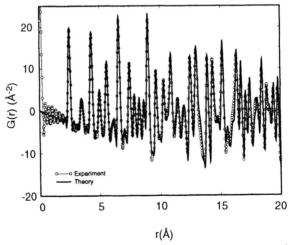

Figure 3. Comparison of theoretical result with experiment in Ni using $q_{max}=21Å^{-1}$ at room temperature (with a synchrotron source).

Example: InAs

Consider a pure binary semiconductor crystal, AC, with A atoms in one sublattice and C in the other. As with Ni, there is only one source of peak broadening, the *thermal* motion, which is characterized by σ_{ij} in Eq. (18). For a pure crystal, the force constants α_{ij} in Eq. (3) assume the same value α for all bonds and we assume that β_{ijl} also takes the same value β for all angles. These parameters can be determined independently from standard experimental data.[6] Since $p = 2$ in the zinc-blende structure, the dynamical matrix is a 6×6 matrix. It can be calculated analytically in a closed form for general $\mathbf{k}$:[8]

$$
D_{\mathbf{K}} = \alpha \begin{bmatrix} \frac{4}{3}\mathbf{1} & \tau_{-\mathbf{k}} \\ -\tau_{\mathbf{k}} & \frac{4}{3}\mathbf{1} \end{bmatrix} + \beta \begin{bmatrix} \frac{4}{3}\mathbf{1} + \frac{1}{4}v_{\mathbf{k}}v_{-\mathbf{k}} & \tau_{-\mathbf{k}} - \frac{2}{3}\gamma_{-\mathbf{k}}\mathbf{1} \\ \tau_{\mathbf{k}} - \frac{2}{3}\gamma_{\mathbf{k}}\mathbf{1} & \frac{4}{3}\mathbf{1} + \frac{1}{4}v_{-\mathbf{k}}v_{\mathbf{k}} \end{bmatrix}
$$
$$
+ \frac{\beta}{9} \begin{bmatrix} \frac{4}{3}\mathbf{1} + \frac{1}{4}v_{-\mathbf{k}}v_{\mathbf{k}} + \frac{4}{3}\gamma_{\mathbf{k}}\tau_{-\mathbf{k}} + \frac{4}{3}\gamma_{-\mathbf{k}}\tau_{\mathbf{k}} & \tau_{-\mathbf{k}} - 2\gamma_{-\mathbf{k}}\mathbf{1} \\ \tau_{\mathbf{k}} - 2\gamma_{\mathbf{k}}\mathbf{1} & \frac{4}{3}\mathbf{1} + \frac{1}{4}v_{\mathbf{k}}v_{-\mathbf{k}} + \frac{4}{3}\gamma_{-\mathbf{k}}\tau_{\mathbf{k}} + \frac{4}{3}\gamma_{\mathbf{k}}\tau_{-\mathbf{k}} \end{bmatrix}. \tag{19}
$$

Here, $\mathbf{1}$ is the 3×3 unit matrix, and the scalar $\gamma_{\mathbf{k}}$ is given by

$$
\gamma_{\mathbf{k}} = \sum_{\delta} e^{-i\mathbf{k}\cdot\mathbf{r}}
$$
$$
= 4\left[\cos\frac{k_x L_e}{\sqrt{3}}\cos\frac{k_y L_e}{\sqrt{3}}\cos\frac{k_z L_e}{\sqrt{3}} + i\sin\frac{k_x L_e}{\sqrt{3}}\sin\frac{k_y L_e}{\sqrt{3}}\sin\frac{k_z L_e}{\sqrt{3}}\right]. \tag{20}
$$

The vector $v_{\mathbf{k}}$ and the tensor $\tau_{\mathbf{k}}$ are defined to be $v_{\mathbf{k}} = (i\nabla_{\mathbf{k}} / L_e)\,\gamma_{\mathbf{k}}$, $\tau_{\mathbf{k}} = (i\nabla_{\mathbf{k}} / L_e)(i\nabla_{\mathbf{k}} / L_e)\,\gamma_{\mathbf{k}}$, respectively. Using Eq. (19), the eigenvalue problem can be solved numerically, and the summation in Eq. (18) can be carried out using Monte Carlo integration over the first Brillouin zone.

There is *no* internal strain for a pure system as discussed previously. Figure 4 presents the result for the width σ for InAs at different temperatures. We used $\alpha = 80$ N/m and $\beta = 10.3$ N/m for InAs.[6] Note that there is a factor of 3 difference in the values of the force constants used here and previously[6] due to different definitions. Also note that the values of α and β are smaller by about 20% from those used in a previous theoretical publication by Chung and Thorpe.[21] We did adjust α and β to get a reasonable fit to the PDF experiments, and then values of α and β used previously[21] were clearly too large. These previous values were derived from measurements of the elastic constants and using the Kirkwood model. In the present set of experiments we have not attempted to optimize α and β, but rather to get them to within a few percent of the optimal values. This was done by noticing that the width of the near neighbor peak is primarily dependent on α alone and so we used this width to fix α. The value of β was then obtained subsequently by fitting the width of the distant neighbors, which is constant beyond about 10Å.

The widths of the peaks are shown in Figure 4 at three separate temperatures. The symbols at 2.62Å depict the width of nearest neighbors, and those at 4.28Å represents that of the second neighbors, etc. Although the highest temperature 1000K is not realistic because it is beyond the melting point, still the width σ is very much less than the interatomic spacing. It is to be noted that the width for the nearest neighbors does not vary as much as the widths for other neighbors as the temperature is increased. This is because nearest neighbor pairs are connected by the strong bond-stretching force. All other neighbors are connected by at least one bond-bending force, which is considerably weaker than the bond-stretching force in covalently-bonded materials. Therefore, further neighbors connected by bond-bending forces become more sensitive to the thermal agitation, while nearest neighbor pairs can remain rather rigid. This has an experimental significance; if one wants to improve the experimental resolution by lowering the temperature, it does not help much for the nearest neighbor peak, which might be the most interesting, as it does for the rest of peaks.

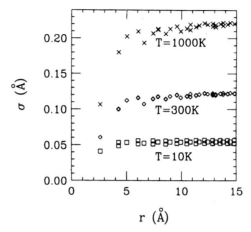

Figure 4. The theoretical width σ for pure InAs crystal as a function of the distance from the origin r at different temperatures ($\times$ at 1000K, $\diamond$ at 300K and $\square$ at 10K). The force constants used are $\alpha = 80$ N/m and $\beta = 10.3$ N/m. The leftmost symbol corresponds to the nearest neighbor. At certain distances, where two types of neighbors (e.g. In-In and As-As) occur, symbols may overlap.

In calculating the PDF, the symmetry of the system can be made use of in the summation of Eq. (18). Since the same type of neighbors have the same distance and the same width, PDF peaks with the same type are simply weighted by the number of neighbors of that type in addition to the weighting factor w_{ij}. For example, nearest neighbor by 4, next nearest neighbor by 12, and so on. The PDF $G(r)$ of InAs at 300K are depicted in Figure 5. The curve at 10K shows much sharper peaks as expected. As the temperature is raised, however, peak widths are increased and hence peak heights are decreased substantially due to the effect of the thermal broadening.

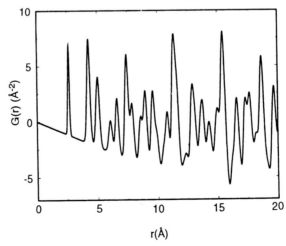

Figure 5. Temperature dependence of the reduced [x-ray] PDF of pure InAs crystal is plotted as a function of the distance from the origin r at 300K. The force constants used in the theory are $\alpha = 80$ N/m and $\beta = 10.3$ N/m.

166

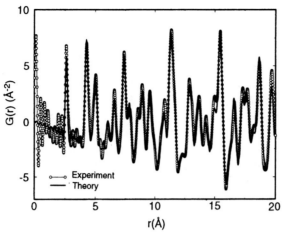

Figure 6. The reduced [x-ray] PDF of InAs from the theoretical calculation is compared with the experiment at 300K. The theoretical curve is convoluted with $q_{max}=22\text{Å}^{-1}$ at room temperature (with synchrotron source). The force constants used in the theory are $\alpha = 80$ N/m and $\beta = 10.3$ N/m.

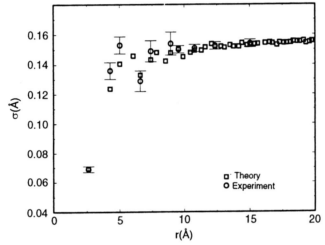

Figure 7. Comparison of theoretical width σ with experimentally determined values for InAs, at room temperature. The force constants used in the theory are $\alpha = 80$ N/m and $\beta = 10.3$ N/m.

In figure 6, the PDF at 300K is compared with an x-ray diffraction experiment.[22] The theoretical curve is convoluted with the experimental resolution function as discussed in the Appendix B. This convolution not only makes small wiggles appear at the bottom of the curves but also lowers and broadens the peaks. The figure shows that the calculation reproduces essentially every feature in the experiment. Our calculation gives better resolution than the experiment, which is not surprising. What is surprising is that our simple model with no adjustable parameters even allows a quantitative comparison with experiment. This confirms that the model is adequate to be used for semiconductor alloys.

The PDF peak widths can be extracted directly from the experimental PDF by fitting Gaussian functions to each peak in the data. This has been done for the InAs data at 300K shown in Fig. 6. The results are shown in Figure 7 compared to the predicted PDF peak widths from the theory.

Example: Ga$_{0.5}$In$_{0.5}$As

Consider a pseudobinary semiconductor alloy in the zinc-blende structure, A$_{1-x}$B$_x$C with A and B atoms in one sublattice and C in the other. Although disorder is introduced only in one sublattice in this study, it would be straight forward to generalize it to both sublattices. As mentioned before, there are two reasons for peak broadening, because the *internal strain* due to bond-length mismatch between A-C and B-C bonds comes into play as well as the *thermal motion*. As in the case of the pure crystal, for simplicity, we take $\alpha_{ij} = \alpha$ for all pairs and $\beta_{ijl} = \beta$ for all angles. We believe that this simplification does not affect the result much because the values of α's and β's do not vary much among III-V and II-VI compounds.[6] This restriction can easily be relaxed to include more general cases. Note that even large changes in the force constants produce only small changes in the internal strains.[5-8]

To realize the alloy, we employ the periodic supercell which consists of $L \times L \times L$ cubic unit cells of the zinc-blende structure, each containing 8 atoms. The dynamical matrix becomes a $3p \times 3p$ matrix where $p = 8L^3$. This method has several advantages over other methods of calculating σ, such as the equation of motion technique.[23] Since this method simply extends the size of the basis, it is conceptually clear and we can closely follow most of the arguments about the pure system given above. Another computational merit of taking a large supercell is that we may sum only over modes at k=0 in Eq. (18). This is because the zone folding in the reduced-zone scheme enables us to sample enough k-points in the original Brillouin zone if we use a big enough supercell. It also reduces computational time since all calculations can be done in real mode rather than in complex mode. For the results presented in this section, we used $L=4$ so that we dealt with 512 atoms and hence a 1536×1536 dynamical matrix. The typical error in σ is estimated to be less than 1% by comparison with the Brillouin zone integration scheme for the perfect crystal. A configurational average is taken over 10 realizations.

The displacement vector **u** in this case contains the distortion due to the *static strain* **u**$_s$ as well as the *thermal motion* **u**$_t$. Since we are interested in the first order correction in the harmonic approximation, the total **u** can be written as a simple sum of these two terms. The static strains **u**$_s$ due to the bond-length mismatch, are found by relaxing the system according to Eq. (6). The thermal fluctuations **u**$_t$ around the relaxed positions enter into Eq. (17). The calculational procedure for alloys is as follows: for a given random number seed a configuration of the system is realized. Then the matrix **M** in Eq. (4) is constructed and the system is relaxed using the conjugate gradient method to find the static equilibrium displacement **u**$_s$ using Eq. (6). From **M**, dynamical matrix **D** is numerically constructed using Eq. (7). The eigenvalue problem for the matrix **D** is solved numerically. The solution is used in the integration (18) to obtain σ_{ij}. This whole procedure is iterated over many realizations to perform a configurational average and finally Eq. (15) gives the PDF.

Figure 8 shows the PDF $G(r)$ for Ga$_{0.5}$In$_{0.5}$As at 10K. Every peak basically consists of many Gaussians as in the pure case. However, due to the internal strains each Gaussian from a particular neighbor is centered at a different distance given by the relaxed positions of each realization. The width σ_{ij} also depends on the particular realization. Therefore, we cannot make use of the symmetry of the system to reduce computational time as in the pure case. The distribution of the pair distance implies that each peak from a particular type of neighbor is already broadened even at very low temperature. Therefore, there is no dramatic change in peak width and height as in the pure system as the temperature is varied.

For a more detailed analysis, the first neighbor peak at 10K is redrawn in Figures 9 and 10 along with the partial bond-length distributions. It is clear that the structure in the first peak results from two different types of bonds (Ga-As and In-As). The lengths of two types of bonds are relaxed to new equilibrium lengths (2.47Å and 2.60Å) from those of pure cases

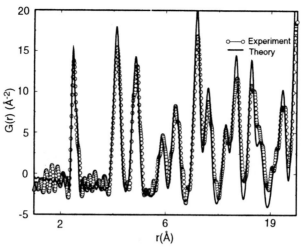

Figure 8. Comparison of theory with experiment for $Ga_{0.5}In_{0.5}As$, using $q_{max}=40Å^{-1}$ (neutron data) at $T=10K$. The force constants used in the theory are $\alpha = 80$ N/m and $\beta = 10.3$ N/m.

(2.45Å and 2.62Å). This change in the bond-lengths has been studied both experimentally[9] and theoretically.[6,24] Our calculation shows that the change in the average length and the width of the distribution of the nearest-neighbor bond lengths can be measured in a PDF experiment at a sufficiently low temperature. A quantitative measurement of the width may not be trivial because the thermal broadening is comparable to the width of the length distribution itself. However, it is this capability of measuring the width that makes a PDF analysis potentially superior to other experimental methods. For example, XAFS experiments only measure the average length of nearest and perhaps also next nearest neighbor peaks. By contrast, a PDF experiment can give the average length *and* the width of the length distribution without any adjustable parameters. The only empirical parameters in our theoretical analysis are the force constants α, β and the lattice constant [See Eq. (3)]. These can be determined independently by standard experiments such as elasticity measurements, optical measurements and Bragg x-ray scattering.

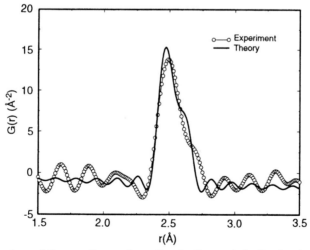

Figure 9. Comparison of theory with experiment for the first peak in $Ga_{0.5}In_{0.5}As$, using $q_{max}=40Å^{-1}$ (neutron data) at $T=10K$. The force constants used in the theory are $\alpha = 80$ N/m and $\beta = 10.3$ N/m..

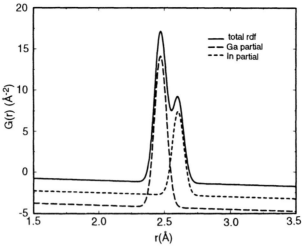

Figure 10. Theoretical calculation of partial PDF of $Ga_{0.5}In_{0.5}As$ at $T=10K$. The partial PDFs are shifted downward for clarity. The force constants used in the theory are $\alpha = 80$ N/m and $\beta = 10.3$ N/m.

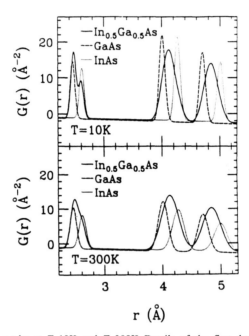

Figure 11. Theoretical results at $T=10K$ and $T=300K$ Details of the first three peaks of the reduced [neutron] PDF of $Ga_{0.5}In_{0.5}As$ are compared with the pure end members, GaAs and InAs, at 10K and 300K. Solid lines are for the alloy, broken lines for GaAs, and dotted lines for InAs. Note that the resolution changes substantially with the temperature. The force constants used in the theory are $\alpha = 80$ N/m and $\beta = 10.3$ N/m.

The first three peaks are plotted again in figure 11 along with those from the pure end members, GaAs and InAs. The internal structure of the first neighbor peak at 10K clearly shows that it retains the characteristics of the pure systems, although it is almost unrecognizable at 300K due to the thermal broadening. This is again because the nearest neighbors are only connected by the strong bond-stretching force. From the second neighbors and beyond, however, there can be many different intermediate configurations connected by the weak bond-bending forces. Hence each peak appears as a distribution of Gaussians centered at the length given by the virtual crystal approximation, with the peak of the alloy tracking the first moment of that peak, which is temperature independent and goes linearly with the composition x between the two pure crystal limits.

CONCLUSIONS

We have developed a method of calculating the PDF of binary semiconductor crystals and pseudobinary alloys having the zinc-blende structure. The PDF reveals the local structure directly and can be compared with experiments. Our approach can be easily generalized for various crystal structures including fcc, diamond and wurtzite structures. To facilitate the calculation, we have used a harmonic Kirkwood potential model with bond-stretching and bond-bending forces. Temperature dependence is treated quantum mechanically using the dynamical matrix and appropriate Bose factors.

The PDF turns out to consist of a series of Gaussians with the weight w_{ij} given by the type of atoms at site i and j and with width σ_{ij} given as a function of phonon properties. In the case of a pure system, each type of neighbor pair has the same width and are further weighted by the number of neighbors of that type. However, in the case of the alloy, each peak from the same type of neighbor is relaxed to a different distance with a different width by the internal strains.

This method is used to calculate the PDF of a InAs pure crystal and a $Ga_{1-x}In_xAs$ alloy, with $x=0.5$. The result for the pure crystal agrees well with the neutron diffraction experiment even though there are no adjustable parameters. The harmonic model we used describes the behavior of the system adequately. The result for the alloy suggests that two different types of nearest neighbors can be resolved experimentally at sufficiently low temperature. The information on the width of the length distribution can be obtained as well as the average length. The resolution of such an experiment, however, is somewhat limited by the zero point motions.

This method does not suffer from possible artifacts which may arise from fitting the experimental data with adjustable parameters, which other methods such as XAFS do. However, it has some limitations in resolutions due to zero-point motions. This makes it difficult to resolve different types of bonds beyond the first neighbor peak. One possible improvement is to explore the partial PDF which measures peaks involving a certain atom.[24] But this would require a large experimental effort, involving isotope substitution or anomalous x-ray scattering techniques.

Despite these limitations, PDF analysis is almost the only method of studying the intermediate range properties of the semiconductor alloys. Therefore it has been important to develop a microscopic model to understand the observed behavior microscopically and provide a clear physical picture. The model and analysis used in this study has proved to be quite versatile and robust.

ACKNOWLEDGMENTS

This work was helped by useful discussions with A. R. Day, R. J. Elliott, R. G. DiFrancesco, and N. Mousseau. This work was supported by the US Department of Energy under the contract DE-FG02-97ER45651 and National Science Foundation grant CHE 9633798. JSC also acknowledges the support from the Basic Science Research Institute Program (BSRI-96-2436), Korean Ministry of Education, and from the KOSEF through a grant to the Center for Theoretical Physics, Seoul National University, Korea. SJB acknowledges financial support from the Alfred P. Sloan Foundation. X-ray experiments were carried out at the National Synchrotron Light Source, Brookahven National Laboratory which is funded by DOE Division of Materials Sciences and Division of Chemical Sciences through contract DE-AC02-76CH00016. Neutron experiments were carried out at the Manuel Lujan, Jr., Neutron Scattering Center which is supported by DOE Offices of Defense Programs and Basic Energy Sciences. A fuller account of the theoretical work described in this paper is given in Chung and Thorpe.[21]

APPENDIX

Derivation of Gaussian Peaks

Let us rewrite Eq. (16) as

$$\rho_{ij}(r) = \frac{1}{2\pi} \int e^{-iqr} C_{ij}(q) dq, \tag{A1}$$

where $C_{ij}(q) \equiv \left\langle e^{iqr_{ij}} \right\rangle$. In the harmonic approximation, the interatomic spacing r_{ij} can be written as

$$r_{ij} = r_{ij}^0 + \mathbf{u}_{ij} \cdot \hat{r}_{ij} , \tag{A2}$$

where r_{ij}^0 is the distance between the atoms i and j in the perfect unstrained lattice. Using the Debye-Waller theorem, we have

$$C_{ij}(q) = e^{iqr_{ij}^0} \left\langle e^{iq\mathbf{u}_{ij} \cdot \hat{r}_{ij}^0} \right\rangle = e^{iqr_{ij}^0} e^{-(1/2)q^2 \left\langle \left(\mathbf{u}_{ij} \cdot \hat{r}_{ij}^0 \right)^2 \right\rangle} . \tag{A3}$$

Putting this back to Eq. (A1), we have

$$\begin{aligned}
\rho_{ij}(r) &= \frac{1}{2\pi} \int_{-\infty}^{\infty} e^{-(1/2)q^2 \left\langle \left[\mathbf{u}_{ij} \cdot \hat{r}_{ij} \right]^2 \right\rangle + iq(r_{ij}^0 - r)} dq \\
&= \frac{1}{\sqrt{2\pi \left\langle \left[\mathbf{u}_{ij} \cdot \hat{r}_{ij} \right]^2 \right\rangle}} e^{-(r_{ij}^0 - r)^2 / 2 \left\langle \left[\mathbf{u}_{ij} \cdot \hat{r}_{ij} \right]^2 \right\rangle} .
\end{aligned} \tag{A4}$$

Therefore, $\rho_{ij}(r)$ is a Gaussian peak centered at r_{ij} with the width

$$\sigma_{ij} = \left\langle \left[\mathbf{u}_{ij} \cdot \hat{r}_{ij} \right]^2 \right\rangle^{1/2} , \tag{A5}$$

which is Eq. (17).

Modeling Finite Data.

In experiments, data can be collected only over a finite range of the scattering momentum q from 0 to q_{max}, although the Fourier transformation in Eq. (12) should be carried out over a range from 0 to ∞. We are interested in how the termination affects the PDF. In fact, the derivation in this appendix may be applied in broader context of modeling finite data. In comparing theory and experiment, it is most convenient to incorporate the effects of truncation into the theory and then compare with the results obtained by Fourier transforming the experimental data. We substitute directly the theoretically calculated $G(r)$ to give

$$G_e(r) = \frac{1}{\pi} \int_0^\infty G(r') \left[\frac{\sin q_{max}(r-r')}{r-r'} - \frac{\sin q_{max}(r+r')}{r+r'} \right] dr' \tag{A6}$$

where $G_e(r)$ is the same quantity that is experimentally. The function in the bracket makes the ideal $G(r)$ broader and produces ripples around the peaks as shown in the many of the figures in the text.

REFERENCES

1. V. Narayanamurti, *Science*, 235:1023 (1987).
2. L. Vegard, *Z. Phys.* 5:17 (1921).
3. L. Nordheim, *Ann. Phys. (Leipz.)* 9:607 (1931); 9:641 (1931).
4. M. F. Thorpe, W. Jin and S. D. Mahanti, in: *Disorder in Condensed Matter Physics*, J. A. Blackman and J. Taguena, ed., Oxford University Press, New York (1991) p. 22.
5. Y. Cai and M. F. Thorpe, *Phys. Rev.* B 46:15872 (1992).
6. Y. Cai and M. F. Thorpe, *Phys. Rev.* B 46:15879 (1992).
7. N. Mousseau and M. F. Thorpe, *Phys. Rev.* B 46:15887 (1992).
8. R. W. Wang, M. F. Thorpe and N. Mousseau, *Phys. Rev.* B 52:17191 (1992).
9. J. C. Mikkelsen, Jr. and J. B. Boyce, *Phys. Rev. Lett.* 49:1412 (1982); *Phys. Rev.* B 28:7130 (1983).
10. B. E. Warren, *X-ray Diffraction*, Addison-Wesley, Reading, (1969).
11. G. S. Cargill III, in: *Solid State Physics*, Vol. 30, H. Ehrenreich, F. Seitz, and D. Turnbull, ed., Academic Press, New York, (1975).
12. S. R. Elliott, *Physics of Amorphous Materials*, Longman, London, (1990).
13. T. Egami, *Mater. Trans.* 31:163 (1990).
14. Different nomenclatures have been used in literature. To minimize confusion we use $G(r)$ for the PDF.
15. J. G. Kirkwood, *J. Chem. Phys.* 7:506 (1939).
16. P. N. Keating, *Phys. Rev.* 145:637 (1966).
17. A. A. Maradudin, E. W. Montroll, G. H. Weiss, and I. P. Ipatova, *Theory of Lattice dynamics in the Harmonic Approximation*, Academic Press, New York, (1971).
18. S. J. L. Billinge, Ph.D. Thesis, U. Pennsylvania (1992).
19. R. J. Birgeneau, J. Cordes, G. Dolling and A. D. B. Woods *Phys. Rev.* 136:A1359 (1964).
20. L. von Heimendahl and M. F. Thorpe *J. Phys.* F 5: L87 (1975).
21. Jean Chung and M. F. Thorpe, *Phys. Rev.* B 55: 1545 (1997).
22. F. Mohiuddin-Jacob and S. J. L. Billinge, private communication; *Bull. Am. Phys. Soc.* 41(No. 1):592 (1996).
23. M. F. Thorpe and S. W. de Leeuw, *Phys. Rev.* B 33:8490 (1986).
24. M. C. Schabel and J. L. Martins, *Phys. Rev.* B 43:11873 (1991).

LOCAL ATOMIC ARRANGEMENTS IN BINARY SOLID SOLUTIONS STUDIED BY X-RAY AND NEUTRON DIFFUSE SCATTERING FROM SINGLE CRYSTALS

J. L. Robertson,[1] C. J. Sparks,[1] G. E. Ice,[1] X. Jiang,[1] S. C. Moss,[2] and L. Reinhard[3]

[1]Oak Ridge National Laboratory, Oak Ridge, TN 37831 USA
[2]University of Houston, Houston, TX 77204 USA
[3]ETH Zentrum, CH-8092 Zurich Switzerland

INTRODUCTION

The concepts of <u>order</u> and <u>disorder</u> are fundamental to understanding the many physical properties exhibited by various materials. In general, these concepts are quite ambiguous, but when applied to a particular circumstance, they often provide needed insight into the relationship between how the atoms are arranged locally in a material and its bulk properties such as phase stability, electrical resistance and magnetism. In addition, the notion of order vs. disorder applies equally to equilibrium and non-equilibrium systems. It is interesting to note that a chemically disordered material can indeed be the equilibrium phase over a large range of temperature, pressure and composition. This phenomenon can be best understood by considering the competition between short-range and long-range ordering tendencies (which can be incompatible with one another) as well as the constant rearrangement of the atoms resulting from thermal diffusion (entropy) at elevated temperatures. Figure 1 shows the FeCr binary alloy phase diagram[1] where the entire phase field denoted as (αFe,Cr) represents a disordered structure as the equilibrium phase.

In this paper, we will be concerned with local atomic arrangements in crystalline binary solid solutions and how information about the local order can be obtained from diffuse x-ray and neutron scattering measurements.[2] For the purposes of this paper, a binary solid solution should be thought of as a crystal lattice decorated by two atomic species, labeled A and B respectively, which occupy the atomic sites in such a way that there is no long-range order. In other words, there is no overall pattern that determines which kind of atom, an A atom as opposed to a B atom, will occupy a particular atomic site. The crystal lattices are assumed to be either simple face-centered cubic (fcc) with four atoms per unit cell or simple body-centered cubic (bcc) with two atoms per unit cell. However, the methods presented here can, with some difficulty, be generalized to include other lattice symmetries.[3] These methods can also be applied to extremely complex crystals in cases where only a small subset of the atomic sites are chemically disordered.

Local Structure from Diffraction
Edited by S.J.L Billinge and M.F. Thorpe, Plenum Press, New York, 1998

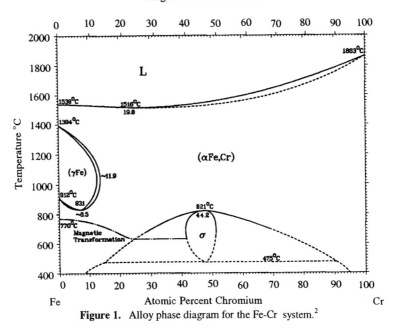

Figure 1. Alloy phase diagram for the Fe-Cr system.[2]

In the development that follows the amount of order present in a system is defined relative to a completely disordered system. In a completely disordered system the probability distribution describing which atomic species will occupy a particular atomic site is simply given by the composition. For example, suppose a lattice is composed of half A atoms and half B atoms. The chance of finding an A atom on a given site is 50% and the chance of finding a B atom there is also 50%. If the composition had been $A_{20}B_{80}$ then there would be a 20% chance for an A atom and a 80% chance for a B atom to occupy a given atomic site. However, such an ideally disordered state is almost never realized in real alloys. The type of atom occupying a particular site invariably influences the distribution of atomic species on neighboring sites due to electronic and/or magnetic interactions, atomic size mismatch, etc. This of course introduces short-range correlations in the chemical order – also referred to as concentration fluctuations in the formalism developed by Krivoglaz.[4] There are two distinct ordering tendencies that can arise from these short-range chemical correlations. The first occurs if the probability of finding unlike pairs of atoms occupying adjacent atomic sites is greater than what would be expected in a completely random alloy with the same composition. This type of order, if extended to include longer-range correlations, would ultimately lead to the formation of a supperlattice structure and is usually referred to as atomic short-range order. If, on the other hand, the atoms on neighboring atomic sites are more likely to be of the same atomic species, then the system will tend toward phase separation. This type of concentration fluctuation is referred to as clustering.

Displacements that arise from the local static, or frozen, deviations of the atoms from their ideal lattice sites usually accompany the concentration fluctuations found in binary solid solutions. These displacements violate the lattice symmetry locally but the symmetry of the lattice as a whole must be preserved. For example, consider the addition of a large atoms, A, to a lattice of smaller atoms, B. The lattice will be expanded around the A atoms which increases the lattice parameter from the value for a lattice of only B atoms. This introduces the concept of the <u>average</u> lattice where every atom in the crystal can be thought of as being displaced from a perfect lattice site by a small amount relative to the lattice parameter. The lattice parameter for the solid solution is then taken to be that of the average lattice, which turns out to be the value one would get by averaging over all of the unit cells in the crystal. What this simple picture implies is that AA near neighbor pairs will have a greater separation than that expected from the lattice parameter and BB near neighbor pairs will have a lesser separation. In general, a nearly linear response of the lattice parameter to the concentration (e.g. the addition of A atoms) throughout the solubility range is observed. This is Vegard's Law.[5] The variation of the lattice parameter with concentration in substitutional alloying is a long-range effect and has been well characterized. The local or near neighbor displacements, however, are not well understood. The local effect of atomic size is crucial to understanding the behavior of substitutional alloys since atomic size disparity between the solvent and solute atoms is known to affect solubility as well as the physical and chemical properties of the alloy. Several theoretical models have been proposed the explain the linear relationship between the lattice parameter and concentration.[5-7] While these models reproduce the almost linear change in lattice parameter with concentration, accurate measurements of the local atomic displacements to test these models on an atomic scale are almost non-existent.

SCATTERING THEORY

As was mentioned above the diffuse scattering from crystalline solid solutions is sensitive to both the local concentration fluctuations and static atomic displacements in crystalline solid solutions. Several methods[8-10] have been developed to extract the desired information from the diffuse scattering data and the development presented here borrows from all of them. Let us begin with intensity at a given scattering vector $\mathbf{Q}$ for a binary alloy expressed by

$$I(\mathbf{Q}) = \sum_p \sum_q f_p f_q e^{i\mathbf{Q} \bullet (r_p - r_q)}. \tag{1}$$

Where $\mathbf{Q} = 4\pi\sin(\theta)/\lambda$, f_p and f_q denote the complex atomic scattering factor for x-rays or the atomic scattering lengths for neutrons, the indices p and q designate lattice sites such that each sum runs over every atom in the crystal, and r_p and r_q are the position vectors for those sites. For crystals where the Bragg reflections are sharp and the average lattice is well-defined, the atomic positions can be represented by $\mathbf{r} = \mathbf{R} + \delta$ where $\mathbf{R}$ is a lattice vector of the average lattice and δ is the displacement of the atom from that lattice site. The exponential in Eq. 1 can then be written as

$$e^{i\mathbf{Q} \bullet (r_p - r_q)} \equiv e^{i\mathbf{Q} \bullet [(\mathbf{R}_p + \delta_p) - (\mathbf{R}_q + \delta_q)]} \equiv e^{i\mathbf{Q} \bullet [(\mathbf{R}_p - \mathbf{R}_q) + (\delta_p - \delta_q)]} \equiv e^{i\mathbf{Q} \bullet (\mathbf{R}_p - \mathbf{R}_q)} e^{i\mathbf{Q} \bullet (\delta_p - \delta_q)} \tag{2}$$

and the exponential involving the displacements, δ, can be expanded as

$$e^{i\mathbf{Q} \bullet (\delta_p - \delta_q)} \equiv 1 + i\mathbf{Q} \cdot (\delta_p - \delta_q) - \frac{[\mathbf{Q} \cdot (\delta_p - \delta_q)]^2}{2!} + \ldots + i^j \frac{[\mathbf{Q} \cdot (\delta_p - \delta_q)]^j}{j!} + \ldots, \tag{3}$$

where j is an integer. This series converges rapidly when $\mathbf{Q} \cdot (\delta_p - \delta_q)$ is sufficiently small.

The total intensity can be separated into the scattering from the average lattice I_{Bragg} and the scattering arises from the deviations from the average lattice $I_{Diffuse}$. The diffuse scattering can further be broken down into contributions from the chemical short-range order and the displacements. Thus by substituting Eqs. 2 and 3 into Eq. 1 we have

$$I_{Total} = I_{Bragg} + I_{SRO} + I_{ISD} + I_{HOT.} \qquad (4)$$

I_{Bragg} and I_{SRO} correspond to the first term in the expansion shown in Eq. 3, I_{ISD} to the second term ($iQ \cdot (\delta_p - \delta_q)$), and I_{HOT} the remaining higher order terms. Following the treatment of Warren and co-workers[8,9,11,12] these terms can be written, in electron units per atom, as follows for a crystal with cubic symmetry

$$I(Q)_{Bragg} = \left| c_A f_A + c_B f_B \right|^2 \sum_p \sum_q e^{iQ \cdot (R_p - R_q)}; \qquad (5)$$

$$\frac{I(Q)_{SRO}}{N} = N c_A c_B \, |f_A - f_B|^2 \sum_{lmn} \alpha_{lmn} e^{-2M\Phi_{lmn}} \cos[\pi(h_1\ell + h_2 m + h_3 n)] \qquad (6)$$

$$\frac{I(Q)_{ISD}}{N} = N c_A c_B \, |f_A - f_B|^2 \sum_{lmn} \gamma_{lmn} \sin[\pi(h_1\ell + h_2 m + h_3 n)] \qquad (7)$$

Here N is the total number of atoms in the crystal, c_A is the concentration of A atoms, c_B is the concentration of B atoms, ℓmn are the Cartesian coordinates[13] the lattice vector $\mathbf{R} = \frac{1}{2}(\mathbf{a}_1\ell + \mathbf{a}_2 m + \mathbf{a}_3 n)$ where a is the cubic lattice parameter) in units of the lattice parameter so that the single sum over ℓmn replaces the double sum over p and q, and h1, h2, and h3 are the Cartesian coordinates of the reciprocal lattice vector ($\mathbf{Q} = \pi/2 \times (\mathbf{b}_1 h_1 + \mathbf{b}_2 h_2 + \mathbf{b}_3 h_3)$ where b is the reciprocal space lattice constant). In the case of a purely random alloy $I_{SRO}(\mathbf{Q})$ would be given by

$$I_{SRO}(\mathbf{Q}) = c_A c_B |f_A - f_B|^2 \qquad (8)$$

which is often referred to as the Laue monotonic scattering. The cosine series is just the Fourier decomposition in reciprocal space of the concentration fluctuations in direct space. The Fourier coefficients α_{lmn}, also know as the Warren-Cowley short-range order parameters,[11] are defined to be

$$\alpha_{lmn} = 1 - \frac{P_{lmn}^{AB}}{c_B} = 1 - \frac{P_{lmn}^{BA}}{c_A} = 1 - \frac{1 - P_{lmn}^{AA}}{c_B} = 1 - \frac{1 - P_{lmn}^{BB}}{c_A} \qquad (9)$$

where P_{lmn}^{AB} is the conditional probability that there will be a B atom at site ℓmn if there is an A atom at the origin. Thus the parameters α_{lmn} represent the pairwise occupation probabilities averaged over all symmetry equivalent pairs separated by the corresponding direct space lattice vector, $\mathbf{R}_{lmn}$. Thus only one parameter is required to describe the occupation probabilities for each neighbor shell, ℓmn.

Figure 2 illustrates the three possible types of concentration fluctuations and their contribution to the total scattering. Three separate computer models were generated of an $A_{50}B_{50}$ bcc alloy. One with the first nearest neighbor $\alpha_{111} = 0.2$ indicating that A atoms are 10% more likely to have A atoms for first nearest neighbors than is expected from the composition and likewise for B atoms; this is referred to as 10% clustering. One where all

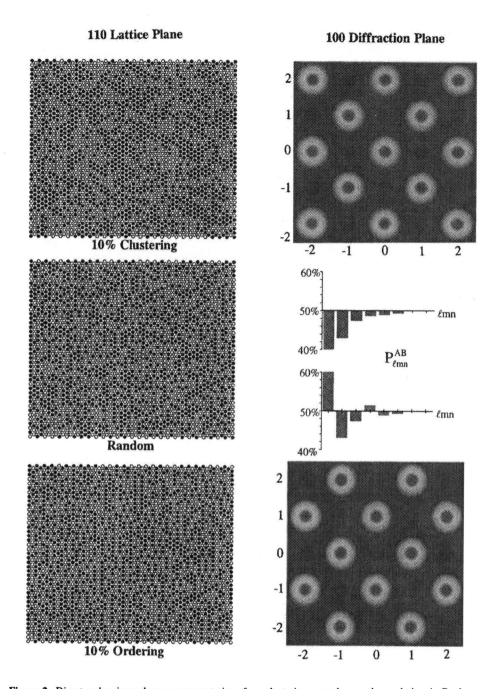

Figure 2. Direct and reciprocal space representations for a clustering, a random, and an ordering $A_{50}B_{50}$ bcc alloy.

the $\alpha_{lmn} = 0.0$, which corresponds to a completely random alloy. And finally one where $\alpha_{111} = -0.2$ indicating that A atoms are 10% more likely to have B atoms as first nearest neighbors and likewise for A atoms as first nearest neighbors of B atoms; this is referred to as 10% ordering. All three models were produced by a reverse Monte Carlo technique.[14] For the 10% clustering and 10% ordering models the α_{lmn} beyond the first neighbor shell were unspecified and thus have whatever value they had when convergence was achieved on α_{111}. The first column in Fig. 2 shows a 110 plane extracted from each model. One can clearly see the difference between the various types of order. The right hand column shows the 100 reciprocal space plane for 10% clustering and 10% ordering models. Only contribution from $I_{SRO}(Q)$ is show. The random model is omitted because it is featureless. The most important feature to notice is that the diffuse scattering peaks at the Bragg peak positions for clustering systems and at the supperlattice positions for ordering systems. This fundamental change in the diffraction pattern means that it is usually readily apparent what type of concentration fluctuations are present. The P_{lmn}^{AB} for the 10% clustering and 10% ordering models are shown in the center of the right had column. Note how the values for the P_{lmn}^{AB} fall off rapidly with increasing distance reflecting the short-range nature of the concentration fluctuations.

The first order term in the displacements, $I(Q)_{1SD}$, is often referred to as the "size effect" scattering. The displacement parameters, γ_{lmn}, are also a pairwise average over all symmetry equivalent pairs. Each γ_{lmn} is a linear combination of the species dependent average pairwise displacements, Δ_{lmn}^{AA} and Δ_{lmn}^{BB}, given by

$$\gamma_{lmn} = Re\left(\frac{f_A}{f_A - f_B}\right)\left(\alpha_{lmn} + \frac{c_A}{c_B}\right)\left(iQ \cdot \Delta_{lmn}^{AA}\right) - Re\left(\frac{f_B}{f_A - f_B}\right)\left(\alpha_{lmn} + \frac{c_B}{c_A}\right)\left(iQ \cdot \Delta_{lmn}^{BB}\right) \qquad (10)$$

where Re() denotes the real part of the ratio of complex scattering factors. It should be noted that it is the individual components of the displacements that are averaged of all the symmetry equivalent pairs in the crystal such that

$$Q \cdot \Delta_{lmn}^{AA} = \frac{2\pi}{a}\left(h_1\left\langle\Delta x_{lmn}^{AA}\right\rangle + h_2\left\langle\Delta y_{lmn}^{AA}\right\rangle + h_3\left\langle\Delta z_{lmn}^{AA}\right\rangle\right) \qquad (11a)$$

and

$$Q \cdot \Delta_{lmn}^{BB} = \frac{2\pi}{a}\left(h_1\left\langle\Delta x_{lmn}^{BB}\right\rangle + h_2\left\langle\Delta y_{lmn}^{BB}\right\rangle + h_3\left\langle\Delta z_{lmn}^{BB}\right\rangle\right). \qquad (11b)$$

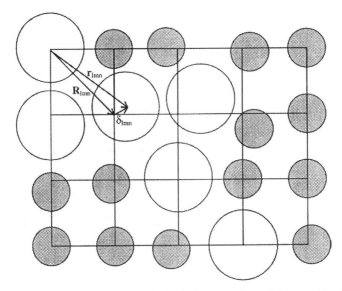

Figure 3. Schematic diagram showing a particular displacement, δ_{lmn}, what is meant by this parameter. It is important to remember that the displacement parameters from Eq. X correspond the to the average of each component of this vector over all symmetry equivalent lmn pairs in the crystal.

The existence of an average lattice requires that the weighted average of the displacements for all AA, AB, BA, and BB pairs for each coordination shell must be zero. Thus, given that $\Delta_{lmn}^{AB} = \Delta_{lmn}^{BA}$ the displacement terms involving AB pairs have been removed from Eq. 10 by applying the average lattice constraint

$$\Delta_{lmn}^{AB} = \frac{1}{2(\alpha_{lmn} - 1)}\left[\left(\frac{c_A}{c_B} + \alpha_{lmn}\right)\Delta_{lmn}^{AA} + \left(\frac{c_A}{c_B} + \alpha_{lmn}\right)\Delta_{lmn}^{BB}\right]. \qquad (12)$$

No assumption has been made as to how the displacements are distributed about the average. This information is contained in $I_{HOT}(Q)$. In order to evaluate $I_{HOT}(Q)$ we make the assumption that either the quadratic and higher order terms in the series expansion of the thermal and static displacements are the same for AA, AB and BB atom pairs or that the different elements have nearly the same atomic scattering factors.[15-18] If either of these assumptions is valid we can write $I_{HOT}(Q)$ as

$$I_{HOT}(Q) = \sum_j \left[\left|c_A f_A + c_B f_B\right|^2 \frac{i^j}{j!}\sum_{lmn}\left\langle(Q\cdot\delta)^j\right\rangle e^{iQ\cdot R_{lmn}} + c_A c_B\left|f_A - f_B\right|^2 \frac{i^j}{j!}\sum_{lmn}\alpha_{lmn}\left\langle(Q\cdot\delta)^j\right\rangle e^{iQ\cdot R_{lmn}}\right]. \qquad (13)$$

The first term in Eq. 13 reduces the intensity of the Bragg peaks and distributes this intensity as thermal and static diffuse scattering. This corresponds to the usual Debye-Waller factor commonly used by crystallographers. The second term in Eq. 13 reduces the intensity associated with the chemical ordering. This term has been treated by Walker and Keating[19] and is included as a Debye-Waller like factor $e^{-2M\Phi_{lmn}}$ in Eq. 6.

EXPERIMENTAL METHODS

In most cases the statistical quality of the data and the presence of systemic the recovery of any information from $I_{HOT}(Q)$. Thus the usual strategy is to try and separate the $I_{SRO}(Q) + I_{ISD}(Q)$ contribution from $I_{Total}(Q)$. For diffuse neutron scattering this can be done when the data are collected by utilizing an energy analyzer to remove all but the elastic scattering from the diffracted beam. The elastic contribution to the scattering from $I_{HOT}(Q)$ is assumed to be small and is simply ignored.

For diffuse x-ray scattering the situation is more complicated. In the past about all one could do was to calculate the thermal diffuse scattering and subtract it from the data. More recently, the availability of x-ray synchrotron sources has made it possible to tune the incident energy of the x-rays so as to vary the scattering contrast between the two atomic species. This technique takes advantage of the anomalous dispersion that occurs when the incident x-ray energy is near an x-ray absorption edge of one of the two atomic species.[10] The two contributions to $I_{Total}(Q)$ of interest, $I_{SRO}(Q)$ and $I_{ISD}(Q)$, are strongly dependent on the scattering contrast, $\Delta f = f_A - f_B$, see Eq. 6 and 7. Thus, one can measure the diffuse intensity at two scattering contrasts; one where Δf is large and one where Δf is small. The data where Δf is small will contain little or no contribution from $I_{SRO}(Q)$ or $I_{ISD}(Q)$ so it can be rescaled to the average scattering per atom at the x-ray energy where Δf is large and then subtracted away leaving only the contrast dependent contribution at that energy.[15] The contrast dependent part of $I_{HOT}(Q)$ also remains, but this is taken to be small and is ignored in much the same way as for diffuse neutron scattering. In both x-ray and neutron diffraction the Bragg intensity is simply omitted since it only occurs at a few points in reciprocal space.

Once $I_{SRO}(\mathbf{Q}) + I_{ISD}(\mathbf{Q})$ has been separated from the total scattering, Eqs. 6 and 7 can be fit to the data whether it comes from x-ray or neutron diffraction. Because $I_{SRO}(\mathbf{Q})$ has even symmetry and $I_{ISD}(\mathbf{Q})$ has odd symmetry the least squares problem is well conditioned so that one should expect little or no interdependence between the α_{lmn}'s and the γ_{lmn}'s. This is in spite of the fact that α_{lmn} appears explicitly in the expression for γ_{lmn} in Eq. 10. The "coupling factors", Φ_{lmn}, in Eq. 6 can be evaluated using various approximations for phonon dispersion in the alloy.[11,20] The leading term $\Phi_{000} = 0$ for x-ray diffuse scattering where the instantaneous correlation function is measured but not for elastic neutron scattering where $\Phi_{000} \approx 1$. Typically, the approximation $\Phi_{000} \approx 1$ is also made for $\ell mn \neq 0$.[21]

With only one data set where the scattering contrast, Δf, is large one can only determine the γ_{lmn}'s but not the species dependent atomic displacement parameters, Δ_{lmn}^{AB}. In order to extract the species dependent parameters, at least one additional high contrast data set is required where the scattering contrast is substantially different from the first. If possible one should attempt to have $f_A > f_B$ for one contrast and $f_B > f_A$ for the other. This can be achieved with x-rays at the synchrotron in exactly the same way as described above for large and small Δf, and by isotopic substitution using neutrons. Quite often at least one of the atomic species will not have an absorption edge within the accessible x-ray energy range available at the synchrotron, and there are no isotopes available (often they are simply too expensive) for use in a neutron diffuse scattering measurement. In this case one should consider using a combination of x-ray and neutron diffuse scattering data to get the required change in scattering contrast. Once the two large scattering contrast data sets are ready, the Δ_{lmn}^{AA} and Δ_{lmn}^{BB}, can be obtained directly from the least squares analysis by substituting Eq. 10 into Eq. 7.

EXAMPLE: FeCr

The FeCr binary system[22] exhibits a bcc solid solution (α-FeCr, see Fig. 1) over a wide temperature and concentration range. At ~1100K, a structural transformation to the σ-phase, a complex close-packed Frank-Kaspar phase, occurs. According to thermodynamic evaluations[22,23] the σ-phase decomposes below ~700K into Fe-rich and Cr-rich bcc phases. Since the bcc to σ transformation is very sluggish, a metastable miscibility gap for α-FeCr is observed well above this decomposition temperature. One might expect the local order in the bcc phase to reveal a tendency toward phase separation. However, there is also the alternative possibility that directly above the σ-phase equilibrium boundary the local atomic arrangements reflect the incipient σ-phase formation through premonitory fluctuations. In alloy systems such premonitory fluctuations can, for example, include atomic short-range order as well as local atomic displacements.

A single crystal of $Fe_{53}Cr_{47}$ was grown at the Materials Preparation Lab, Ames Laboratory, Iowa State University by L. L. Jones using a Bridgeman technique. The purity of the alloying elements was 99.95% and 99.996% for Fe and Cr, respectively, and the Cr concentration was determined by chemical analysis to be 47.2%. The crystal was roughly cylindrical in shape with a diameter of 12mm and a length of ~20mm. After a homogenization anneal at 1600K the crystal was held at 1108K (5K above the σ-phase transition temperature, see Fig. 1) for four days in a sealed quartz tube under a purified argon atmosphere then water-quenched. Extensive small-angle neutron scattering studies[24] of quenched and annealed α-FeCr alloys indicate that such a quench will preserve the high temperature equilibrium configurational order. Small angle neutron scattering was used to verify that this was the case for our sample.

The X-ray scattering experiment[16] was performed on the ORNL beamline X-14A[25] of the National Synchrotron Light Source (NSLS) at Brookhaven National Laboratory. The measurements were done using three different energies for the incident x-rays: (i) E=5.969 keV (20eV below the Cr K absorption edge). This energy was chosen to maximize the scattering contrast between Cr and Fe and thus enhance the contribution from the local order in the crystal. We refer to it as the "Cr edge". (ii) E=7.092keV (20eV below the Fe K absorption edge). At this energy $f_{Cr} > f_{Fe}$, i.e. Cr becomes a stronger scatterer that Fe and we shall refer to it as the "Fe edge". This contrast inversion affects the sign of $I_{ISD}(Q)$, see Eqs. 7 and 10. Therefore, a comparison of the data measured with the "Fe edge" energy with those measured with the "Cr edge" energy highlights the "size effect" scattering. (iii) E=7.600keV. This choice minimizes the scattering contrast whereby the $I_{SRO}(Q) + I_{ISD}(Q)$ contribution to the total intensity are small, thus the measured intensity is predominately due to $I_{Bragg}(Q) + I_{HOT}(Q)$. Figure 4 shows the range of contrast variation obtained during the measurement. Note the greatly enhanced $|\Delta f|^2$ at the "Cr edge" as compared with $|\Delta Z|^2$ = 4 without the anomalous dispersion, and the $|\Delta f|^2$ small for E=7.600keV.

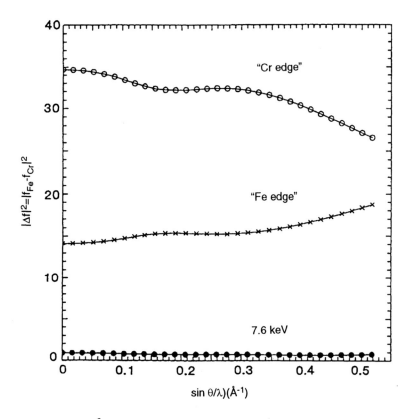

Figure 4. $|\Delta f|^2$ as a function of $\sin\theta/\lambda$ for the three x-ray energies used.

The results of the least-squares analysis are listed in Tables I and II. The first eleven α_{lmn}'s are positive indicating a preference for like neighbors (Fe-Fe and Cr-Cr pairs), i.e. this is a clustering system. For example, $\alpha_{111} = 0.16$ means that the probability of finding an Fe atom in the nearest neighbor shell of another Fe atom is 60.5% as compared to 52.8% ($=c_{Fe}$) for a totally random alloy. Given the high concentration of this alloy, the α_{lmn} are rather small and so the clustering tendency is not very pronounced. According to anomalous x-ray diffraction studies by Yankel,[26] σ-FeCr is partially long-range ordered, i.e. there are sublattices which are preferentially filled with Fe or Cr. For every interatomic vector connecting points in two different sublattices, the corresponding α_{lmn} must be negative. Evidently, the positive α_{lmn}'s of the bcc solid solution above T_σ do not reflect the local chemical order found in the σ-phase.

From Table II we see that most of the components of Δ_{lmn}^{FeFe} are negative. Therefore most of the average Fe-Fe separations (in particular those for the first three neighbor shells) are smaller than the corresponding average lattice distances. This is compatible with the observed decrease of the lattice parameter with increasing Fe concentration. However, the comparatively large negative value of the nearest neighbor Cr-Cr displacement shows that the concentration dependence of the lattice parameter is not necessarily reflected in a simple way in the local atomic distortions, i.e. from Vegard's Law one would expect the first nearest neighbor displacements to be positive! Nevertheless, the Δ_{lmn}^{CrCr} are on the average more positive than the Δ_{lmn}^{FeFe}. Thus the data suggest that taken over a sufficiently large local volume, the Cr atoms are indeed "bigger" than the Fe atoms. The average Cr-Cr nearest neighbor distance is 0.4% smaller and the average Cr-Cr next nearest neighbor distance is 0.3% larger than the corresponding average lattice separations. By comparison, the lattice parameter of pure Cr is 0.6% larger than the lattice parameter of pure Fe. It is also interesting to note that the root-mean-square static displacement amounts to only about ~3% of the root-mean-square thermal displacement.

Table I. Short-range order parameters α_{lmn}.

ℓmn	α_{lmn}	ℓmn	α_{lmn}
0 0 0	1.1806(23)	3 3 3	0.0051(8)
1 1 1	0.1596(14)	5 1 1	0.0025(6)
2 0 0	0.0691(14)	4 4 0	-0.006(7)
2 2 0	0.0455(11)	5 3 1	0.0016(4)
3 1 1	0.0217(10)	4 4 2	0.0022(5)
2 2 2	0.0253(11)	6 0 0	-0.0020(8)
4 0 0	0.0036(11)	6 2 0	0.0009(4)
3 3 1	0.0074(8)	5 3 3	0.0009(4)
4 2 0	0.0074(7)	6 2 2	0.0010(4)
4 2 2	0.0043(7)	4 4 4	0.0007(7)

Table II. Species dependent displacement parameters $\Delta_{lmn} = (\Delta x_{lmn}, \Delta y_{lmn}, \Delta z_{lmn})$ in Å. The Fe-Cr displacements can be obtained from Eq. 12.

ℓmn	Δx_{lmn}^{FeFe}	Δy_{lmn}^{FeFe}	Δz_{lmn}^{FeFe}	Δx_{lmn}^{CrCr}	Δ_{lmn}^{CrCr}	Δ_{lmn}^{CrCr}
1 1 1	-0.00070 (4)	-0.00070(4)	-0.00070(4)	-0.0019(5)	-0.0019(5)	-0.0019(5)
2 0 0	-0.00029(9)	0.00000	0.00000	0.00268(12)	0.00000	0.00000
2 2 0	-0.00022(4)	-0.00022(4)	0.00000	-0.00050(6)	-0.00050(6)	0.00000
3 1 1	0.00018(5)	-0.00022(3)	-0.00022(3)	-0.00007(6)	0.00011(4)	0.00011(4)
2 2 2	-0.00053(5)	-0.00053(5)	-0.00053(5)	0.00039(6)	0.00039(6)	0.00039(6)
4 0 0	0.00009(9)	0.00000	0.00000	0.00063(13)	0.00000	0.00000
3 3 1	-0.00013(4)	-0.00013(4)	-0.00005(4)	0.00016(4)	0.00016(4)	-0.00005(6)

Figure 5 compares the measured intensities in the (h_1, h_2, 0) plane (after subtracting the E=7.600keV data) with those reconstructed from the from the parameters in Tables I and II. The increase in the intensity near the Bragg positions and the details of the intensity in the zone boundary regions are well reproduced. These modulations are largely due to the "size effect" scattering as can be inferred from the systematic differences between the "Cr edge" and the "Fe edge" data. For example, the "dip" near 210 in the "Cr edge" data which becomes a local maximum in the "Fe edge" data, both of which may be related to a measurable zone boundary softness in the [100]$_L$ phonon branch.[27]

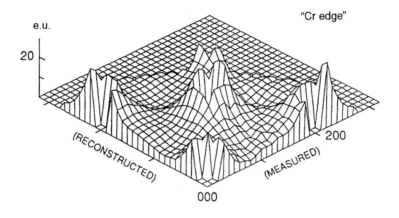

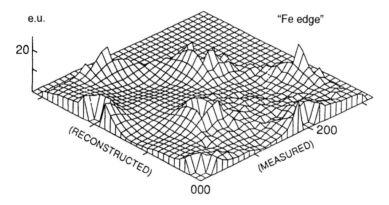

Figure 5. Measured and reconstructed intensities in the (h_1, h_2, 0) plane in electron units for the "Cr edge" and "Fe edge".

The measured intensity is compared with the results of the least-squares refinement along the <hhh> direction in Fig. 6. The intensity minimum around h=0.8 in the "Cr edge" data and the maximum around h=0.7 in the "Fe edge" data are caused by the "size effect" modulation and are related to the dip at 2/3(111) in the [111]$_L$ phonon branch. The difference in there positions can be explained by considering the $I_{SRO}(\mathbf{Q})$ which peaks at the origin, h=0, and therefore will shift a "size effect"-induced minimum towards a higher h value, where as a maximum will be shifted towards lower h. The same arguments apply to the maximum near 4/3(111) in the "Cr edge" data and the corresponding minimum in the "Fe edge" data. This peaking of the static diffuse scattering at the 2/3(111) and 4/3(111) is a direct consequence of the elastic softness of the bcc lattice in response to distortions in these directions as evidenced by the 2/3[111] dip in the [111]$_L$ phonon. Since the restoring force if the lattice to this particular displacement is relatively weak, the atoms are preferentially displaced in these directions. This is clear evidence for the coupling between the static displacements and the elastic response of the lattice.

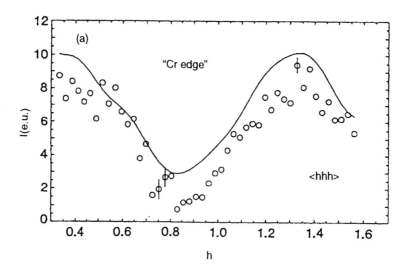

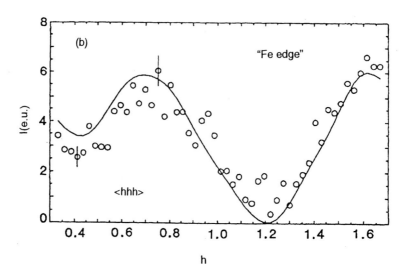

Figure 6. Measured and reconstructed intensities in the <hhh> direction. (a) "Cr edge", (b) "Fe edge".

CONCLUSION

It has been demonstrated that accurate occupational probabilities and first order static displacements can be obtained from diffuse scattering measurements. It is quite remarkable that species dependent atomic displacements on the order of 0.001Å and smaller can be determined from such broad features in the diffraction pattern. The availability of this information will provide theorists with the means to test their models and challenge them to include static displacements in their *ab initio* calculations of phase stability. In even more general terms, knowledge about the local atomic arrangements will help us to understand the connection between local structure and bulk properties.

ACKNOWLEDGEMENTS

The part of this work undertaken at Oak Ridge National Laboratory was supported by the US DOE under Contract No. DE-AC05-96OR22464 with Lockheed Martin Energy Research Corporation. The portion carried out at the University of Houston was supported by the NSF on DMR-9208450 and by the US DOE on DE-FG05-87ER45325.

REFERENCES

1. T. B. Massalski, *Binary Alloy Phase Diagrams*, American Society for Metals, Metals Park, Ohio (1986).
2. Electron diffraction measurements are also very useful for studying local atomic order. However, because the data analysis and interpretation are substantially different from x-ray and neutron scattering measurements, electron diffraction methods will not be discussed in this paper.
3. S. Hashimoto, *Acta Cryst.* A43: 481 (1987).
4. M. A. Krivoglaz, *Theory of X-Ray and Thermal Neutron Scattering from Real Crystals*, Plenum Press, New York (1969).
5. L. Vegard, *Z. Kristallogr.* 67:239 (1928).
6. C. J. Sparks, G. E. Ice, X. Jiang, and P. Zschack, *Mater. Res. Soc. Symp. Proc.* 375:213 (1975).
7. G. E. Ice, C. J. Sparks, J. L. Robertson, J. E. Epperson, and X. Jiang, *Mater. Res. Soc. Symp. Proc.* 437:181 (1996).
8. B. E. Warren, B. L. Averback, and B. W. Roberts, *J. Appl. Phys.* 22:1493 (1951).
9. B. Borie and C. J Sparks, *Acta Cryst.* A17:198 (1971).
10. G. E. Ice, C. J. Sparks, and L. B. Shaffer, *Resonant Anomalous X-Ray Scattering: Theory and Experiment*, ed. G. Materlik, C. J. Sparks, and K. Fischer, North Holland, Amsterdam (1994).
11. B. E. Warren, *X-Ray Diffraction*, Dover, New York (1969).
12. J. M. Cowley, *J. Appl. Phys.* 21:24 (1950).
13. By convention the indices lmn have been multiplied by two so that they are all integers hence the factor of ½ in the expression for **R**.
14. V. Gerold and J. Kern, *Acta Metall.* 35:393 (1987).
15. G. E. Ice, C. J. Sparks, A. Habenschuss, and L. B. Shaffer, *Phys. Rev. Lett.* 68:863 (1992).
16. L. Reinhard, J. L. Robertson, S. C. Moss, G. E. Ice, P. Zschack, and C. J. Sparks, *Phys. Rev. B* 45:2662 (1992).
17. B. Schönfeld, G. E. Ice, C. J. Sparks, H. G. Haubold, W. Schweika, and L. B. shaffer, *Phys. Status Solidi* B183:79 (1989).

18. X. Jiang, G. E. Ice, C. J. Sparks, J. L. Robertson, and P. Zschack, *Phys. Rev. B* 57:3211 (1995).
19. C. B. Walker and D. T. Keating, *Acta Cryst.* 14:1170 (1961).
20. R. O Williams, *Oak Ridge National Laboratory Report No. ORNL-4848* (1972).
21. P. Gerogopoulos and J. B. Cohen, *J. Physique Colloque* 12:C7-191 (1977).
22. O. Kubaschewski, *Iron Binary Phase Diagrams*, Springer, Berlin, (1982).
23. S. Hertzmanand and B. Sundman, *CALPHAD* 6:67 (1982).
24. M. Furusaka, Y. Ishikawa, S. Yamaguchi, and Y. Fujino, *J. Phys. Soc. Jpn.* 55:2253 (1986).
25. A. Habencshuss, G. E. Ice, C. J. Sparks, and R. A. Neiser, *Nucl. Inst. And Meth. Phys. Res.* A266:215 (1988).
26. H. L. Yankel, *Acta Cryst.* B39:0 (1982).
27. J. L. Robertson, L. Reinhard, D. A. Neumann, and S. C. Moss, Mater. Res. Soc. Symp.Proc. 376:689 (1995).

FERMI SURFACE EFFECTS IN THE DIFFUSE SCATTERING FROM ALLOYS

S. C. Moss[1] and H. Reichert[2]

[1] Physics Department
University of Houston
Houston, TX 77204-5506

[2] Universität Wuppertal
Institut für Materialwissenschaften
42285 Wuppertal, Germany

INTRODUCTION

The subject of the diffuse scattering from binary alloys is quite widely studied especially if one includes the consideration of oxide and semiconductor alloys, as one must, since they are often characterized by species disorder on one or more sublattice. For reviews, one may consult the papers and references in the present proceedings of which we mention only a few.[1-4]

The issues confronting X-ray or neutron scatterers are: 1) the local atomic order above a phase transition in the (nominally) disordered state; 2) the attendant (static) atomic displacements associated with a disordered alloy composed of atoms of different sizes; 3) the statistical mechanics/electronic theory of alloys from which we may, on the one hand, derive some of the observed effects and, reciprocally, extract information on alloy energies from the measurement of scattering patterns. Early treatments of the derivation of pair correlations, their Fourier transforms in the scattering patterns and their relation to the energetics of binary alloys include the work of Krivoglaz[5] and Clapp and Moss[6] (together called KCM), Cook and de Fontaine[7,8] and Froyen and Herring.[9] All of these treatments are of the mean-field variety, neglecting, at least at the outset, fluctuations in the several order parameters, and they have as their aim the relationship between the pair correlations and the effective pairwise interaction (EPI's) that govern them and can thus be related to fundamental alloy physics.

More recently, Tokar, Tsatskis (formerly Masanskii) and co-workers have substantially improved these treatments,[10] somewhat in the spirit of the spherical model, (using a γ-expansion method or GEM) and Reinhard and Moss[11] have summarized the situation vis-a-vis experiment and analysis via competing treatments. By competing

Local Structure from Diffraction
Edited by S.J.L Billinge and M.F. Thorpe, Plenum Press, New York, 1998

treatments we mean the KCM, the GEM and the use of an inverse Monte Carlo (IMC) scheme, first proposed by Gerold and Kern[12] and successfully implemented some years ago by Schweika and Haubold.[13] This procedure for extracting the EPI's from measured diffuse scattering is often referred to as exact, accepting the utility of the EPI's.

There is, of course, an experimental problem lurking in all of this; namely the separation of the scattering pattern into the symmetrical short-range-order part, the linear coupling between short-range order and atomic displacements (referred to often as the "size effect scattering") and the quadratic size-effect (atomic displacement) contribution. This last part peaks at the Bragg peaks in the vicinity of which it is called Huang scattering and it has the same functional dependence on distance $|q|$ from the Bragg peak (G_{hkl}) as does the first order thermal diffuse scattering,[1-4] i.e. $\propto q^{-2}$. The entire enterprise of separating these several contributions was first proposed by Borie and Sparks[14] and variants on their analysis, including least-square fits to the several contributions[15] and the use of experimental methods to separate them (three incident wavelengths, 3λ's, for binary alloys with X-rays[16,17], and elastic neutron scattering[18]) have been employed. There is also a more recent, and very complete, treatment of the scattering equations by Dietrich and Fenzl.[19]

It remains true that the physics of disordered alloys, by which we mean the underlying electronic structure and the proper application of statistical mechanics (including Monte Carlo, and cluster variation methods) is the major motivation behind our studies of diffuse scattering. We also hold - with others - that the disordered state presents us with an opportunity to extract this physics in an important way: that the diffuse scattering, carefully analyzed, allows us to make informed statements regarding the electronic structure and bonding in alloys, ultimately leading to the alloy phase stability and phase diagrams. Perhaps the most complete presentation of that program of work is contained in the book of Ducastelle[20] and in several recent conference proceedings.[21-26]

We shall here discuss one aspect of the analysis of the diffuse scattering from alloys, namely the appearance in the scattering pattern of incommensurate features that seem incontestably due to the contribution of the screening electrons, of wavevector $2k_F$, to the EPI's that govern the diffuse scattering (k_F is a Fermi surface wavevector). We give below an abbreviated account of the relevant formalism (it appears elsewhere).[5,27,28] We then concentrate on three alloy systems Cu-Au,[29,30] Cu-Al[30-33] and Cu-Pd[34] on which there has been recent work and which illustrate the effects clearly.

There still remains some disagreement[35] regarding the origin of the putative Fermi-surface effects, especially in their temperature dependence, and it shall also be our intent to describe them within the context of competing interactions in solids (all electronic in our case) which have clearly been shown over the years to be responsible for a hierarchy of incommensurate states. We also wish to discuss these scattering effects not only in q-space but in real space where, by virtue of the incommensurate (diffuse) peaks, we can infer an inhomogeneous disordered state - as suggested by Krivoglaz,[36] and modeled some years ago by Hashimoto[37] through a local anti-phase domain model.

THE PROBLEM

If we confine ourselves solely to the symmetrical short-range order component of the diffuse scattering the pair correlation function α_{oj} may be written as

$$\alpha_{oj} = 1 - \frac{P_{oj}^{A/B}}{c_B} \qquad c_B = \text{concentration of B atoms in the alloy}$$

where $P_{oj}^{A/B}$ is the conditional probability that a site j will be occupied by a B atom of a binary alloy ($c_A + c_B = 1$), given an A atom on the origin site, o. Of course if o=j, $P_{oj}^{A/B} = 0$ and thus $\alpha_{oo} \equiv 1$. If the alloy is random, $P_{oj}^{A/B} = c_B$ and $\alpha_{oj} = 0$. We also may define a set of effective pairwise interactions, V_{ij} for the alloy that must surely depend on the intervening atoms (concentration) and on the temperature. Nonetheless we use, within an Ising formalism,[5,6]

$$H = H_o + 1/2 \sum_{ij} \sigma_i \sigma_j V_{ij}$$

for the Hamiltonian in which the projection operators

$$\sigma_i = 1 \qquad \text{A on i}$$
$$\sigma_i = 0 \qquad \text{B on i} \quad .$$

From these we define deviation operators,

$$\sigma_i - c_A = c_B \qquad \text{A on i}$$
$$\sigma_i - c_A = -c_A \qquad \text{B on i} \quad .$$

The pair correlation functions can then be written as:

$$\alpha_{oj} \propto \langle (\sigma_o - c_A)(\sigma_j - c_A) \rangle$$

to place the formalism on a familiar basis.

The quantities

$$V_{ij} = 1/2 \, [V_{ij}^{AA} + V_{ij}^{BB} - 2V_{ij}^{AB}] \tag{1}$$

are the EPI's for our alloy and we must be careful about both how to use them and how to interpret them. As noted above, there is in principle nothing amiss with a set of V_{ij}'s, given by Eq.(1), that depend on the concentration (and temperature); in a real sense they must. (See the discussion in Schweika and Carlsson.[38])

In the scattering pattern, the symmetrical component due solely to short-range order is given by a Fourier series[1,2] in which the diffuse intensity per atom is,

$$\frac{I_D}{N} = [f_B - f_A]^2 c_A c_B \sum_j \alpha_j e^{iq \cdot r_j} \quad . \tag{2}$$

In Eq.(2) we have dropped the origin prefix (o) because r_j is by definition a vector from an origin site to its jth neighbor where all sites are allowed to be origins. $[f_B-f_A]$ is the scattering contrast which is slightly $|k|$-dependent for X-rays (k is the scattering wavevector; q is the wavevector of the fluctuation). For neutrons we have $[b_B-b_A]$ is independent of $|k|$. In the expression in Eq.(2), we may extract

$$\alpha(q) = \sum_j \alpha_j e^{iq \cdot r_j} \quad , \tag{3}$$

which is simply the transform of the pair correlation function. Within the mean-field (MF) or random phase approximation (RPA), this transform is given by:

$$\alpha(q) = \frac{1}{1 + 2c_A c_B \beta V(q)} \quad , \qquad \beta = 1/kT \tag{4}$$

where in Eq.(4)

$$V(q) = \sum_{ij} V_{ij} e^{iq \cdot r_{ij}} \tag{5}$$

and is the transform of the EPI's. The choice of 1 (unity) in the numerator of Eq.(4) is non-trivial but was initially proposed by Krivoglaz[5] and later shown by Masanskii et al.[39] to be correct and to yield the most reliable values of V_{ij} from measured data (see also Ref.11). Mostly these EPI's are limited to the first few neighbor interactions. There are two cases, particularly in metallic alloys, however, in which longer range interactions must be considered:

1) elastic interactions induced by a "significant" size discrepancy between the atoms; this is also very important in interstitial solutions;[40]

2) conduction electronic screening as with Friedel oscillations which introduces a long-range oscillatory component to the pair potential whose period is given by $2k_F$, where, $2k_F$ is a wavevector spanning parallel flats or nesting pieces of the Fermi surface in order to be observed. The formalism here is rather close to that governing charge-density-wave instabilities in which an atomic displacement or mass density wave is induced with a period of $2|k_F|$. In our case, we have, in the disordered state, an enhanced concentration fluctuation of wavevector $2k_F$.[5,27,28] This can easily be seen as follows:

$$V(q) \rightarrow V_s(q) \equiv V(q)/\varepsilon(q) \quad , \tag{6}$$

where $V_s(\mathbf{q})$ is the screened EPI and $\varepsilon(\mathbf{q})$ is the dielectric function which for the spanning wavevector, i.e. $\varepsilon(2k_F)$, shows a peak whose nature depends on the nesting or flatness of parallel Fermi surface sections (i.e. the number of screening states contributing to the singularity at $2k_F$). The simple formalism[5,27] is borrowed from the magnetic case of a spin density wave[41] but it has since been done, in a much more rigorous fashion, initially for Cu-Pd alloys.[42]

The peak in $\varepsilon(2k_F)$ gives rise to a reduced $V_s(2k_F) < V(2k_F)$ which in turn produces a peak in $\alpha(2k_F)$ and thus in the diffuse scattering. One can then draw the appropriate Kohn construction,[27] as in Fig.1,[29] in which anomaly surfaces intersect giving rise to the observed diffuse satellites which are the subject of this paper.

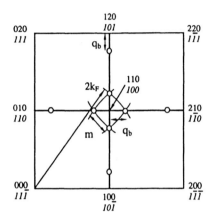

Figure 1. The basic reciprocal space geometry for the observation of diffuse satellites at positions of $\pm q_B$ about the superlattice points. Two equivalent planes of reciprocal space are shown.

That much is old stuff! However, following the initial report of this effect,[27] a series of observations on disordered alloys using electron diffraction was reported in which the position of $2k_F$ clearly confirmed the simple prediction - in a rigid band scheme - of scaling directly with the electron/atom (e/a) ratio of the alloy in question. In all cases, the shifts in the diffuse satellites of Fig.1 were in the "correct" direction - i.e. as predicted. No other scheme would seem to be able both to yield the existence of such a long-range feature in V_{ij} and to predict its change with e/a. These results were summarized by Moss and Walker.[28] For illustration we show below data from three Cu-Al alloys studied by Scattergood et al.[31] whose Fermi surface dimension in the <110> direction was measured directly from the position of the satellites along a symmetry axis in Fig.2 [note in b) the enhancement on the furnace cooling over quenching but without a noticeable shift in position]. Figure 3 presents values of k_F/k_{100} from diffuse scattering and from positron annihilation. Both indicate a deviation from free electron behavior where a flatness in <110> is, as in pure Cu, compensated for by a bulging in <100>. These results all may be understood using Eqs.(4) and (5) and by the construction in Fig.1. A good deal of physics was thereby overlooked in these initial verifications of the basic idea and the calculation of Gyorffy and Stocks,[42] as noted earlier, placed the phenomenon on a firmer theoretical basis; nonetheless neutron studies of the phonon dispersion in pure Cu[43] indicated only an anomaly in the derivative $d\omega(q)/dq$ at $q=2k_F$. This indicated that the actual Kohn anomaly for phonons is considerably less dramatic than the analogous effect on the concentration fluctuations.

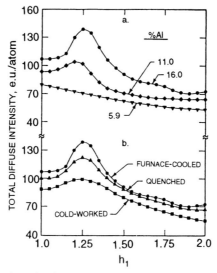

Figure 2. Total diffuse intensity, in electron units per atom, along the [h10] direction where $h_1 = h$; (a) Furnace-cooled single crystals containing 5.9, 11.0 and 16.0 at.% Al. (b) Single crystals containing 16.0 at.% Al in the furnace-cooled, quenched and cold-worked conditions.

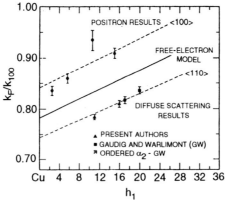

Figure 3. The ratio of the measured Fermi wavevector to the calculated <100> zone boundary vector, k_F/k_{100}, vs. at.% Al. The GW data are from Gaudig and Warlimont and are electron diffraction results. Positron annihilation results are due to Fujiwara et al. The theoretical variation for a free-electron sphere is shown by the solid line; broken lines correspond to a simple scaling of the <100> and <110> dimensions of the Fermi surface of Cu, i.e.: $k_F<110>/0.743 = n(alloy)^{1/3}$; $k_F<100>/0.841 = n(alloy)^{1/3}$, where k_F is in units of $1/a$. (See Ref.31 for additional references.)

There remain, as we can see, interesting questions when one explores this phenomenon in more detail and we shall discuss some of these in the next section.

RECENT EXPERIMENTS

Given the existence of modulations or splittings in the diffuse scattering associated with $2k_F$, it is of interest to explore their dependence on temperature and to extend the measurements of their composition dependence. A good example of the latter has been done by Reinhard and co-workers[44] on fcc Cu-Zn alloys using a ^{65}Cu isotope to enhance the neutron contrast. However here we shall confine ourselves mainly to Cu-Au, Cu-Al and Cu-Pd alloys. (The fact that the phenomenon seems so largely confined to Cu alloys is, in itself, an interesting observation.)

a. Cu₃Au

In Cu_3Au, Reichert et al.[29] examined the temperature dependence of the splittings in a plane of reciprocal space shown in Fig.1 using a [110] single crystal to study the scattering about the *100* position. Scans were made in both the [10ℓ] and [1k0] directions which are through the enhanced spots in Fig. 1 and should reveal diffuse peaks at $\pm q_B$. Scans normal to this plane through the *100* position are along [0.5+h, 0.5-h, 0]. This work was first reported in Ref.29 and recently, with Tsatskis, in Ref.30. Figures 4, 5 and 6 are from Reichert et al.[29,30] in which a fundamental issue is addressed, namely the origin of the temperature-dependent splittings. It is, at first glance, difficult to assign $2k_F$ (at a fixed composition) to a diffuse diffraction feature which changes with T. Certainly $2k_F$ is essentially constant over the narrow range of ~140K. In addition the mean-field expression given in Eq.4, however successful it is in retrieving V_{ij} from α_{ij}, cannot yield a $V_s(2k_F)$ which is temperature dependent. Hence the only temperature dependence enters in $\beta=1/kT$ and this clearly cannot shift the value of q_B in Fig.1. In Ref.30 we discuss the necessary modification to the mean-field equations required to yield the observed shift. It is shown to reside in a proper consideration of the alloy self-energy and this amended treatment of the scattering formalism is given in greater detail by Tsatskis in these proceedings.[45]

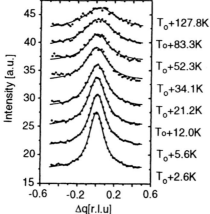

Figure 4. Total diffuse intensity versus momentum transfer Δq through the (100) superstructure position along the line (0.5 + h, 0.5 - h, 0) for a set of temperatures in the disordered phase of Cu_3Au. The lines are fits with single Gaussians. T_0 = ordering temperature. (From Ref.30)

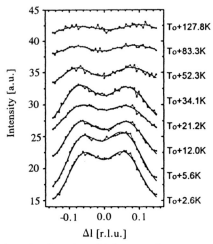

Figure 5. Total diffuse intensity along the (10ℓ) for a set of temperatures in the disordered phase of Cu$_3$Au. The lines are fits with two Gaussians. (From Ref.30)

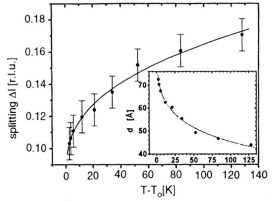

Figure 6. Temperature dependence of the separation distance $\Delta\ell = 2q_b$ (the full line is a fit to a power law.) The inset displays the average separation of domains with the same phase, assuming a correlated anti-phase domain model for the disordered state. (From Ref.30)

Figure 6 is particularly interesting because it shows both the temperature dependence of the splitting ($\Delta\ell \equiv 2q_B$) and its real-space counterpart $d = 2a/\Delta\ell$. Here we have a real-space modulation of the local order of ~70Å falling to ~40Å, whose correlation range in Fig.4 varies from ~15Å near the transition temperature, T_0, to ~6Å at T_0+140K. In other words, the induced splitting indicates a long-range modulation of the short-range order and this, in turn, as noted by Krivoglaz,[36] requires a heterogeneous disordered state of correlated regions of local order. Hashimoto independently wrote a series of papers[37] on a correlated microdomain model for the disordered state to give the splittings a real-space interpretation.

The "true" value of |2k$_F$| for Cu$_3$Au as extracted from these experiments seems ambiguous. However, at higher temperature, Eq.4 becomes exact and thus the value of |2k$_F$| can be taken from the asymptotic value of $\Delta\ell \simeq 0.17$ in Fig.5 and the construction in Fig.1.

196

2. Cu-Al

As noted earlier the Cu-Al system also has shown splittings as in Fig.2 and they have been used to rationalize Fermi surface dimension as a function of composition in Fig.3. We return to Cu-Al alloys because they also reveal a temperature dependence of the split peaks as measured by of Roelofs et al.[32] and reproduced in Fig.7 from Reichert et al.[30] In this case, however, the original data was on a rapidly quenched crystal. The IMC was used to extract a set of V_{ij}'s which were then used in a Monte Carlo calculation to generate the short-range order and hence the diffuse scattering at several temperatures. Within large error bars it would seem that $2q_B$ is leveling off at ~0.55 at ~900K and that would then be an appropriate number to use in calculating $2k_F$. Here it is clear that the splitting originates in the energy but has a value dictated by temperature, i.e. entropy.

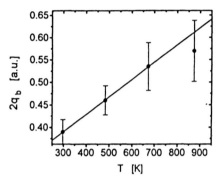

Figure 7. Temperature dependence of the separation distance $2q_b$ for the alloy $Cu_{85.6}Al_{14.4}$ (simulated and measured values extracted from Ref.32.) The straight line indicates a linear increase with temperature, but we believe a leveling should obtain at higher T. (From Ref.30)

Cu-Al alloys so clearly demonstrate concentration-dependent splittings, that they were good candidates for an inelastic neutron scattering study of Kohn anomalies. Chou et al.[33] attempted this only to find that there were no $2k_F$ singularities either in $\omega(q)$ or in $d\omega(q)/dq$ similar to the results on Cu noted earlier.[43] Nonetheless we show in Fig.8 the original Borie-Sparks[46] data for our $Cu_{0.84}Al_{0.16}$ crystal quenched to room temperature. Included is the familiar Kohn construction for peak splittings and the fcc Brillouin zone structure overlaid on the data. This data also shows the strong asymmetry in the diffuse scattering induced by an appreciable atom size disparity. Figure 9 is a set of neutron elastic scattering scans to confirm that the split diffuse satellites are not of dynamic origin. Finally in Table 1 below we present measured phonon frequencies and line widths along the line connecting 100 and 110 in Fig.8, through the strong satellite. The frequencies show no abnormality but the line width of the measured phonons shows an increase at about $\zeta=0.3$ which is the satellite position. Again this has a real space interpretation as the phonons are expected to be damped more strongly by the locally ordered regions; i.e. when the phonon wavevector matches the satellite wavevector, lifetime effects reflect enhanced real space fluctuations and a form of resonant absorption or damping.

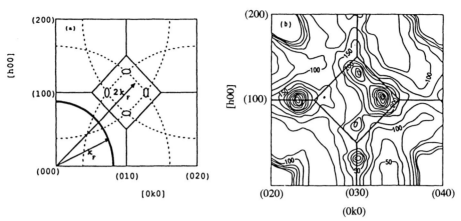

Figure 8. The [001] zone in reciprocal space of a fcc lattice schematically showing (a) Kohn construction of the diffuse satellite positions given k_F in the (100) direction and (b) the diffuse scattering as observed by Borie and Sparks.[46] The asymmetry in the satellite intensity is caused by atomic displacements arising from unequal atom sizes. (From Ref. 33)

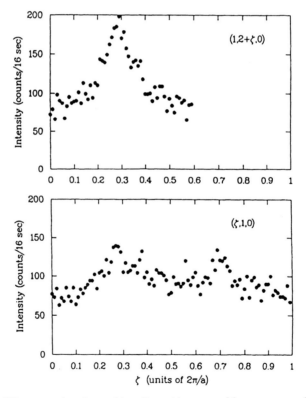

Figure 9. Elastic diffuse scattering observed in a $Cu_{0.84}Al_{0.16}$ crystal by neutrons confirming the satellite positions observed by Borie and Sparks.[46] (From Ref.33)

Table 1. Observed phonon frequencies and linewidths along [1,ζ,0] (linewidths are *not* resolution corrected). (From Ref.33)

ζ (in 2π/a)	ω (THz)	Γ (THz)
0.0	4.98	0.84
0.1	4.94	0.81
0.2	4.88	0.92
0.3	4.72	0.99
0.4	4.64	0.85
0.5	4.64	0.79

3. Cu-Pd

Cu-Pd alloys show a range of long-period superlattices and disordered states. The first self-consistent-field KKR CPA calculation of the Fermi surface of disordered Cu-Pd alloys[42] showed a distinct Fermi surface with enhanced flat portions normal to <110> which in turn gave rise to the composition-dependent diffuse maxima observed by Ohshima and Watanabe.[47] Noda et al.,[34] noting the strong X-ray evidence for the diffuse satellites in Cu-Pd[48] undertook a neutron experiment similar in intent to Chou et al.[33] Figure 10 from Noda et al.[34] shows a neutron elastic intensity profile for $Cu_{0.715}Pd_{0.285}$ with the familiar $2k_F$ "logo" as an insert. There are clear satellites in the quenched disordered alloy with an "m" value (see Fig.1) of 0.095±.01 in good agreement with X-ray data of Saha et al.[48] but somewhat smaller than the estimate of Gyorffy and Stocks.[42] (Perhaps the equilibrium value, at temperature, would increase somewhat as in Cu-Al and Cu-Au.)

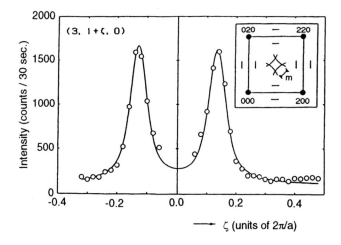

Figure 10. One-dimensional intensity distribution in a $Cu_{0.715}Pd_{0.285}$ crystal along the [010] direction from (3, 0.68, 0) to (3, 1.48, 0) where the λ/2 component was eliminated. (From Ref.34)

The temperature dependence of the peak splitting in Cu-Pd is only now being undertaken in a [110] crystal of $Cu_{0.702}Pd_{0.298}$ by Reichert and co-workers.[49] Their preliminary results for this crystal indicate a value of $q_B \simeq 0.108$ and very little, if any, temperature dependence. This value of q_B differs from Noda et al.[34] who found $q_B \simeq 0.130$ but this discrepancy may well be due to the fact that Reichert and co-workers are studying the diffuse satellite in equilibrium at temperature.

THEORETICAL CONSIDERATIONS

Most recently the original result of Reichert et al.[29] was criticized both for the attribution of the basic effect of the peak splitting to electronic energy terms and for remarking at all on its temperature dependence. This Comment by Ozolins et al.[35] has as its major premise that the ground state of Cu_3Au (and Cu_3Pd) can be calculated from first principles and Monte Carlo methods can then be employed to derive the pair correlations through and above T_c. Their calculations overestimated the transition substantially and gave a value for $\Delta\ell=2q_B$ of 0.09 near T_o compared with the measured value in Ref.30 of 0.11 and a (+100K) higher temperature value of 0.12 which both disagree appreciably with the data. Nonetheless, they did find a temperature-increasing splitting. In Cu-Pd (close to the measured composition of Reichert et al.[49]) they found very little, if any, change with temperature in agreement with the recent data.[49] Their splitting value also seems to disagree with experiment. Inasmuch as the Fermi surface does not explicitly enter their calculations, it is difficult to compare their splitting values to other calculations.

More important perhaps, is their insistence that the peak splitting as discussed is not energetic in origin (the ground state is not modulated) but simply arises from the entropy via Monte Carlo. It is this contention which we would hope to lay to rest, as discussed below.

COMPETING INTERACTIONS

We review briefly the origin of incommensurate modulations in solids which invariably arise from a set of competing interactions. It is important to note that at T=0 the ground state may be a simple commensurate ordered structure; it is often only at $T>T_c$ that the competing interactions reveal themselves. Among the many systems for which calculations have been made are anisotropic magnets, alloys, surface structure of adsorbed species and so forth.

The model that has been quite successfully employed to treat such systems is the axial-next-nearest-neighbor-interaction model (ANNNI). We here quote briefly from a very nice treatment of the 2D ANNNI model in the CVM approximation by Finel and de Fontaine.[50] Figure 11 below shows the basic interaction set where along y the interactions are ferromagnetic while along x the ferromagnetic rows are antiferromagnetically coupled with $J_0 = -J_1$ and $J_2/J_1 = K$ (both antiferromagnetic). A great host of incommensurate phases have been generated in this way (of course 3D treatments abound with planes replacing rows) and we show in Fig.12 the phase diagram of this 2D ANNNI model. All structures have a shorthand such that <1> refers to pure antiferromagnetic ↑↓↑↓ while <2> is a doubled phase ↑↑↓↓, etc. (See Ref.50 for details.) The scales are k_BT/J_0 (temperature) vs. $K = J_2/J_1$. The competing interactions appear along the x-axis. Disregarding the many complications, we may note that for K<0.5, the phases become quite simple. The fully ordered antiferromagnet initially disorders into a commensurate state (diffuse peak at reciprocal lattice point) while above the dashed line the commensurate disorder becomes incommensurate and splits into our familiar satellites. Of course temperature (entropy) is the controlling element, but K (energy ratio) is the determining agent especially as regards the steep "disorder line." Figures 13 and 14 show results for two values of K (0.4 and 0.3). At K=0.4 a very small change in temperature produces the splitting which then increases with T: q is actually q_x and is in units of $2\pi/a$. At K=0.3, the diffuse maximum remains at the commensurate position and then abruptly falls off (the splitting increases) with increasing T as the "disorder line" is crossed.

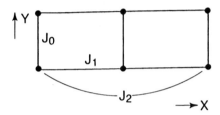

Figure 11. The two-dimensional ANNNI model. (From Ref.50)

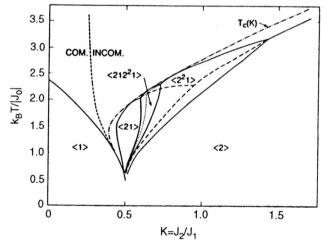

Figure 12. Phase diagram of the two-dimensional ANNNI model. Full lines represent the complete phase diagram with phases <1>, <2>, <21>, <2²1> and <212²1>. The dot-dashed line represents the curve $T_c(K)$ and the dashed line the "disorder" line, above the critical temperature, in the disordered phase. (From Ref.50)

Here then is a clear example of a simple ordered ground state with competing interactions which, upon increasing the temperature, develops a modulated disordered state. It is <u>only</u> in the disordered state for K<0.5, that we are aware of the competing interaction energies. The analogy with our alloys should not be lost; in our case we have local interactions of the Coulombic variety and longer range interactions dictated by the Fermi surface with attendant screening. It is often only in the disordered state that the consequence of these competing interactions appears. But not always, and there are myriad examples of long-period superlattices in which the ground state is indeed also modulated (as in Cu₃Pd - in *two* directions!).

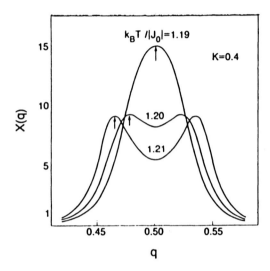

Figure 13. Temperature dependence of the susceptibility $\chi(q)$ for K = 0.4; for susceptibility, one may, as well, read diffuse scattering, $\alpha(q)$. The wavevector q is measured in units of $2\pi/a$ and $\chi(q)$ in arbitrary units. The arrows indicate the location of the maxima of the susceptibility (diffuse scattering). (From Ref.50)

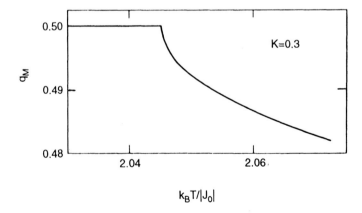

Figure 14. Temperature dependence of q_M, the wavevector where the susceptibility is maximum, for K = 0.3; q_M is measured in units of $2\pi/a$. (From Ref.50)

202

CONCLUSIONS

We have reviewed here the evidence for Fermi-surface-induced modulated scattering in disordered alloys and presented the simplest formalism for understanding - at least - its geometry. We leave for parallel theoretical treatments a more extended discussion that accounts both for the composition and temperature-dependent EPI's and for their proper incorporation into the scattering formalism. Certainly Tsatskis[45] has developed a substantial modification of the original KCM equation that permits temperature-dependent features and yet preserves the utility of the EPI's. There is also the very interesting work of Le Bolloc'h and colleagues at ONERA[51,52] on Pt-V alloys in which there seems to be a pronounced change in the distribution of the diffuse scattering as one goes from Pt_3V to Pt_8V with interactions (EPI's) that seem nearly concentration independent but show a large nearest-neighbor interaction, V_1. Their analysis shows how these large changes can be understood through a large decrease in the influence of V_1 as composition and/or temperature decreases. Their analysis provides a firmer understanding of the meaning and utility of the EPI's and allows us - as does the work of Tsatskis - to retain many of the useful features of a quasi-mean-field treatment by understanding clearly how a renormalized set of interactions can still be employed in handling even highly "unusual" composition or temperature dependencies. The connection with the ideas of competing interactions presented above is particularly relevant here. For a 3D ANNNI model the 2D planes, not the (1D) lines, compete; the result in terms of EPI's is that for fcc lattices, J_2, the second neighbor *planar* interaction includes a large number of EPI's between *atoms* in the respective planes with the *nearest* pair being between two unit cell origin atoms separated by $2a_0$. This is the *8th neighbor* EPI! Even for rather short range competing planar interactions we can see that the EPI's extend to distant neighbors and the competition that stabilizes a modulated state above *or* below the ordering transition is therefore invariably of a long range, oscillatory nature.

In short, then, the use of diffuse scattering as a means of revealing, or retrieving, the basic interactions in alloys seems still to be quite healthy and the parameters one retrieves reflect the basic physics of the alloy.

ACKNOWLEDGMENTS

The authors wish to express their appreciation to Drs. Simon Billinge and Michael Thorpe for organizing a very stimulating conference. We also thank the National Science Foundation for support of this work on DMR-9208450 and the Texas Center for Superconductivity at the University of Houston for partial support. The permission of the authors of Reference 49 to quote work-in-progress is also much appreciated as is the assistance with the manuscript of Ms. Sandi White.

REFERENCES

1. W. Schweika in: *Statics and Dynamics of Alloy Phase Transformations* , Proc. of NATO ASI Series B, Physics Vol. 319, P. E. A. Turchi and A. Gonis, eds., Plenum Press, New York (1994).
2. S. C. Moss, *Mat. Res. Soc. Symp. Proc.* 376:674 (1995).
3. H. Chen, R. J. Comstock and J. B. Cohen, *Ann. Rev. Mater. Sci.* 9:51 (1979).
4. T. R. Welberry and B. D. Butler, *J. Appl. Cryst.* 27:265 (1994), and references therein.
5. M. A. Krivoglaz, *Theory of X-ray and Thermal Neutron Scattering from Real Crystals,* Eng. transl. Plenum Press, New York (1969); note also that English translations of a two-volume update by the late Prof. Krivoglaz are currently available under the titles: *X-ray and Neutron Diffraction in Non-Ideal Crystals;* and *Diffuse Scattering of X-rays and Neutrons by Fluctuations,* Springer, New York (1996).
6. P. C. Clapp and S. C. Moss, *Phys. Rev.* 142:418 (1966); 171:754 (1968); S. C. Moss and P. C. Clapp, *Phys, Rev.* 171:764 (1968).
7. H. E. Cook and D. de Fontaine, *Acta Metall.* 17:915 (1969).
8. D. de Fontaine, *Solid State Phys.* 34:73 (1979).
9. S. Froyen and C. Herring, *J. Appl. Phys.* 52:7165 (1981).
10. V. I. Tokar, I. V. Masanskii and T. A. Grishchenko, *J. Phys.: Cond. Matter* 2:10199 (1990).
11. L. Reinhard and S. C. Moss, *Ultramicroscopy* 52:223 (1993).
12. V. Gerold and J. Kern, *Acta Metall.* 35:393 (1987).
13. W. Schweika and H. G. Haubold, *Phys. Rev. B* 37:9240 (1988).
14. B. Borie and C. J. Sparks, *Acta Cryst. A* 27:198 (1971).
15. R. O. Williams, *Oak Ridge National Laboratory Report* No. ORNL-4828 (1972), unpublished.
16. L. Reinhard, J. L. Robertson, S. C. Moss, G. E. Ice, P. Zschack and C. J. Sparks, *Phys. Rev. B.* 45:2662 (1992).
17. X. Jiang, G. E. Ice, C. J. Sparks, L. Robertson and P. Zschack, *Phys. Rev. B* 54:3211 (1996).
18. B. Schönfeld, L. Reinhard, G. Kostorz and W. Bührer, *Phys. Status Solidi B* 148:457 (1988).
19. S. Dietrich and W. Fenzl, *Phys. Rev. B* 39:8873 (1989).
20. F. Ducastelle, *Order and Phase Stability in Alloys*, North-Holland, New York (1991).
21. Proc. of NATO ASI on *Alloy Phase Stability* , G. M. Stocks and A. Gonis, eds., Kluwer, Boston (1987).
22. Proc. of NATO ASI on *Statics and Dynamics of Alloy Phase Transformations*, P. E. A. Turchi and A Gonis, eds., Plenum Press, New York (1994).
23. MRS Symp. Proc. Vol. 166 on *Neutron Scattering for Materials Science* , S. M. Shapiro, S. C. Moss, J. D. Jorgensen, eds. (1990).
24. MRS Symp. Proc. Vol. 376 on *Neutron Scattering in Materials Science II,* D. A. Neumann, T. P. Russell, B. J. Wuensch, eds. (1996).
25. Proc. of Joint NSF/CNRS Workshop on Alloy Physics in *Computational Materials Science,* Vol. 8 (1997).
26. Fundamental Materials Science Research Series, *Local Structure from Diffraction,* M. F. Thorpe and S. Billinge, eds., Plenum Press, New York (1998).
27. S. C. Moss, *Phys. Rev. Lett.* 22:1108 (1969).
28. S. C. Moss and R. H. Walker, *J. Appl. Cryst.* 8:96 (1975).
29. H. Reichert, S. C. Moss and K. S. Liang, *Phys. Rev. Lett.* 77:4382 (1996).
30. H. Reichert, I. Tsatskis and S. C. Moss, *Comp. Mat. Sci.* 8:46 (1997).
31. R. O. Scattergood, S. C. Moss and M. B. Bever, *Acta Metall.* 18:1087 (1970).
32. H. Roelofs et al., *Scri Mater.,* 34:1393 (1996); B. Schönfeld et al., *Acta Mater.* 44:335 (1996).
33. H. Chou, S. M. Shapiro, S. C. Moss and M. Mostoller, *Phys Rev. B* 42:500 (1990).
34. Y. Noda, D. K. Saha and K. Ohshima, *J. Phys.: Cond. Matter.* 5:1655 (1993).
35. V. Ozolins, C. Wolverton and A. Zunger, *Phys. Rev. Lett. Comment* 79:955; see also Reply by H. Reichert, S. C. Moss and K. S. Liang, 956 (1997).

36. M. A. Krivoglaz, *Sov. Phys. JETP* 57:205 (1983).
37. S. Hashimoto, *Acta Cryst. A* 37:511 (1981); S. Hashimoto, *Acta Cryst. A* 39:524 (1983); S. Hashimoto and H. Iwasaki, *Phys. Status Solidi (a)* 51:673 (1979).; S. Hashimoto and H. Iwasai, *Phys. Status Solidi (a)* 96:435 (1986).
38. W. Schweika and A. E. Carlsson, *Phys. Rev. B* 40:4990 (1989).
39. I. V. Masanskii, V. I. Tokar and T. A. Grishchenko, *Phys. Rev. B* 44:4647 (1991).
40. A. G. Khachaturyan, *The Theory of Structural Transformations in Solids*, Wiley, New York (1983).
41. L. M. Roth, H. J. Zeiger and T. A. Kaplan, *Phys. Rev.* 149:519 (1966).
42. B. L. Gyorffy and M. Stocks, *Phys. Rev. Lett.* 50:374 (1983).
43. G. Nilsson and S. Rolandson, *Phys. Rev. B* 9:3278 (1974).
44. L. Reinhard, B. Schönfeld, G. Kostorz and W. Bührer, *Phys. Rev. B* 41:1727 (1990).
45. I. Tsatskis, these proceedings.
46. B. Borie and C. J. Sparks, *Acta Cryst.* 17:827 (1964).
47. K. Ohshima and D. Watanabe, *Acta Cryst. A* 29:520 (1973).
48. D. K. Saha, K. Koga and K. Ohshima, *J. Phys.: Cond. Matter* 4:10093 (1992).
49. H. Reichert, H. H. Hung, V. Jahns, K. S. Liang, D. Zehner and H. Dosch (work in progress).
50. A. Finel and D. de Fontaine, *J. Stat. Phys.* 43:645 (1986).
51. D. Le Bolloc'h, T. Cren, R. Caudron and A. Finel, *Comp. Mat. Sci.* 8:241 (1997).
52. D. Le Bolloc'h, R. Caudron and A. Finel, *Phys. Rev. B* (in press).

NON-MEAN-FIELD THEORIES OF SHORT RANGE ORDER AND DIFFUSE SCATTERING ANOMALIES IN DISORDERED ALLOYS

Igor Tsatskis[*]

Department of Earth Sciences
University of Cambridge
Downing Street
Cambridge CB2 3EQ
United Kingdom

INTRODUCTION

Local, or short-range, order in disordered alloys is an important and exciting phenomenon which is quantified in electron, X-ray and neutron scattering experiments. It is discussed in many excellent reviews and books,[1-6] as well as in the multitude of original research papers.

This relatively short review of the subject does not attempt to discuss all aspects of the problem of local correlations in alloys. In particular, we will not touch such issues as multiatom (cluster) interactions, static displacements and vibrations of alloy atoms, partially ordered, multicomponent or amorphous alloys. As a result, we will concentrate on the Hamiltonian traditional for the considered problem, that of the Ising model on a rigid ideal lattice with pair, but otherwise arbitrary (i.e., of any range) interatomic interactions.

The central object of the paper is the pair correlation function of the corresponding dynamical variables of the model, the occupation numbers or spin variables, the Fourier transform of which is proportional to the intensity of diffuse scattering caused by atomic short-range order. The main aim is to show that the expression for this quantity has certain internal structure analogous, e.g., to that of the averaged Green's function used in the electronic theory of disordered alloys. This structure is independent of the approximation used for the quantitative description of correlations. As will be seen, this structure alone, without further specification of a particular theory of short-range order, allows us to see new possibilities in diffuse scattering, some of which have recently

---

[*]Former name: I.V. Masanskii

Local Structure from Diffraction
Edited by S.J.L Billinge and M.F. Thorpe, Plenum Press, New York, 1998

been observed experimentally.

The present contribution is organized as follows. First two sections are auxiliary and serve mainly for reference purposes; the former introduces relation between the phenomenological Hamiltonian of binary solid solutions and the Ising model, as well as necessary definitions and formulae, while the latter describes briefly standard approaches to the theory of electronic structure of disordered alloys. Readers familiar with the material contained in these sections can skip it and proceed directly to the third section where the key expression (75) for the diffuse intensity is introduced. In the fourth section its derivation and relation with alloy thermodynamics are discussed. Without any doubt, many readers would quickly realize that Eq. (75) is simply one of the possible forms of the famous Dyson equation. Such readers can then focus on the two last sections. The fifth section reviews, from the point of view adopted in this paper, existing theoretical approaches – both traditional and relatively new – to the problem of calculation of the pair correlation function and the diffuse-scattering intensity. Finally, in the sixth section based on the recent author's work it is shown how the internal structure of Eq. (75) leads to understanding existing and predicting new diffuse-scattering anomalies.

LIST OF ABBREVIATIONS

AE   alpha-expansion
ATA   average t-matrix approximation
CPA   coherent potential approximation
GEM   gamma-expansion method
HTE   high-temperature expansion
KCM   Krivoglaz-Clapp-Moss (approximation)
MC   Monte Carlo (simulation)
MFA   mean-field approximation
PCF   pair correlation function
RPA   random-phase approximation
SM   spherical model
SRO   short-range order
SSA   single-site approximation
VCA   virtual crystal approximation

DESCRIPTION OF A BINARY ALLOY

We will consider the standard model of a binary alloy used in the statistical theory of ordering.[1-6] In this model two sorts of atoms (A and B) are distributed over N sites of a rigid lattice; there are no vacancies or other lattice defects. For simplicity reasons the consideration is confined to the case of the lattice with one site per unit cell. All sites of the lattice are equivalent, i.e., there are no distinct sublattices; this situation corresponds to disorder or ferromagnetic ordering. A particular alloy configuration is fully described by the set of occupation numbers p_i^α,

$$p_i^\alpha = \begin{cases} 1, & \text{atom of type } \alpha \text{ at the lattice site } i, \\ 0, & \text{otherwise.} \end{cases} \tag{1}$$

The occupation numbers satisfy obvious relations

$$\sum_{\alpha} p_i^{\alpha} = 1 , \tag{2}$$

$$\sum_{i} p_i^{\alpha} = N^{\alpha} , \tag{3}$$

$$\sum_{\alpha} N^{\alpha} = N , \tag{4}$$

where N^{α} is the total number of atoms of type α, and the last equation holds because the total number of alloy atoms is equal to the total number of lattice sites. Indices α and β take only two values, A or B. It is assumed that the interatomic interactions are pairwise; an atom of type α at site i interacts with an atom of type β at site j with a potential $V_{ij}^{\alpha\beta}$. Then the configurational part of the alloy Hamiltonian is

$$H = \frac{1}{2} \sum_{ij} \sum_{\alpha\beta} V_{ij}^{\alpha\beta} p_i^{\alpha} p_j^{\beta} . \tag{5}$$

The main aim of statistical mechanics is the calculation of the partition function[7]

$$Z = \operatorname{Tr} \exp(-\beta H) , \tag{6}$$

where $\beta = 1/k_B T$, k_B is the Boltzmann constant, T the absolute temperature, and Tr denotes the trace of a matrix. It is convenient to avoid restriction (3) on the total number of atoms of each type and work in the grand-canonical ensemble, calculating the grand partition function $\tilde{Z}$,

$$\tilde{Z} = Z \exp \left(\beta \sum_{\alpha} \mu^{\alpha} N^{\alpha} \right) , \tag{7}$$

where μ^{α} is the chemical potential of atoms α. In this approach the number of atoms is a function of the corresponding chemical potential, and after having done the calculation the chemical potential is adjusted to get the required number of atoms. To unify canonical and grand-canonical treatments, i.e., to get the same formula (6) for $\tilde{Z}$, an effective Hamiltonian $\tilde{H}$ is often introduced:

$$\tilde{H} = H - \sum_{\alpha} \mu^{\alpha} N^{\alpha} . \tag{8}$$

The expression for the effective Hamiltonian is then

$$\tilde{H} = \frac{1}{2} \sum_{ij} \sum_{\alpha\beta} V_{ij}^{\alpha\beta} p_i^{\alpha} p_j^{\beta} - \sum_{i} \sum_{\alpha} \mu^{\alpha} p_i^{\alpha} . \tag{9}$$

As is well-known, the statistical-mechanical problem defined by the Hamiltonian (9) can be transformed into the equivalent problem of an Ising magnet in a magnetic field. The occupation numbers are not yet all independent, due to Eq. (2). Dependent variables can be eliminated by introducing a spin variable s_i,

$$s_i = \begin{cases} +1, & \text{spin up (atom } A\text{) at the site } i, \\ -1, & \text{spin down (atom } B\text{) at the site } i. \end{cases} \tag{10}$$

The relations between the occupation numbers and the spin variables are

$$p_i^A = \frac{1}{2}(1 + s_i) , \tag{11}$$

$$p_i^B = \frac{1}{2}(1 - s_i) . \tag{12}$$

Inserting Eqs. (11) and (12) into the expression (9) for the effective Hamiltonian $\tilde{H}$, we find that, apart from the configuration-independent term, the Hamiltonian is that of the Ising model,

$$\tilde{H} = -\frac{1}{2}\sum_{ij} J_{ij}s_i s_j - h\sum_i s_i \ , \tag{13}$$

where the effective exchange integral J_{ij} and the magnetic field h are given by

$$J_{ij} = -\frac{1}{2}V_{ij} \ , \tag{14}$$

$$h = \frac{1}{2}\left(\mu^A - \mu^B\right) - \frac{1}{4}\sum_j \left(V_{ij}^{AA} - V_{ij}^{BB}\right) \ , \tag{15}$$

$$V_{ij} = \frac{1}{2}(V_{ij}^{AA} + V_{ij}^{BB}) - V_{ij}^{AB} \ , \tag{16}$$

and V_{ij} is the pair ordering potential. The second term in Eq. (15) does not, in fact, depend on the index i because all interatomic potentials are functions of only the distance between interacting atoms.

For the following we define concentrations c^α and magnetization m which are statistical averages of the occupation numbers (1) and the spin variables (10), respectively,

$$c^\alpha = \langle p_i^\alpha \rangle \ , \tag{17}$$

$$m = \langle s_i \rangle \ , \tag{18}$$

getting from Eqs. (11) and (12)

$$c = \frac{1}{2}(1 + m) \ , \tag{19}$$

where $c = c^A = 1 - c^B$. Here the symbol $\langle \ldots \rangle$ denotes the statistical averaging with the effective Hamiltonian $\tilde{H}$, Eq. (9) or (13),

$$\langle \ldots \rangle = \frac{\text{Tr}\left[\ldots \exp\left(-\beta\tilde{H}\right)\right]}{\text{Tr}\exp\left(-\beta\tilde{H}\right)} \ . \tag{20}$$

We also introduce irreducible PCFs for the occupation numbers,

$$G_{ij}^{\alpha\beta} = \langle p_i^\alpha p_j^\beta \rangle - \langle p_i^\alpha \rangle\langle p_j^\beta \rangle \ , \tag{21}$$

and for the spin variables,

$$G_{ij} = \langle s_i s_j \rangle - \langle s_i \rangle\langle s_j \rangle \ . \tag{22}$$

The notion of irreducibility comes from the diagram technique for the Ising model, in the framework of which it can be shown that the irreducible PCF does not contain so-called split diagrams which are present in the expansion for the correlation function $\langle s_i s_j \rangle$.[8] Using relations (11) and (12) between the occupation numbers and the spin variables, it is easy to obtain the correspondence between the PCFs (21) and (22),

$$G_{ij}^{AA} = G_{ij}^{BB} = -G_{ij}^{AB} = \frac{1}{4}G_{ij} \ , \tag{23}$$

and, therefore, there exists only one independent PCF for the occupation numbers. In the theory of alloys the Warren-Cowley parameters α_{ij}, instead of the PCFs (21), are traditionally used for the description of SRO. They are defined as

$$\alpha_{ij} = 1 - \frac{\langle p_i^A p_j^B \rangle}{c(1 - c)} \ , \tag{24}$$

and, according to Eqs. (17), (19), (21) and (23),

$$G_{ij} = 4c(1-c)\alpha_{ij} = (1-m^2)\alpha_{ij} \; . \tag{25}$$

The diagonal matrix elements of the PCFs (21) and (22) can be expressed in terms of the averages (17) and (18). Taking into account identities $(p_i^\alpha)^2 = p_i^\alpha$ and $s_i^2 = 1$ which follow from the definitions (1) and (10), we have

$$G_{ii}^{AA} = G_{ii}^{BB} = -G_{ii}^{AB} = c(1-c) \; , \tag{26}$$

$$G_{ii} = 1 - m^2 \; , \tag{27}$$

$$\alpha_{ii} = 1 \; . \tag{28}$$

The last equation leads to the important sum rule in the reciprocal space. Let us introduce the lattice Fourier transformation,

$$f(\mathbf{k}) = \frac{1}{N} \sum_{ij} f_{ij} \exp(i\mathbf{k}\mathbf{r}_{ij}) \; , \tag{29}$$

$$f_{ij} = \frac{1}{\Omega} \int d\mathbf{k} \; f(\mathbf{k}) \exp(-i\mathbf{k}\mathbf{r}_{ij}) \; , \tag{30}$$

where f_{ij} is an arbitrary lattice function, $\mathbf{k}$ the wavevector, $\mathbf{r}_{ij} = \mathbf{r}_i - \mathbf{r}_j$, $\mathbf{r}_i$ the radius-vector of the site i, and the integration in Eq. (30) is carried out over the Brillouin zone of volume Ω. Eq. (28) can then be rewritten as

$$\frac{1}{\Omega} \int d\mathbf{k} \; \alpha(\mathbf{k}) = 1 \; . \tag{31}$$

Recalling that the Fourier transform $\alpha(\mathbf{k})$ of the SRO parameters α_{ij} is the SRO part of the diffuse-scattering intensity measured in Laue units,[5] we conclude that the sum rule (31) expresses the property of conservation of the integrated SRO intensity.

ELECTRONIC THEORY OF DISORDERED ALLOYS

In what follows we will occasionally turn to ideas which were developed in the theory of electronic structure of disordered alloys.[2,6,9–11] With this in mind, we deviate now from the main theme of this paper and consider briefly (and rather formally, without discussing the physical meaning of derived formulae) methods of calculating the Green's function of electrons averaged over possible alloy configurations. There is some overlapping of notations used in this section with those in the rest of the paper; however, the present section is quite isolated from the statistical-mechanical part of the discussion, and hence this overlapping should not lead to any confusion.

The central role in the one-electron theory of disordered alloys is played by the electronic Green's function G which may be defined as a resolvent of the alloy Hamiltonian H,

$$G = (E - H)^{-1} \; , \tag{32}$$

where E is the energy of an electron. The Hamiltonian of a disordered binary alloy is usually assumed to be a sum of two operators,

$$H = H_0 + V \; . \tag{33}$$

The first of these two terms, H_0, is translationally invariant, while the second, the random one-electron potential V, depends on particular alloy configuration and is usually assumed to be diagonal in the site representation. The operator V is a sum of individual potentials V_i centred at each site i and acquiring two possible values, V^A and V^B, in accordance with the type of atom occupying the considered site.

Propagator expansion

The potential V is often viewed as a perturbation, though not necessarily small, of the initial unperturbed Hamiltonian H_0. Defining the unperturbed Green's function,

$$G_0 = (E - H_0)^{-1} , \tag{34}$$

one can easily construct the Dyson equation,

$$G = G_0 + G_0 V G . \tag{35}$$

Iterating this equation and averaging over all possible alloy configurations (the averages are denoted by brackets), we get the propagator expansion

$$\langle G \rangle = G_0 + G_0 \langle V \rangle G_0 + G_0 \langle V G_0 V \rangle G_0 + \dots . \tag{36}$$

The averaged Green's function satisfies another Dyson equation,

$$\langle G \rangle = G_0 + G_0 \Sigma \langle G \rangle , \tag{37}$$

where the operator Σ is called the self-energy. The self-energy is, in terms of the diagrammatic expansion for the averaged Green's function $\langle G \rangle$ generated by Eq. (36), the irreducible part of $\langle G \rangle$, i.e., the sum of all graphs for $\langle G \rangle$ which cannot be separated into two parts by cutting a single bare-propagator line G_0. From Eq. (37) it follows that

$$\langle G \rangle = (G_0^{-1} - \Sigma)^{-1} = (E - H_0 - \Sigma)^{-1} , \tag{38}$$

the second equation comes from Eq. (34).

Locator expansion

There exists another perturbation series, the locator expansion, which is the expansion in powers of the unperturbed Hamiltonian H_0 rather than the potential V. In this case the unperturbed Green's function is that of the localized atomic states,

$$g = (E - V)^{-1} , \tag{39}$$

and the corresponding Dyson equation has the form

$$G = g + g H_0 G . \tag{40}$$

As in the case of the propagator expansion, we iterate this equation and average over the ensemble of configurations term by term, obtaining the following expansion:

$$\langle G \rangle = \langle g \rangle + \langle g H_0 g \rangle + \langle g H_0 g H_0 g \rangle + \dots . \tag{41}$$

By introducing the so-called fully renormalized interactor U,

$$U = H_0 + H_0 G H_0 , \tag{42}$$

full formal analogy between the propagator (Eq. (36)) and locator expansions is achieved:

$$\langle U \rangle = H_0 + H_0 \langle g \rangle H_0 + H_0 \langle g H_0 g \rangle H_0 + \dots . \tag{43}$$

The corresponding sum of all irreducible graphs σ in the case of the locator expansion is called the locator; similarly to Eq. (37), we have

$$\langle U \rangle = H_0 + H_0 \sigma \langle U \rangle . \tag{44}$$

From Eqs. (42) and (44) it immediately follows that in terms of the locator the Dyson equation and the expression for $\langle G \rangle$ are

$$\langle G \rangle = \sigma + \sigma H_0 \langle G \rangle , \tag{45}$$
$$\langle G \rangle = (\sigma^{-1} - H_0)^{-1} , \tag{46}$$

respectively. The relationship between the locator σ and the self-energy Σ, according to Eqs. (38) and (46), is

$$\sigma = (E - \Sigma)^{-1} . \tag{47}$$

Renormalization of one-electron potential

We now renormalize the potential V subtracting a configuration-independent, site-diagonal operator S and adding it to the unperturbed Hamiltonian H_0:

$$H = H_0 + V = (H_0 + S) + (V - S) = \tilde{H} + \tilde{V} . \tag{48}$$

If one defines the unperturbed Green's function with respect to the renormalized Hamiltonian $\tilde{H}$,

$$\tilde{G} = (E - \tilde{H})^{-1} , \tag{49}$$

then the Green's functions G and $\tilde{G}$ are related by the Dyson equation analogous to Eq. (35),

$$G = \tilde{G} + \tilde{G} \tilde{V} G . \tag{50}$$

As follows from Eqs. (38) and (48),

$$\langle G \rangle = (E - \tilde{H} - \tilde{\Sigma})^{-1} , \tag{51}$$

where $\tilde{\Sigma}$ relates to $\tilde{H}$ and $\tilde{V}$ in the same way as Σ to H_0 and V, and

$$\Sigma = S + \tilde{\Sigma} . \tag{52}$$

The operator S may be regarded as an initial approximation for the exact self-energy Σ.

The next step is to introduce the total scattering operator T and express the exact Green's function G and the self-energy Σ in terms of T. The total scattering operator is defined by the relation

$$T = \tilde{V} + \tilde{V} \tilde{G} T , \tag{53}$$

which gives

$$T = (\tilde{V}^{-1} - \tilde{G})^{-1} . \tag{54}$$

Excluding $\tilde{V}$ from Eqs. (50) and (54) and averaging the result over realizations of the random potential, we get the relation between $\langle G \rangle$ and $\langle T \rangle$:

$$\langle G \rangle = \tilde{G} + \tilde{G} \langle T \rangle \tilde{G} . \tag{55}$$

Taking into account the relation (38) between the averaged Green's function and the self-energy and using Eq. (55), we finally obtain

$$\Sigma = S + ((\langle T \rangle^{-1} + \tilde{G})^{-1} \, . \tag{56}$$

Thus, having calculated the averaged total scattering operator $\langle T \rangle$ one can determine the averaged Green's function $\langle G \rangle$ and the self-energy Σ according to Eqs. (55) and (56). The problem of the description of the disordered alloy is therefore reduced to the problem of finding reasonable approximate expression for the operator $\langle T \rangle$. The operator T can be expressed in terms of operators describing scattering on individual atomic potentials V_i. To do this, it is convenient to decompose the Green's function $\tilde{G}$ into two parts which are diagonal and off-diagonal in the site representation, respectively:

$$\tilde{G} = \tilde{G}_d + \tilde{G}_{od} \, . \tag{57}$$

Inserting this equation into Eq. (54), we have after some straightforward algebra,

$$T = (t^{-1} - \tilde{G}_{od})^{-1} \, , \tag{58}$$

where the operator

$$t = (\tilde{V}^{-1} - \tilde{G}_d)^{-1} \tag{59}$$

is, similarly to the potential V, site-diagonal and represents a sum of individual scattering operators t_i corresponding to atomic potentials.

Single-site approximations

At any level of approximation there exist two main approaches to the problem of calculation of the averaged total scattering operator $\langle T \rangle$ and, consequently, the averaged Green's function $\langle G \rangle$ and the self-energy Σ. First, it is possible to choose the operator S from the very beginning, and then calculate these three quantities using Eqs. (54)-(56); this is the non-self-consistent scheme. In most cases the VCA choice $S = \langle V \rangle$ is used. Second, one can consider S as an operator variable and, noticing that the scattering operator T is a function of S, determine the latter as a solution of the equation

$$\langle T[S] \rangle = 0 \, . \tag{60}$$

This equation gives, according to Eqs. (55) and (56),

$$\langle G \rangle = \tilde{G} \, , \tag{61}$$
$$\Sigma = S \, , \tag{62}$$

and such an approach is called self-consistent. To make the calculation of the averaged total scattering operator practically possible, the following decoupling of the configurationally-averaged Eq. (58), called the SSA, is usually adopted:

$$\langle T \rangle = \langle (t^{-1} - \tilde{G}_{od})^{-1} \rangle \overset{SSA}{\longrightarrow} (\langle t \rangle^{-1} - \tilde{G}_{od})^{-1} \, . \tag{63}$$

In the framework of the SSA Eq. (56) for the self-energy takes the form

$$\Sigma^{SSA} = S + (\langle t \rangle^{-1} + \tilde{G}_d)^{-1} \, , \tag{64}$$

which means that in this approximation the self-energy is diagonal in the site representation. Consider now the two approaches in combination with the SSA. The non-self-consistent one, as is seen from Eqs. (59) and (64), gives the self-energy as a sum of the operator S and the effective scattering potential $\tilde{V}_{eff}$ corresponding to the average scattering operator $\langle t \rangle$,

$$\tilde{V}_{eff} = (\langle t \rangle^{-1} + \tilde{G}_d)^{-1} \; ; \tag{65}$$

accordingly, this method of calculation is known as the ATA. The application of the SSA to the self-consistent scheme (Eq. (60)) gives the following equation for the evaluation of the self-energy:

$$\langle t[S] \rangle = 0 \; . \tag{66}$$

This equation determines the CPA.

EXPRESSION FOR SHORT-RANGE ORDER DIFFUSE INTENSITY

We now derive a formally exact expression for the SRO part $\alpha(\mathbf{k})$ of the diffuse-scattering intensity. A key observation here is that the spin PCF (22) satisfies the Dyson equation,[8]

$$G = \sigma + \sigma \Delta G \; , \tag{67}$$

where $\Delta = \beta J$ and, like in the previous section, matrix notations are used. Here σ is again the sum of all irreducible graphs in the diagrammatic expansion for G, but irreducibility is now defined with respect to the interaction-to-temperature ratio Δ. Eq. (67) has the same form as the Dyson equation (45) for the Green's function of electrons (32) averaged over alloy configurations. From Eq. (67) it follows that

$$G = \left(\sigma^{-1} - \Delta \right)^{-1} \; . \tag{68}$$

This expression for the spin PCF is analogous to Eq. (46).

The irreducible part σ of the PCF G is sometimes called the self-energy.[8] However, to maintain the analogy with the electronic theory of alloys (i.e., with the terminology of the previous section) we will refer to this quantity as the locator, and reserve this term for another object, defining the PCF self-energy Σ by the relation similar to Eq. (47):

$$\sigma = -\Sigma^{-1} \; . \tag{69}$$

In terms of Σ Eq. (68) becomes

$$G = (-\Sigma - \Delta)^{-1} \; . \tag{70}$$

We also introduce the locator $\tilde{\sigma}$ and the self-energy $\tilde{\Sigma}$ for the occupation-number PCFs $G^{\alpha\beta}$ (Eq. (21)):

$$\tilde{\sigma} = \frac{1}{4}\sigma \; , \tag{71}$$

$$\tilde{\Sigma} = 4\Sigma \; . \tag{72}$$

Then matrices $\tilde{\sigma}$ and $\tilde{\Sigma}$ are related by the same Eq. (69),

$$\tilde{\sigma} = -\tilde{\Sigma}^{-1} \; , \tag{73}$$

and we have

$$G^{AA} = G^{BB} = -G^{AB} = c(1 - c)\alpha = (-\tilde{\Sigma} + 2\beta V)^{-1} \; . \tag{74}$$

215

Written in k-representation, this equation leads to the following expression for the diffuse-scattering intensity:[12,13]

$$\alpha(\mathbf{k}) = \frac{1}{c(1-c)\left[-\tilde{\Sigma}(\mathbf{k}) + 2\beta V(\mathbf{k})\right]} \, . \tag{75}$$

The central quantity of interest for us here is the PCF self-energy $\tilde{\Sigma}(\mathbf{k})$; apart from the wavevector, it depends also on two other variables, temperature and concentration. Later on we will be focusing on existing approximations for the PCF self-energy.

SELF-ENERGY AND THERMODYNAMICS

In the last section the Dyson equation (67) and the related expression (70) for the spin PCF were simply postulated. However, it would be useful to know how Eq. (70) could be derived and how the self-energy is related to the thermodynamics of the system.

Variational formulation of statistical mechanics

As was mentioned before, the main task of the statistical-mechanical treatment is to calculate the partition function (6) of a system described by the Hamiltonian H. The Hamiltonian usually is a linear combination,

$$H = \sum_n x_n a_n \, , \tag{76}$$

of some operators a_n with coefficients x_n. Variables α_n conjugated to the parameters x_n are defined as averages (20) of the operators a_n,

$$\alpha_n = \langle a_n \rangle = \frac{\partial F}{\partial x_n} \, , \tag{77}$$

where

$$F = -k_B T \ln Z \tag{78}$$

is the free energy of the system, and the second equation in (77) comes from the definition (6) and Eq. (76). Our aim now is the calculation of the averages α_n and the free energy F as functions of the parameters x_n. This problem can be formulated as a variational one, if the Legendre transform $\Gamma(\alpha)$ of the free energy $F(x)$ is introduced:[14,15]

$$\Gamma(\alpha) = F(x(\alpha)) - \sum_n \alpha_n \, x_n(\alpha) \, . \tag{79}$$

Here the averages α_n are the independent variables, and $x_n(\alpha)$ are solutions of Eqs. (77). Differentiating $\Gamma(\alpha)$ with respect to α_n and using Eqs. (77), we get

$$\frac{\partial \Gamma}{\partial \alpha_n} = \sum_m \frac{\partial F}{\partial x_m} \frac{\partial x_m}{\partial \alpha_n} - \sum_m \left(\frac{\partial \alpha_m}{\partial \alpha_n} x_m + \alpha_m \frac{\partial x_m}{\partial \alpha_n} \right) = -x_n \, . \tag{80}$$

Finally, introducing another function $\Phi(x, \alpha)$,

$$\Phi(x, \alpha) = \Gamma(\alpha) + \sum_n \alpha_n x_n \, , \tag{81}$$

216

which depends on both α_n and x_n, we find from Eq. (80) that it is stationary with respect to variations of α_n at fixed x_n:

$$\frac{\partial \Phi}{\partial \alpha_n} = 0 \ . \tag{82}$$

At the stationary point $\alpha_n = \alpha_n(x)$, where $\alpha_n(x)$ are solutions of Eqs. (80), $\Phi(x, \alpha)$ coincides with the free energy $F(x)$:

$$\Phi(x, \alpha(x)) = F(x) \ . \tag{83}$$

Function $\Phi(x, \alpha)$ is usually called the variational free energy. Noting that the internal energy is the statistical average (20) of the Hamiltonian, $E = \langle H \rangle$, $\Phi(x, \alpha)$ can be written in the standard thermodynamic form:

$$\Phi(x, \alpha) = E(x, \alpha) - TS(\alpha) \ , \tag{84}$$

$$E(x, \alpha) = \sum_n x_n \alpha_n \ , \tag{85}$$

$$S(\alpha) = -\beta \Gamma(\alpha) \ , \tag{86}$$

where $E(x, \alpha)$ and $S(\alpha)$ are the variational internal energy and configurational entropy, respectively. Eq. (82) now becomes

$$T \frac{\partial S}{\partial \alpha_n} = x_n \ . \tag{87}$$

From Eqs. (77) and (80) it follows that

$$\sum_l \frac{\partial^2 \Gamma}{\partial \alpha_n \partial \alpha_l} \frac{\partial^2 F}{\partial x_l \partial x_m} = -\sum_l \frac{\partial x_n}{\partial \alpha_l} \frac{\partial \alpha_l}{\partial x_m} = -\frac{\partial x_n}{\partial x_m} = -\delta_{nm} \ . \tag{88}$$

Eq. (88) shows that matrices of second derivatives of the free energy $F(x)$ and its Legendre transform $\Gamma(\alpha)$ (or the variational free energy $\Phi(x, \alpha)$ which differs from $\Gamma(\alpha)$ only by the bilinear term $\sum_n x_n \alpha_n$) are mutually inverse up to a sign.

First Legendre transformation for the Ising model

We will now apply the general technique of the Legendre transformations outlined above to the particular case of the Ising model in the inhomogeneous magnetic field. The Hamiltonian of the model is a straightforward generalization of Eq. (13):

$$H = -\frac{1}{2} \sum_{ij} J_{ij} s_i s_j - \sum_i h_i s_i \ . \tag{89}$$

From comparison of Eqs. (76) and (89) it follows that the latter contains two kinds of operators a_n - single spin variables s_i and products $s_i s_j$ of two spin variables. Corresponding parameters x_n, except for sign, are h_i and J_{ij}, and variables α_n conjugated to these parameters are $\langle s_i \rangle = m_i$ (Eq. (18)) and $\langle s_i s_j \rangle = m_i m_j + G_{ij}$ (Eq. (22)). In the general case considered before the Legendre transformation was carried out with respect to all parameters x_n; as a result, all conjugated variables α_n were calculated using the variational principle. Here, however, this approach will be applied only to the magnetic field h_i, and only the magnetization m_i will be calculated by means of the variational procedure. The resulting partial transformation is called the first Legendre transformation.[15-17] In this case h_i plays the role of x_n, and comparison with Eq. (76) shows that a_n corresponds to $-s_i$. Eqs. (77), (79)-(83) and (88) now become

$$\Gamma(m, J) = F(h(m), J) + \sum_i m_i h_i(m) , \qquad (90)$$

$$\Phi(h, m; J) = \Gamma(m, J) - \sum_i m_i h_i , \qquad (91)$$

$$\frac{\partial F}{\partial h_i} = -m_i , \qquad (92)$$

$$\frac{\partial \Gamma}{\partial m_i} = h_i , \qquad (93)$$

$$\frac{\partial \Phi}{\partial m_i} = 0 , \qquad (94)$$

$$\Phi(h, m(h); J) = F(h, J) , \qquad (95)$$

$$\sum_k \frac{\partial^2 \Gamma}{\partial m_i \partial m_k} \frac{\partial^2 F}{\partial h_k \partial h_j} = \sum_k \frac{\partial^2 \Phi}{\partial m_i \partial m_k} \frac{\partial^2 F}{\partial h_k \partial h_j} = -\delta_{ij} , \qquad (96)$$

where, in the same manner as earlier, $h_i(m)$ and $m_i(h)$ are solutions of Eqs. (92) and (93), respectively.

Derivation of the Dyson equation and meaning of the self-energy

First of all, we note that from the definitions of the partition function, the statistical average, the PCF and the free energy (Eqs. (6), (20), (22) and (78)) it follows that for the Hamiltonian (89)

$$\frac{\partial^2 F}{\partial h_i \partial h_j} = -\beta G_{ij} . \qquad (97)$$

Then, combining this equation with Eq. (96), we obtain

$$G_{ij} = k_B T \left(\frac{\partial^2 \Phi}{\partial m \partial m} \right)^{-1}_{ij} . \qquad (98)$$

This notation means that the real-space matrix element of the PCF is proportional to the corresponding matrix element of the inverse of the matrix whose matrix elements are second derivatives $\partial^2 \Phi / \partial m_i \partial m_j$ of the variational free energy with respect to the corresponding magnetizations.

The variational free energy can always be written as a sum of its mean-field (Bragg-Williams) part and the non-mean-field correction:

$$\Phi = \Phi^{MFA} + \delta\Phi , \qquad (99)$$

$$\Phi^{MFA} = E^{MFA} - T S^{MFA} . \qquad (100)$$

The expressions for the mean-field internal energy E^{MFA} and the configurational entropy S^{MFA} are well-known:[1-6]

$$E^{MFA} = -\frac{1}{2} \sum_{ij} J_{ij} m_i m_j - \sum_i h_i m_i , \qquad (101)$$

$$S^{MFA} = -k_B \sum_i \left(\frac{1+m_i}{2} \ln \frac{1+m_i}{2} + \frac{1-m_i}{2} \ln \frac{1-m_i}{2} \right) . \qquad (102)$$

Substituting Eqs. (99)-(102) into Eq. (98) and noticing that

$$\frac{\partial^2 E^{MFA}}{\partial m_i \partial m_j} = -J_{ij} \, , \tag{103}$$

we recover the result (70) for the PCF (and, therefore, the Dyson equation (67)) with the following expression for the self-energy:

$$\Sigma_{ij} = \frac{\partial^2 S^{MFA}}{\partial m_i \partial m_j} - \beta \frac{\partial^2 (\delta \Phi)}{\partial m_i \partial m_j} \, . \tag{104}$$

It is seen that the self-energy is the sum of the second derivatives, with respect to the magnetizations, of the two terms contributing to the expression for the variational free energy Φ: the mean-field configurational entropy and the non-mean-field part of Φ. Noting further that

$$\frac{\partial^2 S^{MFA}}{\partial m_i \partial m_j} = -\frac{\delta_{ij}}{1 - m_i^2} \, , \tag{105}$$

we finally obtain

$$\Sigma_{ij} = -\frac{\delta_{ij}}{1 - m_i^2} - \beta \frac{\partial^2 (\delta \Phi)}{\partial m_i \partial m_j} \, . \tag{106}$$

The first term in this expression is diagonal in the direct space. This means that, back to the homogeneous case in which all the lattice sites are equivalent, this term is k-independent in the reciprocal space, and all the wavevector dependence of the self-energy comes from the second term, i.e., is the result of the non-mean-field corrections to the MFA variational free energy.

APPROXIMATIONS FOR THE SELF-ENERGY

In this section our attention will be focused on the formally exact result (75) for the SRO diffuse intensity $\alpha(\mathbf{k})$, which is the Fourier transform of the Warren-Cowley SRO parameters α_{ij} (24). Available theories of SRO will now be considered, in the light of the structure of Eq. (75), as different approximations for the self-energy $\tilde{\Sigma}$.

Random-phase (Krivoglaz-Clapp-Moss) approximation

The simplest and by far the most popular theory of SRO is the RPA (or, in the alloy language, the KCM approximation):[5,18,19]

$$\alpha^{RPA}(\mathbf{k}) = \frac{1}{1 + 2c(1 - c)\beta V(\mathbf{k})} \, . \tag{107}$$

Eq. (107) is usually derived using mean-field-like arguments. Comparing Eqs. (75) and (107), we see that

$$\tilde{\Sigma}^{RPA}(\mathbf{k}) = -\frac{1}{c(1 - c)} \, . \tag{108}$$

The RPA self-energy (108) is thus wavevector- and temperature-independent; it depends only on alloy composition. Returning via Eqs. (19) and (72) to the magnetic language used in the previous section, we conclude that the RPA for the self-energy corresponds precisely to neglecting the second term in the right-hand side of Eq. (106). The non-mean-field contribution to the self-energy is therefore ignored in the RPA, and the self-energy in the direct space is simply the second derivative of the mean-field configurational entropy with respect to the magnetization:

$$\Sigma_{ij}^{RPA} = \frac{\partial^2 S^{MFA}}{\partial m_i \partial m_j} = -\frac{\delta_{ij}}{1 - m_i^2} \ . \tag{109}$$

From the point of view of the terminology used in the electronic theory of disordered alloys, the RPA for the self-energy could be referred to as the SSA; indeed, both the RPA (Eq. (104)) and the SSA (Eq. (64)) self-energies are diagonal in the site representation. The RPA resembles the non-self-consistent SSA, i.e., the ATA, in the sense that both approximations define the corresponding self-energies explicitly.

The RPA reduces to the well-known Ornstein-Zernike description of correlations,[7] when only those wavevectors which are close to the position $\mathbf{k}_0$ of the absolute minimum of the interaction $V(\mathbf{k})$ are considered (this approximation corresponds to the long-wave limit in the case of ferromagnetic ordering). Let us expand $V(\mathbf{k})$ in powers of $\mathbf{q} = \mathbf{k} - \mathbf{k}_0$ and retain only the lowest-order (quadratic) term,

$$V(\mathbf{k}) \approx V(\mathbf{k}_0) + \frac{1}{2} \sum_{ij} g_{ij} q_i q_j \ , \tag{110}$$

where g is the 3×3-matrix of second derivatives of $V(\mathbf{k})$ at $\mathbf{k}_0$. We will take only the simplest example of cubic symmetry, when $g_{ij} = g \delta_{ij}$; in this case $\sum_{ij} g_{ij} q_i q_j = g q^2$, where $q \equiv |\mathbf{q}|$. Substituting the result into Eq. (107), we obtain

$$\alpha^{RPA}(\mathbf{k}) = \frac{k_B T}{k_B(T - T_c) + c(1 - c) g q^2} \ , \tag{111}$$

where it is taken into account that the mean-field result for the instability temperature T_c is

$$T_c = 2c(1 - c)|V(\mathbf{k}_0)|/k_B \ ; \tag{112}$$

at the position of the absolute minimum the interaction value is negative, since the average of the interaction over the Brillouin zone is zero (see Eq. (116) below). In real space we get asymptotically (i.e., at large distances)

$$\alpha_{ij}^{RPA} \propto \frac{1}{r_{ij}} \exp(-i \mathbf{k}_0 \mathbf{r}_{ij} - r_{ij}/\xi) \ , \tag{113}$$

where $r_{ij} \equiv |\mathbf{r}_{ij}|$, and the correlation length ξ is

$$\xi = \sqrt{\frac{c(1 - c) g}{k_B(T - T_c)}} \ . \tag{114}$$

Eqs. (111)-(114) represent the Ornstein-Zernike result for the PCF.

However, the RPA expression (107) has a serious disadvantage: it is unable to satisfy the sum rule (31). Using the identity

$$\frac{1}{1 + 2c(1 - c)\beta V(\mathbf{k})} = 1 - 2c(1 - c)\beta V(\mathbf{k}) + \frac{4c^2(1 - c)^2 \beta^2 V^2(\mathbf{k})}{1 + 2c(1 - c)\beta V(\mathbf{k})} \ , \tag{115}$$

it can be shown that the RPA formula (107) always leads to the overestimation of the value of the integral over the Brillouin zone (31). Since

$$\frac{1}{\Omega} \int d\mathbf{k} \, V(\mathbf{k}) = V_{ii} = 0 \ , \tag{116}$$

from Eq. (115) it follows that

$$\frac{1}{\Omega} \int d\mathbf{k} \; \alpha^{RPA}(\mathbf{k}) = 1 + \frac{1}{\Omega} \int d\mathbf{k} \; \frac{4c^2(1-c)^2\beta^2 V^2(\mathbf{k})}{1 + 2c(1-c)\beta V(\mathbf{k})} \; , \tag{117}$$

and, therefore,

$$\frac{1}{\Omega} \int d\mathbf{k} \; \alpha^{RPA}(\mathbf{k}) > 1 \; . \tag{118}$$

This integral is close to unity only at sufficiently high temperatures. As temperature decreases, the deviation from the value prescribed by the sum rule becomes more and more significant, and the integral finally diverges at the instability point.[20,21]

Spherical model

Another analytical expression for $\alpha(\mathbf{k})$ is given by the SM,[6,19,22-24] also known as the Onsager cavity field theory,[25,26]

$$\alpha^{SM}(\mathbf{k}) = \frac{1}{c(1-c)\left[-\tilde{\Sigma}^{SM} + 2\beta V(\mathbf{k})\right]} \; , \tag{119}$$

where $\tilde{\Sigma}^{SM}$ is, at fixed temperature and concentration, a number determined from the sum rule (31). Therefore, the sum rule is satisfied in the SM by definition, contrary to the case of the RPA. From the definition of the SM it also follows that the self-energy depends not only on concentration, like its RPA counterpart (108), but also on temperature. Nevertheless, the SM self-energy is still wavevector-independent. The explicit expression for $\tilde{\Sigma}^{SM}$ can be derived from the sum rule (31):

$$\tilde{\Sigma}^{SM} = \tilde{\Sigma}^{RPA} + \delta\tilde{\Sigma} \; , \tag{120}$$

$$\delta\tilde{\Sigma} = 2\beta \sum_j \alpha_{ij} V_{ij} = \frac{2\beta}{\Omega} \int d\mathbf{k} \; \alpha(\mathbf{k}) V(\mathbf{k}) \; . \tag{121}$$

Similarly to the RPA, the SM is analogous to the SSA, since the SM self-energy is diagonal in the direct space. However, the SM is rather the self-consistent SSA, like the CPA, because the sum rule here plays the role of the CPA self-consistency condition. In fact, the sum rule is the self-consistency condition: it simply means that the diagonal matrix elements of the approximate and exact PCFs are the same. More surprisingly, it was shown[27,28] that both the CPA and the SM could be obtained by summation of the same sets of diagrams in the corresponding perturbation expansions. We can conclude, therefore, that the SM is the CPA for the Ising model.

Cluster variation method

The CVM[29] is at present the standard technique for quantitative calculation of thermodynamic properties of alloys. It is discussed in great detail in almost every book or review on the subject.[1-3,6] We do not attempt to do this here; instead, we will consider only those features of the CVM which are relevant to our discussion, without going into technical aspects of the method.

The CVM is essentially a procedure which allows us to derive an approximate expression for the variational configurational entropy $S(\alpha)$ of the system. The CVM entropy is a function of probabilities of various atomic configurations on lattice clusters which belong to the so-called basic, or maximal, cluster. A particular CVM approximation for $S(\alpha)$ is therefore defined by the choice of the basic cluster. Combined with the variational internal energy $E(x, \alpha)$, the CVM configurational entropy gives the expression (84) for the variational free energy $\Phi(x, \alpha)$. The operators a_n in the alloy Hamiltonian are products of the spin variables or of the occupation numbers, and the averages α_n (Eq. (77)) are thus related to, or coincide with, cluster probabilities entering the expression for the CVM configurational entropy. The variational free energy Φ is then minimized with respect to all cluster probabilities, taking into account various self-consistency constraints.

The self-energy obtained in the framework of the CVM depends, in general, on all three parameters – temperature, concentration and wavevector. The problem with the CVM, however, is that this method is intrinsically numerical and does not lead to analytical approximations for pair correlations. The reason for this is that in most cases the number of cluster variables used to get a reasonably accurate formula for the configurational entropy is far too large. From the point of view of the general technique of the Legendre transformations, the CVM corresponds to the high-order transformation with respect to all coefficients x_n conjugated to cluster variables which are involved in the expression for the variational configurational entropy $S(\alpha)$. Therefore, as far as SRO is concerned, general Eq. (88) is still valid, as is the Ising model-specific Eq. (97). However, in practical sense this case is very different from that of the first Legendre transformation. In the latter, the inversion of the matrix $\partial^2 \Phi / \partial \alpha \partial \alpha$ is carried out trivially using the Fourier transformation. In the CVM this object in the reciprocal space is still a sufficiently large matrix, and in all realistic situations it is necessary to resort to numerics. Analytical formulae were obtained only for such simple model cases as the pair (also known as quasichemical, or Fowler-Guggenheim, or Bethe) approximation, or the square (Kramers-Wannier) approximation for the nearest-neighbour Ising model on the square lattice.[2,20] Besides, the CVM suffers from the same drawback as the RPA, though to a lesser extent: the integrated intensity is not conserved, and its behavior with temperature is similar to the case of the RPA.[2,20]

High-temperature expansion

As we have seen, the RPA and the SM are the analogs of the SSAs in the electronic theory of disordered alloys and thus fail to take account of the wavevector dependence of the self-energy. On the other hand, the CVM leads to the k-dependent self-energy, but loses the simplicity of the former two approximations by not providing analytical expressions for correlations. At the semiquantitative level of the RPA this problem can be cured by using the HTE for the self-energy.[30]

To do this, we return to Eq. (104) which, in combination with Eq. (109), can be written as

$$\Sigma = \Sigma^{RPA} + \delta\Sigma,$$ (122)

$$\delta\Sigma_{ij} = -\beta \frac{\partial^2 (\delta\Phi)}{\partial m_i \partial m_j}.$$ (123)

It is known that the RPA is exact to first order in $1/T$ (i.e., in Δ);[2] it means that $\delta\Phi$ and $\delta\Sigma$ are of order Δ^2. It is also known how to construct the first Legendre transformation Γ (and, therefore, $\delta\Phi$) for the Ising model.[15-17] Sorting all contributions to $\delta\Phi$ (diagrams) according to the number of lines Δ, we obtain the HTE; first eight orders in Δ are available in the literature.[17] For simplicity reasons, the discussion here is confined to two first orders,

$$-\beta \delta\Phi = \frac{1}{4} \bigcirc + \frac{1}{12} \ominus + \frac{1}{6} \triangle + O(\Delta^4),$$ (124)

where a line corresponds to Δ, and a vertex with n attached lines represents the function $u_n(m_i)$. For the diagrams in Eq. (124) n is equal to either 2 or 3:

$$u_2(m_i) = 1 - m_i^2 , \tag{125}$$

$$u_3(m_i) = -2m_i\left(1 - m_i^2\right) . \tag{126}$$

The expressions for the diagonal and off-diagonal parts of $\delta\Sigma$ thus are, according to Eq. (123),

$$\delta\Sigma_{ii} = \frac{1}{2}\,\bigcirc\!\!\!| + \frac{1}{6}\,\bigcirc\!\!\!\bigcirc + \frac{1}{2}\,\bigtriangledown + O(\Delta^4) , \tag{127}$$

$$\delta\Sigma_{ij} = \frac{1}{2}\,\bigcirc\!\!\!\bigcirc + \frac{1}{6}\,\bigcirc\!\!\!\bigcirc + \bigtriangleup + O(\Delta^4) , \quad i \neq j . \tag{128}$$

Here a vertex with n internal lines and k external legs corresponds to the kth derivative of the function $u_n(m)$. The corresponding analytical expressions have the form

$$\delta\Sigma_{ii} = -\sum_l(1 - m_l^2)\Delta_{il}^2 - 4m_i\sum_l m_l(1 - m_l^2)\Delta_{il}^3$$
$$- \sum_{kl}(1 - m_k^2)(1 - m_l^2)\Delta_{ik}\Delta_{il}\Delta_{kl} + O(\Delta^4) , \tag{129}$$

$$\delta\Sigma_{ij} = 2m_im_j\Delta_{ij}^2 + \frac{2}{3}(1 - 3m_i^2)(1 - 3m_j^2)\Delta_{ij}^3$$
$$+ 4m_im_j\Delta_{ij}\sum_l(1 - m_l^2)\Delta_{il}\Delta_{jl} + O(\Delta^4) , \quad i \neq j . \tag{130}$$

In the homogeneous case $m_i = m$ for all sites i. Defining constants

$$a_1 = \sum_l J_{il}^2 , \quad a_2 = \sum_l J_{il}^3 , \quad a_3 = \sum_{kl} J_{ik}J_{il}J_{kl} \tag{131}$$

and lattice functions

$$(f_1)_{ij} = J_{ij}^2 , \quad (f_2)_{ij} = J_{ij}^3 , \quad (f_3)_{ij} = J_{ij}\sum_l J_{il}J_{jl} , \tag{132}$$

which depend only on the interaction J, one can finally write the expression for $\delta\Sigma$ in the k-space:

$$\delta\Sigma(\mathbf{k}) = -(1 - m^2)\beta^2 a_1 + 2m^2\beta^2 f_1(\mathbf{k}) - 4m^2(1 - m^2)\beta^3 a_2$$
$$-(1 - m^2)^2\beta^3 a_3 + \frac{2}{3}(1 - 3m^2)^2\beta^3 f_2(\mathbf{k})$$
$$+ 4m^2(1 - m^2)\beta^3 f_3(\mathbf{k}) + O(\beta^4) . \tag{133}$$

In alloy notations this result reads

$$\delta\tilde{\Sigma}(\mathbf{k}) = -4c(1 - c)\beta^2\tilde{a}_1 + 2(1 - 2c)^2\beta^2\tilde{f}_1(\mathbf{k}) + 8c(1 - c)(1 - 2c)^2\beta^3\tilde{a}_2$$
$$+ 8c^2(1 - c)^2\beta^3\tilde{a}_3 - \frac{4}{3}[1 - 6c(1 - c)]^2\beta^3\tilde{f}_2(\mathbf{k})$$
$$- 8c(1 - c)(1 - 2c)^2\beta^3\tilde{f}_3(\mathbf{k}) + O(\beta^4) , \tag{134}$$

where

$$\tilde{a}_1 = \sum_l V_{il}^2 , \quad \tilde{a}_2 = \sum_l V_{il}^3 , \quad \tilde{a}_3 = \sum_{kl} V_{ik}V_{il}V_{kl} , \tag{135}$$

$$(\tilde{f}_1)_{ij} = V_{ij}^2 , \quad (\tilde{f}_2)_{ij} = V_{ij}^3 , \quad (\tilde{f}_3)_{ij} = V_{ij}\sum_l V_{il}V_{jl} . \tag{136}$$

Alpha- and gamma-expansions

The approximate expressions (133), (134) for the self-energy given by the first orders of the HTE satisfy almost all requirements: they are analytical, relatively simple and take into account dependence on all relevant parameters, including the wavevector. However, the limits of applicability of the HTE are essentially the same as those of the RPA; the HTE is quantitatively correct only at reasonably high temperatures, as can be concluded already from the name of the expansion. What is needed, therefore, is an approximation which would combine all the advantages of the HTE self-energy with the applicability at moderate temperatures, including the range not far away from the transition or instability points.

The theory of SRO which will now be discussed[12,13] is based on the fairly general procedure[27] of self-consistent renormalization of the bare propagator Δ^{-1} in the functional-integral representation of the generating functional for correlation functions. The resulting expansion for the matrix elements of the self-energy is in powers of the matrix elements of the fully dressed propagator, which in the case of the Ising model coincides with the PCF (70). Two first non-zero orders of this expansion for the off-diagonal part of the self-energy were calculated,[12,13]

$$\Sigma_{ij} = aG_{ij}^2 + bG_{ij}^3 + O(G^4) , \quad i \neq j , \tag{137}$$

$$a = \frac{2m^2}{(1-m^2)^4} , \tag{138}$$

$$b = \frac{2[(1-3m^2)^2 - 12m^4]}{3(1-m^2)^6} , \tag{139}$$

in terms of the alloy variables

$$\tilde{\Sigma}_{ij} = \tilde{a}\alpha_{ij}^2 + \tilde{b}\alpha_{ij}^3 + O(\alpha^4) , \quad i \neq j , \tag{140}$$

$$\tilde{a} = \frac{(1-2c)^2}{2[c(1-c)]^2} , \tag{141}$$

$$\tilde{b} = \frac{[1-6c(1-c)]^2 - 3(1-2c)^4}{6[c(1-c)]^3} . \tag{142}$$

Eq. (140) is the expansion in powers of the SRO parameters, and is therefore referred to as the AE,[31] though initially it was obtained in the framework of the GEM.[12,13] The difference between the AE and GEM is discussed below. In the two calculated orders the AE preserves the sum rule (31), and the expression for the diagonal part of the self-energy then comes from Eqs. (31) and (140), similarly to the case of the SM (Eqs. (120), (121)):

$$\tilde{\Sigma}_{ii} = \tilde{\Sigma}^{RPA} + \delta\tilde{\Sigma}_{ii} , \tag{143}$$

$$\delta\tilde{\Sigma}_{ii} = 2\beta \sum_j \alpha_{ij} V_{ij} - \sum_{j(\neq i)} \left(\tilde{a}\alpha_{ij}^3 + \tilde{b}\alpha_{ij}^4 \right) + O(\alpha^5) . \tag{144}$$

Note that the first term in Eq. (144) corresponds to the SM (cf. Eq. (121)) which is the zero-order approximation for both the AE and the GEM; in the SM the self-energy is diagonal ($\tilde{a} = \tilde{b} = 0$).

The difference between the GEM and the AE is in the way of selection of leading terms in the expansion (140). The GEM, originally proposed by Tokar[27] and further developed by Tokar, Grishchenko and the author,[12,13,32-35] is based on the assumption that correlations decrease rapidly with distance, and the GEM expansion parameter is

$$\gamma = \exp(-1/\xi) , \qquad (145)$$

where ξ is the dimensionless correlation length. The leading terms in the diagrammatic expansion for the self-energy are selected in the framework of the GEM according to the total length of all lines in the diagrams, where the line connecting lattice sites i and j represents α_{ij}. For example, in the case of three Bravais lattices belonging to the cubic system taking into account several first terms of the perturbation theory leads to the result[12,13]

$$\tilde{\Sigma}_s = \tilde{a}\alpha_s^2 + \tilde{b}\alpha_s^3 , \quad s = 1 , \qquad (146)$$

$$\tilde{\Sigma}_s = \tilde{a}\alpha_s^2 , \quad s = 2,3 , \qquad (147)$$

$$\tilde{\Sigma}_s = 0 , \quad s > 3 , \qquad (148)$$

where subscript s denotes the matrix elements corresponding to the sth coordination shell. However, the GEM assumption about the rapid decay of correlations is not always valid; e.g., it is incorrect in the cases where distant interactions are essential. The AE abandons this assumption and uses instead the SRO parameters α_{ij} themselves as the expansion parameters; the leading terms are chosen according to the number of lines in the diagrams (i.e., the powers of α_{ij}), since all α_{ij} are relatively small. A particular AE approximation is defined by neglecting higher-order terms and including only finite number of coordination shells in the AE expression (140) for the off-diagonal part of the self-energy. The GEM was successfully applied to both the direct and inverse problems of alloy diffuse scattering,[12,13,36-38] leading to reliable results everywhere except in the vicinity of the instability point, while the AE was used in the analysis of some of the diffuse-scattering anomalies discussed in the next section.[31,39]

ANOMALIES IN ALLOY DIFFUSE SCATTERING

In this last section we will show how expression (75) for the intensity $\alpha(\mathbf{k})$ and, in particular, the wavevector dependence of the self-energy $\tilde{\Sigma}(\mathbf{k})$ lead to straightforward explanation of recently observed unusual features (anomalies) of diffuse scattering from disordered alloys and to prediction of some new effects.

Temperature-dependent Fermi surface-induced peak splitting

This curious effect (the temperature dependence of the splitting) was discovered in 1996 by Reichert, Moss and Liang[40] in the first *in situ* experiment to resolve the fine structure of the equilibrium diffuse scattering intensity from the disordered Cu_3Au alloy. The separation of the split maxima changed reversibly, increasing with temperature. The same behavior of the splitting was also found[41] by analysing results of the MC simulations for the $Cu_{0.856}Al_{0.144}$ alloy.[42] The fourfold splitting of the intensity peaks located at the (110) and equivalent positions (Figure 1) is attributed to the indirect interaction of atoms via conduction electrons in an alloy whose Fermi surface has flat portions; the effective interatomic pair interaction $V(\mathbf{k})$ itself has split minima in the reciprocal space, and their location is determined by the wavevector $2k_F$ spanning these flat portions of the Fermi surface.[5,43] It is usually assumed that $V(\mathbf{k})$ is temperature-independent. This assumption is justified at least as far as positions of the $V(\mathbf{k})$ minima are concerned, since the $2k_F$ value is unlikely to change over the considered temperature range. Besides, the MC calculations[42] in which the increase of the splitting with temperature was found[41] were carried out for the temperature-independent pair interaction parameters. The standard RPA (KCM) treatment (Eq. (107)) predicts that positions of the intensity peaks coincide with those of the corresponding minima of $V(k)$, and, therefore, the splitting does not depend on temperature.

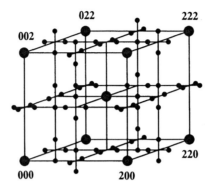

Figure 1. 3D reciprocal-space picture of scattering from the FCC alloys discussed in the text. Large dots represent the Bragg reflections. Characteristic crosses formed by small dots correspond to the split diffuse intensity peaks.

As will be demonstrated below, this phenomenon can be easily understood by employing the notion of the k-dependent self-energy.[31] Let us consider the $\alpha(\mathbf{k})$ profile along one of the lines containing split peaks, e.g., the (h10) line, and concentrate on the two peaks around the (110) position. The peak positions k_α (k is the deviation of the wavevector from the (110) position along the (h10) line) are determined by the condition $\partial\alpha/\partial k = 0$ which gives

$$2\frac{\partial V}{\partial k} = T\frac{\partial\tilde{\Sigma}}{\partial k} .$$

(149)

Eq. (149) means that the k-dependence of $\tilde{\Sigma}$ leads to the shift $\delta k = k_\alpha - k_V$ of the peak position with respect to the position k_V of the corresponding minimum of $V(\mathbf{k})$ (the latter is defined by the condition $\partial V/\partial k = 0$). Furthermore, the right-hand side of Eq. (149) is a function of T, while its left-hand side is T-independent. The $\alpha(\mathbf{k})$ peaks will therefore change their positions with temperature. The "local" temperature behavior of the splitting is reflected in the sign of the derivative $\partial_T k_I$ which can be calculated by expanding the derivatives in Eq. (149) in powers of small changes of T and k_I and retaining only linear terms:

$$\frac{\partial k_\alpha}{\partial T} = \left[\left(\frac{\partial\tilde{\Sigma}}{\partial k} + T\frac{\partial^2\tilde{\Sigma}}{\partial k\partial T}\right)\bigg/\left(2\frac{\partial^2 V}{\partial k^2} - T\frac{\partial^2\tilde{\Sigma}}{\partial k^2}\right)\right]_{k=k_\alpha} .$$

(150)

Our aim now is to develop a kind of minimal, i.e., simplest possible, theory which would be able to describe essential features of the considered effect (and, apropos, two other

226

anomalies discussed in this section). Interestingly, the approach formally rather similar to the Landau theory of second-order phase transitions[7] could be used. Indeed, in the case of not very large splitting the expansion of $V(k)$ and $\tilde{\Sigma}(k)$ in powers of k can be used. To describe the split minimum of $V(k)$, only the second- and fourth-order terms are necessary; since the (110) position serves as the origin, the expansions do not contain odd powers of k. We therefore assume that in the area of the splitting $V(k)$ and $\tilde{\Sigma}(k)$ have the following approximate form,

$$V(k) = V(0) + \frac{1}{2}A_V k^2 + \frac{1}{4}B_V k^4 \,, \tag{151}$$

$$\tilde{\Sigma}(k) = \tilde{\Sigma}(0) + \frac{1}{2}A_\Sigma k^2 + \frac{1}{4}B_\Sigma k^4 \,, \tag{152}$$

where $A_V < 0$, $B_V > 0$, the signs of A_Σ and B_Σ are arbitrary (there are no apparent restrictions on the behavior of the self-energy), and $k = 0$ corresponds to the (110) position. The resulting inverse intensity $\alpha^{-1}(k)$ has exactly the form of the Landau free energy functional in the low-symmetry phase where the latter possesses a double minimum; this implies $2|A_V| + TA_\Sigma > 0$ and $2B_V - TB_\Sigma > 0$. The wavevector k plays the role of the order parameter. Substituting approximations (151) and (152) into general Eqs. (149) and (150), we get

$$k_\alpha = \pm\sqrt{\frac{2|A_V| + TA_\Sigma}{2B_V - TB_\Sigma}} \,, \tag{153}$$

$$k_\alpha^{-1}\frac{\partial k_\alpha}{\partial T} = \frac{1}{2}\left[\frac{A_\Sigma + T\,\partial A_\Sigma/\partial T}{2|A_V| + TA_\Sigma} + \frac{B_\Sigma + T\,\partial B_\Sigma/\partial T}{2B_V - TB_\Sigma}\right] \,, \tag{154}$$

while $k_V = \pm\sqrt{|A_V|/B_V}$. It is seen that the shifts of the two peaks and their temperature derivatives have opposite signs and the same absolute values. Eq. (154) clearly shows two scenarios for the temperature behavior of the splitting, depending on the sign of its right-hand side which can be either positive or negative. The first one is the increase of the splitting with temperature discussed above; this takes place when the right-hand side of Eq. (154) is negative. Apart from that, the theory predicts that the decrease of the splitting with increasing temperature is also possible. This regime corresponds to the case of positive right-hand side of Eq. (154), and such temperature dependence has not yet been observed experimentally. Thus, the behavior of the self-energy determines whether the splitting increases or decreases with temperature. At high temperatures the correction $\delta\Sigma$ to the wavevector-independent Σ^{RPA} (Eq. (122)), and, therefore, A_Σ and B_Σ, are of order Δ^2 (i.e., T^{-2}). From Eq. (153) it then follows that the absolute value of the shift δk decreases as T^{-1} with temperature, unless the corresponding coefficient identically vanishes.

Coalescence of Fermi surface-related intensity peaks

The analogy with the Landau theory of phase transitions, though rather formal, immediately leads to the following question: in the considered case, what would correspond to the transition point? The answer to this question is fairly obvious; there exists a possibility for the splitting to disappear at some point as temperature decreases, before the transition to the low-symmetry phase occurs.[44] This anomaly was neither observed experimentally nor correctly described theoretically, though the coalescence of intensity peaks (unrelated to any Fermi surface effects) was found for the exactly solvable 1D Ising model[45] and in the CVM calculations for the 2D ANNNI model.[46]

As we will show in this subsection, Eq. (75) provides clear understanding of how such effect takes place.

In the treatment given above it is, in fact, implicitly assumed that the wavevector dependence of the interaction term $2\beta V(\mathbf{k})$ in Eq. (75) dominates, at least in the area of the splitting, i.e., near the (110) position. In this case the shape of the diffuse intensity closely follows that of $V(\mathbf{k})$; in particular, there exists one-to-one correspondence between the split minima of the interaction and the split intensity peaks. The variation of the self-energy with $\mathbf{k}$ in this part of the reciprocal space is comparatively weak, though qualitatively important for the description of the temperature-dependent splitting. This assumption is certainly correct at sufficiently high temperatures, where the RPA (KCM) approximation in which the self-energy is $\mathbf{k}$-independent works reasonably well. Meanwhile, as temperature starts to decrease, the correction to the RPA self-energy ($\propto T^{-2}$) grows faster than the interaction term $2\beta V(\mathbf{k})$ ($\propto T^{-1}$). We can then encounter a situation when the variations of $\tilde{\Sigma}(\mathbf{k})$ and $2\beta V(\mathbf{k})$ with the wavevector are of the same order of magnitude. With temperature further decreasing, the wavevector dependence of the self-energy can even become dominant. In this regime positions of the diffuse intensity peaks would be determined by the maxima of $\tilde{\Sigma}(\mathbf{k})$.

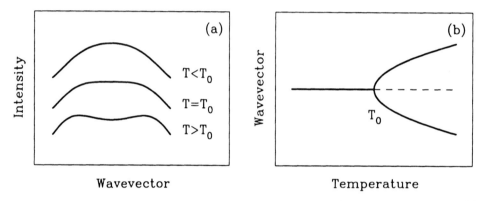

Figure 2. Schematical temperature dependence of the intensity profile (a) and the peak positions (b) in the coalescence case.

The behavior of the self-energy in the $\mathbf{k}$-space is, in general, qualitatively different from that of the interaction. In particular, there is no special reason to expect that the self-energy would have any extrema away from the special points. Let us assume that the self-energy does not have such extrema and that the variation of the self-energy in the reciprocal space becomes more and more important in comparison with the corresponding variation of $2\beta V(\mathbf{k})$ as temperature decreases. Then the qualitative picture of the temperature behavior of the splitting is as follows. At high temperatures the self-energy is almost $\mathbf{k}$-independent, and the intensity peak positions deviate little from those of the corresponding minima of $V(\mathbf{k})$. As temperature decreases, the wavevector dependence of $\tilde{\Sigma}$ becomes more pronounced; the peak positions move farther away from the positions of $V(\mathbf{k})$ minima and towards those special points $\mathbf{k}_0$ at which $\tilde{\Sigma}(\mathbf{k})$ has maxima. Eventually, as temperature reaches certain value T_0, the intensity peaks coalesce at these special points and the splitting disappears (Figure 2).

The coalescence temperature T_0 can be found from the condition of vanishing second derivative $\alpha'' = \partial^2\alpha/\partial k^2$ of the intensity with respect to the wavevector at k_0; the sign of this derivative controls the presence or absence of the splitting. At the special points all the first derivatives are equal to zero, and from Eq. (75) we obtain

$$\alpha''(k_0) = c(1-c)\alpha^2(k_0)\left[\tilde{\Sigma}''(k_0) - 2\beta V''(k_0)\right] . \tag{155}$$

Therefore, the splitting disappears when the second derivatives, or curvatures, of the self-energy and of the interaction term $2\beta V(k)$ at the special point k_0 compensate each other, i.e., when

$$\tilde{\Sigma}''(k_0) = 2\beta V''(k_0) . \tag{156}$$

To analyse the behavior of the splitting close to the coalescence point, it is convenient to use the same Landau-type approach as in the previous case. In this temperature range the splitting above T_0 is small, and expansions (151) and (152) are valid. Substituting them into Eq. (75), we get

$$\alpha^{-1}(k) = \alpha^{-1}(0) + \frac{1}{2}Ak^2 + \frac{1}{4}Bk^4 , \tag{157}$$

where second-order coefficient

$$A = c(1-c)(-A_\Sigma + 2\beta A_V) \tag{158}$$

vanishes at $T = T_0$ (see Eq. (156)), while the fourth-order coefficient B remains positive at that temperature. We can then, at temperatures close to T_0, regard A as linear in $T - T_0$ with a negative coefficient and B as temperature-independent. Thus, the inverse intensity $\alpha^{-1}(k)$ behaves almost in the same way that the Landau free energy. The only difference here is that the role of temperature is reversed; $\alpha^{-1}(k)$ has a double minimum above the coalescence temperature T_0 and a single minimum below it. Therefore, at small positive values of $T - T_0$ the splitting increases with temperature as $(T - T_0)^{1/2}$. Contrary to the corresponding result of the genuine Landau theory, obtained bifurcation exponent is exact, since the intensity is an analytical function of the wavevector and can legitimately be expanded into the Taylor series. At higher temperatures behavior of the splitting changes, and sufficiently far away from T_0 it starts to approach the value dictated by the interaction $V(k)$.

"Thermal" splitting of intensity peaks

We have just considered the situation when the double-well profile of the interaction in the vicinity of a special point is compensated by the wavevector dependence of the self-energy which has a simple maximum at this position. The competing curvatures $2\beta V''(k_0)$ and $\tilde{\Sigma}''(k_0)$ are both negative. As a result, the second derivative (155) of the intensity vanishes at some temperature, and the splitting induced by the Fermi surface effects disappears.

It is not very difficult to realize that another kind of curvature compensation is possible; this is the case in which both curvatures are positive.[39] This is, perhaps, the most physically interesting situation: here the interaction with a single minimum produces the intensity peak with no fine structure at higher temperatures (in full accordance with the RPA-like considerations), but then, as temperature decreases and the wavevector dependence of the self-energy becomes more and more significant, the compensation takes place and the intensity peak splits. This is especially probable when the minimum of $V(k)$ is shallow (i.e., $V''(k_0)$ is small); in particular, in the limiting case of vanishing $V''(k_0)$ it is the curvature of $\tilde{\Sigma}(k)$ that controls the fine structure (single- vs.

double-peak) of the maximum of $\alpha(k)$. The application of the Landau-type description gives essentially the same results as before. However, the coefficient in front of $T - T_0$ in A is now positive; the inverse intensity $\alpha^{-1}(k)$ has a single minimum above the splitting temperature T_0 and a double minimum below this point, and the splitting increases as $(T_0 - T)^{1/2}$ with decreasing temperature at small negative $T - T_0$ values (Figure 3) . The bifurcation exponent is again exact, for the reasons mentioned above.

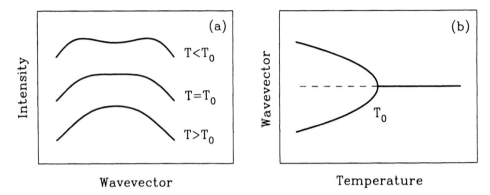

Figure 3. Schematical temperature dependence of the intensity profile (a) and the peak positions (b) in the case of the "thermal" splitting.

This type of behavior was recently experimentally observed (and subsequently reproduced in the MC simulations) for the Pt-V alloy system by Le Bolloc'h et al.[37] In this system the splitting of the (100) intensity peak along the (h00) line occured with decreasing concentration of vanadium rather then temperature. The explanation of this anomaly was proposed by the author,[39] who also pointed out that similar splitting should take place when temperature decreases at fixed composition. The predicted effect was then discovered in the MC simulations by the same group.[47] Experimental confirmation of its existence remains to be seen.

REFERENCES

1. D. de Fontaine, *Solid State Physics* 34:73 (1979); 47:33 (1994).
2. F. Ducastelle. *Order and Phase Stability in Alloys*, North-Holland, Amsterdam (1991).
3. G. Inden and W. Pitsch, in: *Phase Transformations in Materials*, P. Haasen, ed., VCH Press, New York, (1991).
4. A.G. Khachaturyan. *Theory of Structural Transformations in Solids*, Wiley, New York, (1983).
5. M. A. Krivoglaz. *Theory of X-Ray and Thermal Neutron Scattering by Real Crystals*, Plenum, New York (1969); *Diffuse Scattering of X-Rays and Neutrons by Fluctuations*, Springer, Berlin (1996).
6. J. M. Ziman. *Models of Disorder*, Cambridge University Press, Cambridge (1979).
7. L.D. Landau and E.M. Lifshitz. *Statistical Physics*, Part I, Pergamon, Oxford (1980).
8. Yu.A. Izyumov and Yu.N. Skryabin. *Statistical Mechanics of Magnetically Ordered*

Systems, Consultants Bureau, New York and London (1988).

9. H. Ehrenreich and L.M. Schwartz, *Solid State Phys.* 31:149 (1976).

10. R.J. Elliott, J.A. Krumhansl, and P.L. Leath, *Rev. Mod. Phys.* 46:465 (1974).

11. J.S. Faulkner, *Prog. Mater. Sci.* 27:1 (1982).

12. V.I. Tokar, I.V. Masanskii, and T.A. Grishchenko, *J. Phys. Condens. Matter* 2:10199 (1990).

13. I.V. Masanskii, V.I. Tokar, and T.A. Grishchenko, *Phys. Rev. B* 44:4647 (1991).

14. D.J. Amit. *Field Theory, the Renormalization Group and Critical Phenomena*, World Scientific, Singapore (1984).

15. A.N. Vassiliev. *Functional Methods in Quantum Field Theory and Statistics*, Leningrad State University Press, Leningrad (1976).

16. A.N. Vassiliev and R.A. Radzhabov, *Teor. Mat. Fiz.* 21:49 (1974); 23:366 (1975).

17. N.M. Bogoliubov et al., *Teor. Mat. Fiz.* 26:341 (1975).

18. P.C. Clapp and S.C. Moss, *Phys. Rev.* 142:418 (1966); 171:754 (1968).

19. R. Brout. *Phase Transitions*, Benjamin, New York (1965).

20. J. M. Sanchez, *Physica A* 111:200 (1982).

21. T. Mohri, J.M. Sanchez and D. de Fontaine, *Acta Metall.* 33:1463 (1985).

22. G.S. Joyce, in: *Phase Transitions and Critical Phenomena*, Vol. 2, C. Domb and M.S. Green, eds., Academic Press, New York (1972).

23. J. Philhours and G.L. Hall, *Phys. Rev.* 170:496 (1968).

24. D.W. Hoffmann, *Metall. Trans.* 3:3231 (1972).

25. L. Onsager, *J. Am. Chem. Soc.* 58:1468 (1936).

26. J.B. Staunton and B.L. Gyorffy, *Phys. Rev. Lett.* 69:371 (1992).

27. V.I. Tokar, *Phys. Lett. A* 110:453 (1985).

28. V. Janiš, *Czech. J. Phys. B* 36:1107 (1986); *Phys. Rev. B* 40:11331 (1989).

29. R. Kikuchi, *Phys. Rev.* 81:988 (1951).

30. I. Tsatskis (unpublished).

31. I. Tsatskis (submitted).

32. V.I. Tokar and I.V. Masanskii, *Fiz. Metal. Metalloved.* 64:1207 (1987).

33. I.V. Masanskii and V.I. Tokar, *Teor. Mat. Fiz.* 76:118 (1988).

34. T.A. Grishchenko, I.V. Masanskii, and V.I. Tokar, *J. Phys. Condens. Matter* 2:4769 (1990).

35. I.V. Masanskii and V.I. Tokar, *J. Phys. I France* 2:1559 (1992).

36. L. Reinhard and S.C. Moss, *Ultramicroscopy* 52:223 (1993).

37. D. Le Bolloc'h et al., in: *Proceedings of the Joint NSF/CNRS Workshop on Alloy Theory*, Mont Sainte Odile Monastery, Strasbourg, France, October 11-15, 1996, *Comput. Mater. Sci.* 8:24 (1997).

38. M. Borici-Kuqo and R. Monnier, Ref. 37, p. 16.

39. I. Tsatskis (submitted).

40. H. Reichert, S.C. Moss and K.S. Liang, *Phys. Rev. Lett.* 77:4382 (1996).

41. S.C. Moss and H. Reichert (private communication); H. Reichert, I. Tsatskis and S.C. Moss, Ref. 37, p. 46.

42. H. Roelofs et al., *Scripta Mat.* 34:1393 (1996).

43. S.C. Moss, *Phys. Rev. Lett.* 22:1108 (1969); S.C. Moss and R.H. Walker, *J. Appl. Crystallogr.* 8:96 (1974).

44. I. Tsatskis (in preparation).

45. J. Kulik, D. Gratias, and de D. Fontaine, *Phys. Rev. B*, 40:8607 (1989).

46. A. Finel and D. de Fontaine, *J. Statist. Phys.* 43:645 (1986).

47. D. Le Bolloc'h et al. (private communication and in preparation).

DIFFUSE SCATTERING BY CRYSTALS WITH
DEFECTS OF COULOMB DISPLACEMENT FIELD

R.I. Barabash

National Technical University of Ukraine
Metal Physics Department, pr. Pobedy 37,
Kiev 252056, Ukraine

INTRODUCTION

Distortions in crystals caused by impurity atoms, point defects, appearing during radiation, their groups, small dislocation loops, particles of new phase result in the formation of diffuse scattering. Peculiarities of the diffuse scattering intensity distribution (isodiffuse surface shape) in the reciprocal lattice space essentially depend on the defect type, its position in crystal lattice and on the interaction between defects with matrix atoms. That is why by means of diffuse X-ray or neutron scattering data it is possible to analyze defect types and there characteristics. This possibility was shown theoretically [1,2] and observed experimentally in a number of papers [3-6]. The development of new experimental techniques allows much more than just the investigation of diffuse scattering.

By analyzing experimental diffuse scattering data in every case, we must simulate intensity distributions for each defect type in a certain crystal and choose the best one to fit the experimental results. Therefore it is interesting to simulate isodiffuse surfaces for typical defects in a number of different crystals and to determine important parameters of such surfaces near different reciprocal lattice points, depending on the characteristics of the defects. Comparison of experimental data with simulated diffuse scattering distributions will enable us to define defects type and parameters. The general analysis of such changes was made by Krivoglaz [1] in the framework of the kinematical theory of scattering. According to his defect classification, for the first type of defect the value of the exponent of static Debye-Waller factor $\exp(-2M)$ is finite, and for the second type of defects $2M$ is infinite.

The presence of the first type of defects in crystals causes the appearance of diffuse scattering, besides the usual regular reflections. The positions of both diffuse and regular intensity maximums are displaced relatively to the ideal positions of intensity maximums corresponding to the crystal in the absence of defects. The analysis of diffuse scattering in crystals with the first type of defects is different in two limiting cases of weakly and strongly distorted crystals. These cases can be distinguished by the value $2M$. If $2M \le 1$

crystals are considered to be weakly distorted, if $2M \gg 1$, they are strongly distorted. The value of the parameter $2M$ depends on the displacement fields $\vec{U}$ caused at the point s of the matrix by the defect situated in point t, on the defects concentration, and on the diffraction vector $\vec{Q}$.

DIFFUSE SCATTERING BY SOLID SOLUTIONS WITH ISOLATED DEFECTS

Detailed analysis of isodiffuse surfaces types in the vicinity of reciprocal lattice sites for cubic crystals was carried out in [1,7]. According to it, in principle the defect type is connected with the directions or planes of zero intensity near reciprocal lattice points (h00), (hh0), (hhh). Really it is rather difficult to observe the regions of zero intensity because of other factors which influence on the diffuse scattering. Therefore, to determine the defect type and parameters, it will be useful to analyze other directions in reciprocal lattice space besides those chosen in [7]. Below we shall describe such computer simulations for defects of cubic, tetragonal, rombohedron and orthorhombic symmetry in different cubic crystals. Assume the defect concentration to be not high, so that crystals can be considered as weakly distorted. The diffuse scattering intensity distribution is analyzed in the vicinity of reciprocal lattice points. In this case it is influenced mainly by static distortions at large distances from the defect.

Diffuse Scattering Intensity

According to [1,8] at small defect concentration c (c is determined as the ratio of defect quantity to the number of crystal elementary cells N) diffuse scattering intensity in the crystals with point defects can be written as

$$I_1 = cN \left| \Phi_{\vec{q}} \right|^2$$

$$\Phi_{\vec{q}} = \sum_{s=1}^{N} \exp(i\vec{q}\vec{R}_s) \left[f\left(e^{i\vec{q}\delta\vec{R}_s} - 1 \right) + \varphi_s(\vec{Q}) \right].$$

(1)

Here $\vec{Q}$ is the diffraction vector ($\vec{Q} = \vec{k} - \vec{k}_0$), $\vec{k}$ and $\vec{k}_0$ are wave vectors of the scattered and initial waves; $q = \vec{Q} - \vec{G}$, where $\vec{G}$ is reciprocal lattice vector nearest to $\vec{Q}$, $\vec{R}_s$ is the radius vector of the first atom of the s elementary cell for the ideal crystal, $\delta\vec{R}_s$ static displacement of this atom due to the defect, f is the matrix structure amplitude, $\varphi_s(\vec{Q})$ is the change of structure amplitude in the s elementary cell due to the defect.

Let us introduce Fourier components for static displacements

$$\delta\vec{R}_s = \frac{i}{N} \sum_k \vec{A}_{\vec{k}} \exp(i\vec{k}\vec{R}_s)$$

$$\vec{A}_{\vec{k}} = -i \sum_s \delta\vec{R}_s \exp(i\vec{k}\vec{R}_s),$$

(2)

Equation (1) can be written as

$$\Phi_{\vec{q}} = -f\vec{Q}\vec{A}_{\vec{q}} + \Delta f(\vec{Q}); \Delta f(\vec{Q}) = \sum_s e^{i\vec{q}\vec{R}_s} \left[f\left(e^{i\vec{Q}\delta\vec{R}_s} - 1 - i\vec{Q}\delta\vec{R}_s \right) + \varphi_s(\vec{Q}) \right].$$

(3)

234

As the static displacement decreases slowly with the distance to the defect, $\vec{A}_{\vec{q}}$ increase as $1/q$ as $q \to 0$. At the same time $\Delta f(\vec{Q})$ tends to the final limit. That is why in the vicinity of reciprocal lattice points at relatively small q, we can neglect $\Delta f(\vec{Q})$ in (3) compared with $f\vec{Q}\vec{A}_{\vec{q}}$. Then the diffuse scattering intensity approximately will coincide with Xuang scattering I_1' given by

$$I_1(\vec{Q}) \approx I_1'(\vec{Q}) = cNf^2(\vec{Q}\vec{A}_{\vec{q}})^2 \qquad (q \to 0) \qquad (4)$$

This intensity increases as q^{-2} with the approach to the reciprocal lattice point. More accurately I_1' can be determined by measuring the sum of the diffuse scattering intensity in two points being symmetrically situated relatively to the reciprocal lattice point:

$$\frac{1}{2}\left(I_1(2\pi\vec{K}_n + \vec{q}) + I_1(2\pi\vec{K}_n - \vec{q})\right) \approx I_1'(\vec{q}) + cN|\Delta f(\vec{q})|^2 \qquad (q \to 0) \qquad (5)$$

If the static displacements in the neighborhood of defect are small ($|\vec{Q}\vec{U}_{st}| \langle\langle 1$) or known, the quantity $\Delta f(\vec{Q})$ may also be considered to be known. In the case of small displacements, $\Delta f(\vec{Q})$ can be also determined experimentally by defining the intensity in the vicinity of two points (h k l) and (2h 2k 2l). Below we shall assume that this additional term is small and already taken into account. So we shall analyze I_1' increasing as q^{-2} with the approach to the point.

Non-cubic defects in cubic crystals usually are oriented in several directions λ with the equal probability $1/\lambda$. The resulting intensity in this case will be a sum over different defects orientations δ

$$I_1' = cNf^2 \frac{1}{\lambda} \sum_{\delta=1}^{\lambda} \left(\vec{Q}\vec{A}_{\vec{q}}^{\delta}\right)^2 \qquad (6)$$

In the macroscopic approximation (small q) Fourier components for atom static displacement around δ type defects can be determined by the following simple equation system

$$\sum_{j=1}^{3} Q_{\vec{q}_{ij}} A_{\vec{q}_i}^{\delta} = P_{\vec{q}_i}^{\delta}$$
$$Q_{\vec{q}_{ij}} = C_{44} + (C_{11} - C_{44})n_i^2$$
$$Q_{\vec{q}_{ij}} = (C_{12} + C_{44})n_i n_j \qquad (7)$$
$$qP_{\vec{q}_i}^{\delta} = C_{11}n_1 L_{11}^{\delta} + C_{12}n_1(L_{22}^{\delta} + L_{33}^{\delta}) + 2C_{44}(n_2 L_{12}^{\delta} + n_3 L_{13}^{\delta}))$$

Here $\vec{n} = \vec{q}/q$. Let us introduce force dipole tensor P_{ij}^{δ} in terms of which system (7) can be written as following

$$qP_{\vec{q}_i}^{\delta} = P_{ij}^{\delta} n_j$$
$$P_{ij}^{\delta} = \lambda_{ijlm} L_{lm}^{\delta}$$
$$P_{11}^{\delta} = C_{11} L_{11}^{\delta} + C_{12}(L_{22}^{\delta} + L_{33}^{\delta}) \qquad (8)$$
$$P_{12}^{\delta} = 2C_{44} L_{12}^{\delta}$$

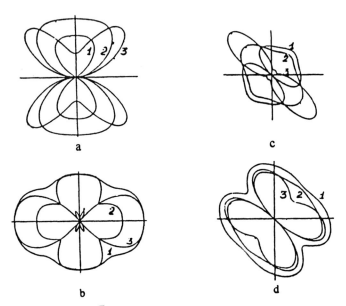

Figure 1. Isodiffuse curves for $\alpha - Fe$ crystals around (00h),(h0h) reciprocal lattice points(a,b),and (0hh), (hhh) points (c,d) with different type defects:a,c-cubic (1),tetragonal with $b=0$ (2),$b=-0,5$ (3); b,d-rombohedron defects:b)- point (h0h),$b=-0,5$ (1), point (00h) ,$b=0$ (2),point (00h),$b=-0,5$ (3);d- point (hhh), $b=-0,5$ (1),point (0hh),$b=0$ (2),point (0hh),$b=-0,5$ (3)

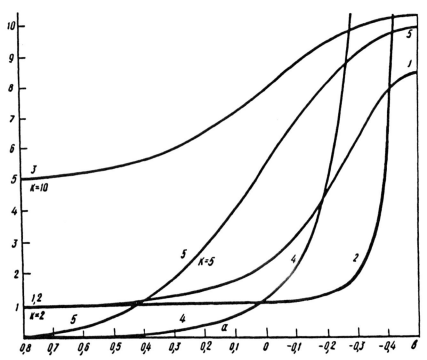

Figure 2. Intensity ratios along different directions in the reciprocal lattice points to the intensity along [00h] direction for crystal with W elastic modulus.

These equations become simpler for certain defect types. For example for tetragonal defects with the tetragonal axis along the [001] direction they look like:

$$L_{11}^{\delta} = L_{22}^{\delta} = L_1^{\delta} \neq L\delta_{33}^{\delta} = L_2$$

$$L_{ij}^{\delta} = P_{ij}^{\delta} = 0 \qquad (i \neq j)$$

$$P_{11}^{\delta} = P_{22}^{\delta} = (C_{11} + C_{12})L_1 + C_{12}L_2$$

$$P_{33}^{\delta} = C_{11}L_2 + 2C_{12}L_1 \tag{9}$$

$$qP_{qi} = ((C_{11} + C_{12})L_1 + C_{12}L_2)n_i \qquad (i = 1,2);$$

$$qP_{\bar{q}_\delta}^{\delta} = (C_{11}L_2 + 2C_{12}L_1)n_3$$

For orthorhombic defects along [110] direction :

$$L_{11}^{\delta} = L_{22}^{\delta} = \frac{1}{2}(L_1 + L_2)$$

$$L_{33}^{\delta} = L_3^{\delta}$$

$$L_{12}^{\delta} = L_{21}^{\delta} = \frac{1}{2}(L_1 - L_2)$$

$$L_{13}^{\delta} = L_{23}^{\delta} = 0$$

$$P_{11}^{\delta} = P_{22}^{\delta} = \frac{1}{2}(C_{11} + C_{12})(L_1 + L_2) + C_{12}L_3$$

$$P_{33}^{\delta} = C_{12}(L_1 + L_2) + C_{11}L_3 \tag{10}$$

$$P_{12}^{\delta} = P_{21}^{\delta} = C_{44}(L_1 - L_2)$$

$$P_{13}^{\delta} = P_{23}^{\delta} = 0$$

$$qP_{\bar{q}_1} = P_{11}^{\delta}n_1 + P_{12}^{\delta}n_2$$

$$qP_{\bar{q}_2} = P_{12}^{\delta}n_1 + P_{22}^{\delta}n_2$$

$$qP_{\bar{q}_3} = P_{33}^{\delta}n_3$$

Here L_1, L_2, L_3 define deformations caused by defects along [110], [1$\bar{1}$0], [001] directions. So for tetragonal or orthorombic defects, isodiffuse surfaces depend not on the elastic modulus and L_{ij}^{δ}, but on their ratios: $P_1 = C_{12}/C_{11}$, $P_2 = C_{44}/C_{11}$, $b = L_1/L_2$. Two of them characterize the crystal and b is the extent of the defect deviation from the cubic form (for cubic defects it is equal 1). The value of b determines the shape of the isodiffuse surface and the ratio $I_1(\vec{Q})$ for different directions. For tetragonal defects according to (9), these ratios coincide at two different meanings x' and x'' of the parameter $x = P_{33}^{\delta} / P_{11}^{\delta}$ or for two different values b', b'' of the parameter b :

$$(1+2x')(1+2x'')=9$$

$$\frac{2 + P_1 + (1 + 5P_1)b'}{P_1 + (1 + P_1)b'} = \frac{2 + P_1(1 + 5P_1)b''}{P_1 + (1 + P_1)b''} \tag{11}$$

$$b = \frac{1 - P_1 x}{(1 + P_1)x - 2P_1}$$

For rombohedron defects along the axis [111]:

$$L_{ii}^{\delta} = \frac{1}{3}(L_3 + 2L_1)$$

$$P_{ii}^{\delta} = (C_{11} + 2C_{12})L_{ii}^{\delta} \qquad (i = 1,2,3)$$

$$L_{ij}^{\delta} = \frac{1}{3}(L_3 - L_1) \tag{12}$$

$$P_{ij}^{\delta} = 2C_{44}L_{ij}^{\delta}, \qquad (i = j)$$

$$qP_{\bar{q}}^{\delta} = \frac{1}{3}(L_3 + 2L_1)(C_{11} + 2C_{12})n_1 + \frac{2}{3}(L_3 - L_1)C_{44} + (n_2 + n_3)$$

Here (L_3, L_1) determine deformations along and perpendicular to the [111] direction. In this case isodiffuse surfaces coincide for the pair b', b'' being connected by the equation:

$$(4b'-1)(4b''-1) = 9 \tag{13}$$

Isodiffuse Curves Computer Simulations

For constant intensity value, and using (1)-(13), we can analyze isodiffuse surfaces shape for different defects types. In Fig.1a there are presented the results of computer simulations of isodiffuse curves in the planes (h00) (or (0h0)) in the vicinity of the (00h) point for cubic and tetragonal defects. For tetragonal defects, the isodiffuse curves shape depends essentially on the parameter b value, being the extent of defect tetragonality. With $(-b)$ increasing, the intensity becomes very small in the direction [00h] (at any b, the intensity is equal to zero in the plane (00h)). Isodiffuse curves are the same in the vicinity of the (hoh) point for these defects. Isodiffuse curves shape changes for [111] defects. At large $(-b)$ there are small lamellae of intensity in the direction [001] near the point (00h). The isodiffuse curves differ essentially in the vicinity of the point (0hh). To characterize the isodiffuse curve shape more briefly depending on the value of parameter b for cubic (C), tetragonal (T), rombohedron (R) and orthorhombic (O) defects, there were calculated intensity ratios for different directions around reciprocal lattice points to the intensity along the direction [00h] (Fig. 2). Numbers near the curves correspond to the following directions: 1 - [011] near (00h),(h0h) points for (T) defects; 2 - [011] near (00h), 3 - [011] near (h0h), 4 - [010] near (00h), 5 - [010] near (h0h), 2 - 5 for (R) defects. Intensity distributions are equal for the pairs of x' and x'' or b' and b''. For (T) defects, if x belongs to the interval $-2 < x' < 1$, then the parameter x'' lies in the interval $\infty > x'' > 1$ and $(-\infty < x'' < -2)$. Therefore, it is enough to determine I_1' only in the above x'' intervals . If x'' changes in the intervals $[\infty, 1]$ and $[-\infty, -2]$, then b changes I the interval $-1/2 < b < 1$. So it is sufficient to calculate the intensity for the latter interval of $b = b''$ values. Out of this interval for $b = b'$ intensities $I_1'(b') = I_1'(b'')$ and b' is connected with b'' by the equation (11). For (R) defects, the interval $-1/2 < b < 1$ is also sufficient for intensity calculations, because $b = b'$ are connected with b'' by equation (13) and gives the same intensity. For some reciprocal lattice directions intensity I_1' is equal to zero due to the lattice symmetry. For (C) defects, in the vicinity of (00h), (0hh), (hhh) points this takes place in the planes perpendicular to the directions [001], [011] and [111]. For (T) defects near the point (00h), zero intensity corresponds to the whole plane (001) and near the point (0hh) only to the direction [100]. For (R) defects, zero intensity corresponds only to the directions

[$\bar{1}$ 10], or [1 $\bar{1}$ 0] near the point (hh0). For the arbitrary defect type isodiffuse curves in the vicinity of (h00), (hh0), (hhh) points the directions [100], [110], [111] are symmetry axes. The parameter $b=L_1/L_3$ characterizes the influence of the defect force on the crystal with the subsequent change in the shape of isodiffuse curves. When $b\sim1$ defects are almost cubic and isodiffuse curves for different type defects are very similar. So their type analysis is very difficult in this case. But for noncubic defects b differs essentially from 1 and is often negative. In this case for (T) and (R) defects, isodiffuse surface shape strongly differs from the one for cubic defects and depends on the b value. It enables us to determine the defect type and the b value from the data on diffuse scattering intensity.

To make such analysis reliable, we must use the data on several curves simultaneously (if we use only one curve for one direction we can obtain the same intensity values for different defect type corresponding to different b values). Curves corresponding to different directions will never cause such uncertainty. To avoid this ambiguity, we also may take into account that in weakly distorted crystals defects result in the relative displacement of intensity maximums equal to $-(c/3)\sum l_{ii}^{\delta}$. If crystals contain simultaneously different types of defects (such as vacancies, interstitial atoms, other point defects, small dislocation loops) such analysis becomes much more accurate. In the case of (O) defects, the shape of isodiffuse curves depend not on one, but on two parameters $b_1=L_3/L_1$ and $b_2=L_2/L_1$. For final defect type analysis it will be useful to restore the whole isodiffuse surfaces for corresponding b value by means of the above equations and to compare them with the experimental ones.

DIFFUSE SCATTERING BY CRYSTALS WITH PRECIPITATES

Precipitates with Weak Distortions

In weakly distorted crystals, the Debye-Waller factor connected with distortion fields around the new particles $\exp(-2M)$ is of the order of unity. For simplicity let us consider the case when similar precipitates are formed in the cubic matrix and correlation between their positions is not essential. Even in this simple model we can get a large majority of intensity distributions.

The presence of new phase particles causes local lattice distortions due to the existence of lattice misfit between matrix and forming phase. These distortions are described by the law: $|\bar{U}(\bar{r})|\sim Cr^{-2}$ (C is the constant that characterizes the strength of defects; its value being of the order of absolute value of the volume change $|\Delta V|$ caused by defect). They cause the appearance of intensive X-ray diffuse scattering in the vicinity of reciprocal lattice points. These intensity distributions may differ essentially for different materials after different treatment.

If the distortions caused by precipitates are negligible, symmetric intensity distributions appear around reciprocal lattice points. In previous papers [9-12], the intensity distributions were analyzed for the case when diffuse scattering is mainly influenced by distortions. The presence of distortions makes the distributions asymmetric and results in hyperbolic intensity increase, when we approach the reciprocal lattice points. There was carried out the general analysis of intensity distributions and computer simulation of certain intensity distributions in elastically anisotropic crystals. Often the interference of scattering connected with the difference between structure amplitude and distortions is very essential. We have analyzed the effects of such mutual influence of these factors on diffuse scattering for elastically anisotropic crystals. When describing the formation of precipitates we have to distinguish two of the following cases. The first and most simple one is a case when precipitate with diffuse boundary is formed in a uniform matrix. Usually it occurs after relatively large aging time when depletion layers around different particles interflow and create almost constant concentration level in the matrix. In this case it can be described simply by one parameter for example Gauss function and corresponds to continuous concentration distribution. Usually on the early stages of aging when diffusion length is much less than the distance between the particles the second case is realized when the particle of new phase is surrounded by the depletion layer of impurity atoms. To analyze the influence of depletion layer existence on diffuse scattering we have analyzed simple model of sphere particles with the same structure as that of matrix and with diffuse boundaries. In these cases intensity distribution tends to zero in reciprocal lattice points and often is asymmetric and concentrates on one side of the point being crescent-shaped. In contrast to the case of homogeneous matrix in the last case isodiffuse surfaces do not pass through reciprocal lattice points but surround their neighborhood. The picture may become still more complicated due to the correlation between the precipitate distribution.

Diffuse Scattering Intensity for Weakly Distorted Crystals

Let us analyze diffuse scattering I_1 by the crystal containing Np equal particles of new phase with the same orientation being chaotically distributed in the matrix. If the particles have sharp boundary with matrix and their structure differs from that of matrix, then according to [1,9,10,13] intensity I_1 in the vicinity of matrix reciprocal lattice points is defined by the equation:

$$I_1 = N_p v^{-2} \left| \Delta f - f \vec{Q} \vec{A}_{\vec{q}} \right|^2 \left| s(\vec{q}) \right|^2. \tag{14}$$

Here v is unit cell volume; Δf is the difference between new phase and matrix structure amplitudes (if precipitate and matrix have the same structure and are coherent); or $\Delta f = -f$ if their structures are different; $s(\vec{q})$ is the Fourier component of shape function of the precipitate; $\vec{A}_{\vec{q}}$ determines Fourier component of displacement field $\vec{U}(\vec{r})$ in the matrix around the new phase particle. At small q they may be found from the equations of macroscopic elasticity theory.

The solution of the corresponding system of equations for cubic particles in cubic matrix has the following form([1,13])

$$\vec{A}_{\vec{q}} = \frac{3L\vec{\alpha}_{\vec{q}}}{q}$$
$$\alpha_{\vec{q}x} = \frac{C_{11} + 2C_{12}}{3D} \left(1 + \xi n_y^2\right)\left(1 + \xi n_z^2\right) n_x, \tag{15}$$

where $L\delta_{ij}$ is the tensor of self deformation due to the formation of precipitates; C_{11}, C_{12}, C_{44} - elastic modulus and

$$D = C_{11} + \xi\left(C_{11} + C_{12}\right)\left(n_x^2 n_y^2 + n_x^2 n_z^2 + n_y^2 n_z^2\right) + \xi^2\left(C_{11} + 2C_{12} + C_{44}\right)n_x^2 n_y^2 n_z^2$$

$$\xi = \frac{C_{11} - C_{12} - 2C_{44}}{C_{44}} \qquad \vec{n} = \vec{q}/q. \tag{16}$$

In (14) we suppose that the local distortions are small and $\left|\vec{Q}\vec{U}\right| \langle\langle 1$ or $q \langle\langle (LR^3Q)^{-1/2}$, (R-effective particle size). If precipitates have the same structure as matrix has and are formed as a result of continuous diffusive change $\delta c(\vec{r})$ in concentration c of atoms A in binary solution A-B, then they do not form sharp boundary with matrix. In this case I_1 can be again determined by formula (14), if we put:

$$\Delta f = f_A - f_B$$

$$S(\vec{q}) = \int \delta c(\vec{r}) \exp(i\vec{q}\vec{r}) d\vec{r} \tag{17}$$

$$L = \frac{1}{3v} \frac{\partial v}{\partial c}.$$

From (14)-(17), it follows that the intensity distributions around reciprocal lattice points are influenced by the dependence of $\vec{q}$ on two factors: $|\Delta f - f\vec{Q}\vec{A}_q|^2$ and $|S(\vec{q})|^2$. The first of them is connected with the displacement field in elastically anisotropic matrix and the second one describes the influence of concentration distribution on the diffuse scattering. To analyze the influence of the last factor let us consider as in [1,13] two simple functions $\delta c(\vec{r})$ for precipitates with diffuse boundary.

Precipitates in Homogeneous Matrix. Concentration distributions become the most simple after relatively large annealing time when depletion layers around the particles unite and create almost constant concentration level. In this case concentration distribution around separate spherical particles can be described by, for instance, Gaussian function

$$\delta c(\vec{r}) = \delta c_m \exp\left(-\frac{r^2}{R^2}\right) \tag{18}$$

where δc_m is the maximal change in the concentration, R characteristic radius of the distribution. According to (17), $S(\vec{q})$ is the "form function" for the distribution (18) and looks like

$$S(\vec{q}) = nv \exp\left(-\frac{1}{2} R^2 q^2\right)$$
$$n = \frac{(2\pi)^{3/2} R^3}{v} \tag{19}$$

where n is the effective number of atoms in the region where the composition is changed. According to (14), (19) in this case

$$I_1 = K \left| C - \frac{\vec{m}\vec{\alpha}_q}{qR} \right|^2 \exp(-R^2 q^2)$$
$$K = N_p n^2 f^2 \left(\frac{1}{v} \frac{\partial v}{\partial c} QR\right)^2 \tag{20}$$
$$\vec{m} = \vec{Q}/Q$$
$$C = \frac{\Delta f}{f} \frac{1}{QR} \left(\frac{1}{v} \frac{\partial v}{\partial c}\right)^{-1}. \tag{21}$$

241

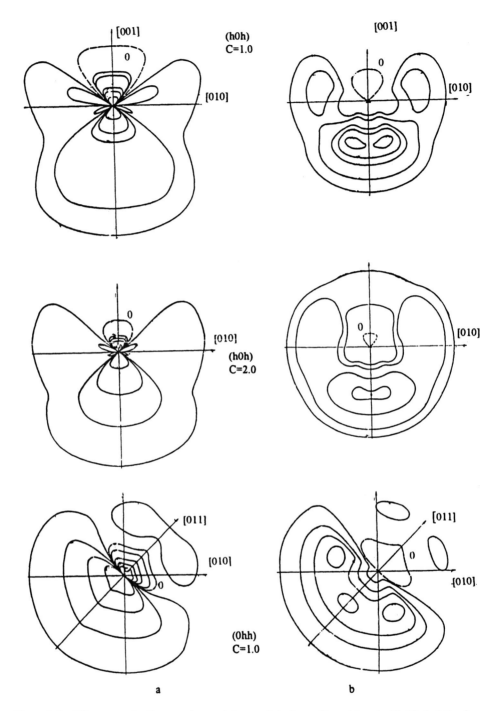

Figure 3. Isodiffuse curves for Cu crystals containing precipitations without (a) and with (b) depletion layer around reflection (h0h) in the plane (100) for different values of parameter C

The parameter C describes the relative influence of the effects connected with the difference between atomic scattering factors of the alloy and distortions. Last exponent factor in (20) restricts the region of intensive diffuse scattering by the condition $q \leq 1/R$.

For small distortions and not very large particles when $|C| \gg 1$ at $q \sim 1/R$ in the brackets in (20) we may retain only the term C. In that case isodiffuse surfaces are spheres and symmetrically surround reciprocal lattice points. Dependence of I_1 on q is bell shaped. By studying the width of this dependence one can determine the size of particles R. If particle size and misfit between this particle and matrix is large, then I_1 is mainly influenced by distortions caused by these particles in the matrix. It is described by the second term in the brackets in (20). In this case the dependence of I_1 on q is not bell shaped but hyperbolic one and isodiffuse surfaces are not spheres, but are stretched along the diffraction vector. In this case reciprocal lattice points are still the inversion centers for isodiffuse surfaces.

In the parameter region with $|C| \sim 1$, the interference of scattering is essential on both the changes in structure amplitude and the distortions, so that both terms in brackets (20) should be taken into account. In this case isodiffuse surfaces differ essentially from the two of above-mentioned limiting cases. Reciprocal lattice points are no longer inversion centers and isodiffuse surfaces become asymmetric. Their intensity is much higher to one side of the point than to another.

Isodiffuse curves - sections of isodiffuse surfaces by certain planes in the reciprocal lattice space - were numerically calculated for different values of parameter C in the latter case according to formulas (15), (18), (20). Elastic moduli of the matrix were set equal to those of Cu as an example of the crystal with "average" anisotropy. We have analyzed cross-sections by planes (100) near the points (00h) (Fig.3). The numbers near the curves show the values of I_1/K for the corresponding curves and on the rest of figures they show $\alpha\, I_1/K$ (α is shown at each figure). At a relatively small value $C = 0.4$ the intensity distribution is mostly influenced by distortions . Asymmetry of the isodiffuse curves is very weak and can be observed only for the curves corresponding to very small intensity level. For $C = 0$ scattering takes place only on the distortions and isodiffuse curves are strictly symmetric and zero intensity curve coincides with the line [010]. If C differs from zero, then this curve deviates from straight line and is shown in dash. When C increases (for $C=1$, $C=2$), the curves become essentially asymmetric. But with further increase in C, asymmetry of the curves decreases again, although the shape of the curves changes. At C < 0 isodiffuse curves have the same shape, but are inversed relatively to reciprocal lattice point. It means that the asymmetry sign is determined by the signs of Δf and $\partial v / \partial c$. Isodiffuse curves in the plane (100) near the points (h0h) are analogous to those near (00h). Symmetry axis coincides with [001]. Isodiffuse curves near the points (0hh) are similar to those near (hhh). Their symmetry axis coincides with [011]. Such axial symmetry will be spoilt for arbitrary point not belonging to the high symmetry direction.

Influence of Depletion Layer. At early stages of ageing the effective diffusion length is essentially less than the particle separation. Around each precipitate there will appear the depletion layer with less concentration of one of the components. To analyze the influence of depletion layer we shall consider a simple model of spherical particles having the same structure with the matrix. We shall suppose that both the increase in concentration inside the particle and its decrease in the depletion layer can be described by simple distribution

$$\delta c(\vec{r}) = \delta c_m \left[\exp\left(-\frac{r^2}{2R^2} \right) - \left(\frac{R}{R'} \right)^{3/2} \exp\left(-\frac{r^2}{2R'^2} \right) \right]$$

$$\int \delta c(\vec{r}) d\vec{r} = 0 \tag{22}$$

At the same time the total atom concentration near each type of precipitates is constant. If we determine $S(\vec{q})$ using formulas (17), (22) and substitute this expression into (14), we find that the intensity distribution I_1 can be written as following

$$I_1 = K \left| C - \frac{\vec{m}\vec{\alpha}_{\vec{q}}}{qR} \right|^2 \left[\exp(-R^2 q^2) - \exp(-R'^2 q^2) \right]^2. \tag{23}$$

In this case $S(\vec{q}) \to 0$ and $I_1(\vec{Q}) \to 0$ at $\vec{q} \to 0$. It means that in the reciprocal lattice points intensity I_1 tends to zero. Therefore intensity distributions have minima in the reciprocal lattice points and are symmetric in the limiting cases when one of the items in (20) is much larger than another. It results in the spherically symmetrical halo at $|C| \gg 1$, $qR \sim 1$ and in the appearance of two maxima with almost equal intensity at $|C| \ll 1$. If both items are of the same order of magnitude, then the resulting intensity distribution is essentially asymmetric and is concentrated mainly on one side of the reciprocal lattice point having crescent shape. Contrary to the above-considered case of a homogeneous matrix, isodiffuse surfaces do not pass through the reciprocal lattice points, but surround them in their vicinity. This crescent shape of isodiffuse surfaces is caused by vanishing of displacement field due to the mutual compensation of effects from the precipitate and the depletion layer. Such compensation takes place upon linear dependence v on c (Vegard's law) and if the condition $\int \delta c(\vec{r}) d\vec{r} = 0$ is true. The deviations from Vegard's law will result in the not complete compensation of displacement fields at large distances from the particle and intensity $I_1(q \to 0)$ being not zero. But still the value of coefficient near q^{-2} in the expression for I_1 will be much less comparing with the case of homogeneous matrix. Its value will characterize the extent of distortion penetration through the depletion layer into the matrix. Such transformation of isodiffuse curves for concentration distribution (9) was illustrated by numerical calculations (Fig. 3) upon the linear dependence of v on c for different meanings of parameter C. Calculations were fulfilled for the case $R' = 2R$ for crystals with elastic moduli of Cu and NaCl (asymmetry parameter ξ for Cu and NaCl has opposite signs). Isodiffuse curves in the plane (100) around the reciprocal lattice points (00h) for Cu (Fig. 3) differ from the case of homogeneous matrix. Presence of the depletion layer makes them essentially asymmetric even upon the value $C = 0.4$ (in the case $C = 0$ which was considered earlier [8,9] asymmetry disappears). They consist of different intensity maxima displaced from their reciprocal lattice points the distance approximately equal to the reverse particles size. The positions of the displaced peaks are indicated by the cross. Besides those peaks that are displaced along the direction $\vec{Q} \approx \vec{G}$, there exist also weak isolated maxima in the direction at certain angle to $\vec{G}$. When $C = 1.0$ and 2.0 intensity distribution becomes crescent like. In the region of large q isodiffuse curves tend to the circle. For AgCl precipitates in NaCl crystals isodiffuse curves were calculated in two different sections (100) and ($1\bar{1}0$) near the point (hhh). Section of the point (hhh) by the plane (100) has axial symmetry. But its section by the plane ($1\bar{1}0$) has no axial symmetry. In this crystal that has $\xi > 0$ crescent shape of isodiffuse curves at $C = 0.5$ and $C = 2.0$ is much more distinct than for Cu crystals. As can be seen from Fig. 2 at $C = 2.0$ the edges of isodiffuse curves with lower intensity value unite almost into a circle

244

and form asymmetric halo. Computer simulations were compared with interesting experiments [14, 15] for decomposition in the system NaCl-AgCl. The interference of scattering due to the distortions and depletion layer gives the same intensity picture as observed in [14, 15] even without correlation between precipitates.

Correlation between Precipitates. The influence of correlation between particles in the case of $2M \ll 1$ and small volume fraction of the particles for equal and eqiaxial precipitates results in the additional factor in the expression for intensity I_1

$$I_1 = I_1^{(0)} \frac{N}{c} \langle |C_{\vec{q}}|^2 \rangle = I_1^0 \left[1 + \frac{1}{c} \sum_{\vec{\rho} \neq 0} \varepsilon(\vec{\rho}) \cos \vec{q}\vec{\rho} \right] \tag{24}$$

$$\varepsilon(\vec{\rho}) = \langle (C(\vec{R}_t) - C)(C(\vec{R}_t - \vec{\rho}) - C) \rangle,$$

where I_1^0 is defined by formulas (14), (20) or (23) and radius vector $\vec{R}_t$ determines possible positions of precipitate centers in different crystals cells t, N is the number of cells. $C(\vec{R}_t) = 1$ if precipitate center is in the point t and $C(\vec{R}_t) = 0$ in the opposite case. $\varepsilon(\vec{\rho})$ is the correlation parameter of precipitates divided by vector $\vec{\rho}$; and <...> means averaging over statistical ensemble. So correlation results in the additional factor $\langle |\tilde{N}_{\vec{q}}|^2 \rangle$ in the expression for intensity. The above results correspond to diffuse scattering intensity:

$$I_1^{(0)} = I_1 \frac{c}{N} \left[\langle |C_{\vec{q}}|^2 \rangle \right]^{-1} \tag{25}$$

They can be used if the dependence $\langle |\tilde{N}_{\vec{q}}|^2 \rangle$ on $\vec{q}$ is not too strong. If it is possible to calculate function $\langle |\tilde{N}_{\vec{q}}|^2 \rangle$ using certain model for decomposition kinetics these additional correlation effects may be simply taken into account by formula (24).

Precipitates Causing Strong Distortions

In strongly distorted crystals defects power C and the average size if new particles R_0 corresponds to the condition $QC \gg R_0^2$. Precipitates of new phase also cause the appearance of diffuse scattering besides the usual regular reflections. They shift the positions of both diffuse and regular intensity maxima relatively to their positions for crystal without defects and also decrease the intensity of regular reflections. The character of the appearing intensity distribution essentially depends on the defect power C and their concentration $c = nv$, where n is a number of particles per unit volume and v is volume of the matrix unit cell.

In this case, intensity I_1 is given

$$I_1(\vec{Q}) = |f|^2 \sum_{s,s'} e^{i\vec{Q}(\vec{R}_s^0 - \vec{R}_{s'}^0)} e^{-T}$$

Function I_1 depends on displacement fields $\vec{u}_{st}$ in s^{th} cell by defect in t^{th} cell. Defect function T has the following real and imaginary parts: $T = T' + iT''$

$$T' = n \int d\vec{R}_t (1 - \cos \vec{Q}\vec{u}_{ss't}); T'' = -n \int d\vec{R}_t \sin \vec{Q}\vec{u}_{ss't}$$

For defects with one orientation and for not large ρ that is the distance between cells s and s', we have $T = nCQ\rho(\varphi' + i \ \varphi'') + i\delta G\rho$; $\varphi'' = \varphi_1'' + \varphi_2''$. For strongly distorted crystals($2M \gg 1$) with cubic symmetry of matrix being retained after the formation of new phase particles diffuse scattering intensity $I_1(\vec{Q})$ according to [1, 16, 17] can be described as following:

$$I_1 = \frac{N}{v}|f|^2 \int d\vec{\rho} \exp(i\vec{q}'\vec{\rho}) \exp(-nCQ\rho(\varphi' + i\varphi_1'')), \tag{26}$$

$$\vec{q}' = \vec{Q} - \vec{G}^0 - \delta\vec{G}' = \vec{Q} - \left(1 - \frac{1}{3}n\Delta V_{im}\right)\vec{G}^0 = \vec{q} - \frac{1}{3}n\Delta V_0 \vec{G}^0, \vec{q} = \vec{Q} - \vec{G}^0 - \delta\vec{G},$$

Maximum $\vec{q}_{0m}$ displacement of the intensity distribution $I_1(\vec{Q})$ relatively to the ideal positions of intensity maxima for crystal without defects, $\vec{q}_{0m} = \delta\vec{G}' + \vec{q}'_m \neq \vec{q}_{0m}^0 = \delta\vec{G}$ essentially differs from the maximum $\vec{q}_{0m}^0$ displacement of regular peak even after taking into account the additional term $\vec{q}'_m$. Taking into account that ratio $|\varphi_1''|/\varphi'$ is very small, we can simply find the value of maximal intensity I_{1max} corresponding to the maximum of intensity distribution (26) at $\varphi_1'' = 0$ and $\vec{q}' = 0$.

Diffuse scattering intensity distribution I_{1r} in the reciprocal space in radial direction ($\| \vec{Q}$), can be written as following:

$$I_{1r}(q_{0r}) = Iri \ \frac{1}{2\pi} \int\limits_{-\infty}^{+\infty} d\rho \exp(iq_{0r}) \exp[-T(\vec{\rho}\vec{m})]. \tag{27}$$

Expression (27) describes intensity distribution as a function of q_{0r}. For radial direction it can be also expressed in terms of scattering angle 2θ. Both of these distributions correspond to Lorenz function. Integral width of this curve $2\delta\theta$ and their maximum displacement $2(\theta_m - \theta_0^0)$ relatively to the ideal positions of intensity maximums can be determined for the case of cubic crystal by the expressions

$$2\delta\theta = 2\pi nC\varphi'(\vec{m}, \vec{m})tg\theta,$$

$$\theta_m - \theta_0^0 = nC\left[\frac{\vec{m}\delta\vec{G}}{nCQ} + \varphi''(\vec{m}, \vec{m})\right]tg\theta = \left(-\frac{\Delta V_{im}}{3} - C\varphi_1''\right)ntg\theta = -\frac{\Delta d}{d}tg\theta$$

Here $\Delta d/d$ is the relative change of lattice parameter corresponding to diffuse maximum displacement. Displacement of regular maximum $2(\theta_m - \theta_0^0)$ (or maximum in weakly distorted crystal) is quite different from $(\theta_m - \theta_0^0)$ and is equal to

$$(\theta_m - \theta_0^0) = -\frac{\Delta d}{d}tg\theta = \frac{1}{G}\delta\vec{G}\vec{m}tg\theta = -\frac{1}{3}n\Delta Vtg\theta. \tag{28}$$

It is essential that both half-width and maximum displacement depend on $n\Delta Vtg\theta$

Both regular and diffuse maxima are displaced relatively to their ideal positions but diffuse maximum displacement is approximately six times less.

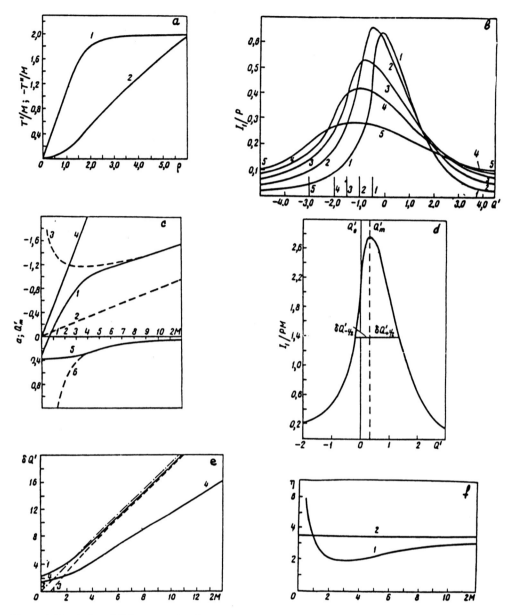

Figure 4. Simulation of diffuse scattering for strongly distorted scrystals with precipitations:
a - real (T'/M), and imaginary (T''/M) parts of function T($\vec{p}$); b -intensity distributions in radial directions for 2M equal -1 (1), -2(2), -3(3),- 4(4), -6 (5); c - dependence of maximums positions q_{Dm}' (1-computer simulation, 2-asymptotic value 0,08 $2M(CQ)^{-1/2}$, 3-asymptotic value with certain additional terms,4-regular reflection q_{D0}', 5- asymmetry parameter,6- asymptotic asymmetry parameter; d - diffuse scattering intensity distribution for small 2M; e - halfwidth intensity dependence in radial direction on 2M: 1-asymptotic value 1,89 * $2M(CQ)^{-1/2}$,2-computer simulation, 3-asymptotic value q_{Di} with additional terms, 4-numerically simulated halfwidth; f- computer simulation of η dependence on 2M

247

Analytical expressions for half-width and maximum displacements can be used only if $2M \gg 1$, but much more frequently experimental data correspond to the condition $2M \sim 1$. In this case we can observe simultaneous existence of both diffuse and regular peaks of intensity. In the region $2M \leq 1$ analytical analysis of intensity distribution is impossible and numerical calculations were used to obtain diffuse scattering intensity distributions.

At Fig. 4a there are shown the results of numerical calculations for real T' and imaginary T'' parts of the parameter T that is necessary to calculate diffuse scattering intensity. The results of computer simulations of diffuse scattering intensity distributions for small $2M$ values are presented at Fig. 4d for the case $\Delta V > 0$. It is displaced from the ideal position that is shown by a straight line. The value of this displacement depends on the value of $2M$. If $V < 0$ these intensity distributions will be situated on the other side of ideal peak position. As it can be seen from the Fig.4, at $2M \sim 1$ intensity distribution is essentially asymmetric and does not correspond to Lorenz shape. While $2M$ increases, curves asymmetry gradually decreases their shape tending to Lorenz one. Half-width and maximum displacement also decrease. The positions of quasiline maxima in radial direction

q'_{Dm} and their integral widths $qr_i = 4\pi\lambda^{-1}cos\theta\delta\theta$, half-widths $q_{r1/2}$, being defined as the distance between points $q'_{D\frac{1}{2}} + \delta^+_{\frac{1}{2}}$ and $q'_{D\frac{1}{2}} - \delta^-_{\frac{1}{2}}$, in which intensity is equal to the half

of maximal values and also asymmetry parameter $a = \left(\delta^+_{\frac{1}{2}} - \delta^-_{\frac{1}{2}} \right) / \left(\delta^+_{\frac{1}{2}} + \delta^-_{\frac{1}{2}} \right)$ were

determined. The results of numerical calculation at $2M = 2...4$ differ essentially from the analytical one. When $2M \to 0$ ($CQ \gg R_0^2$) intensity distribution $I_{1r}(qr)$ becomes symmetric with parameters $q_{rm}' = 0.315(CQ)^{-1/2}$; $q_{r1/2} = 1.525(CQ)^{-1/2}$; $q_{ri} = 2.27(CQ)^{-1/2}$; $Q = 0.38$ ($2M \ll 1$) which do not depend on the defect density, but only on their power. Distance between maxima I_1 and I_0 in the crystals with precipitate is equal to

$$q_{rm} - q_{rm}^{\,0} = \eta_0 nCQ \frac{\Delta V}{|\Delta V|} . \tag{29}$$

Parameter η_0 for the region $2M \sim 1$ was calculated numerically (Fig. 4f). For large values of $2M$ it tends to the constant value 3.515. It makes the distance $q_{Dm} - q_{Dm}^{\,0}$ to depend only on the product nC. If during the aging this product remains the same, then the relative difference between maxima $q_{Dm} - q_{Dm}^{\,0}$ will not change. In the region of not very large $2M \leq 4$ both quasi (diffuse) and regular maxima can be observed simultaneously and η_0 strongly depends on $2M$. So when $2M$ is not large the displacement of quasi-reflections depends not only on the product nC, but also on n and C separately.

Experimental Diffuse Scattering Analysis. Such intensity redistribution changes with the increase of defect power were observed experimentally in the region $2M \sim 1$ for Ni-based aging single crystals corresponding to the negative sign ΔV. X-Ray intensity distributions were received for the samples after solidification, quenching from the temperature 1240^0 C and different time of aging at 900^0 C [18]. Both structure intensity distributions (002), (004), (111), (222), (220) for γ-phase and superstructure -(001), (003), (110) for γ'-phase in the $(1\bar{1}0)$ reciprocal lattice plane were restored. Intensity distributions were examined both in radial ($\parallel \vec{G}$) and azimuth (perpendicular to $\vec{G}$) directions around (222) reflection.. In Fig. 5 there are shown space intensity distributions around (222) reflections and corresponding isodiffuse curves for different aging time. This intensity distribution always consists of two maxima. The azimuth section ($\perp \vec{G}$) of the reflection is the same for both I_0 and I_1 (Fig. 5). Intensity distribution is stretched along $\vec{G}$ and has complicated shape in radial direction. In azimuth direction it is relatively narrow and does not change along the whole reflection. Parameter $2M$ depends on $(h_i^2)^{3/4}$ (Miller indices).It means that in one sample different reflections may correspond to different $2M$ values. It will result in different contribution of I_0 and I_1 to the final intensity distribution. Therefore the resulting intensity distributions for the first (111) and the second (222) orders of the same diffraction maximum are not similar. The angle position of I_0 and I_1 depends on the sign of ΔV being the characteristics of volume change between matrix and precipitates. In our experiments I_1 was always situated at the smaller $\vec{Q}$ values relatively to I_0 . It corresponds to the negative $\Delta V < 0$.

Parameter η_0 in the region $2M \approx 1$ was determined using the results of numerical calculations (Fig. 4f). In this $2M$ region it had values $\eta_0 \approx 2...2,2$. This allows us to determine the product nC. Taking into account the dependence of M on ΔV we determined the main parameter for dispersively hardened alloys $n\Delta V = p\Delta v / v = -(2-2.3) \cdot 10^{-2}$. The dependence of width and displacement of reflections on the modulus and orientation of vector $\vec{Q}$ is in good agreement with theoretical dependences. The ratio $2\delta\theta/tg\theta$ also depends on the direction of $\vec{Q}$ and is equal to for (420), (331) and (222) reflections: 1.92:1.66:1.22:1. The dependence of reflection integral width on $\vec{Q}$ orientation is mainly a result of anisotropy of matrix and precipitates.. Aging results first of all in the displacement of ideal peak θ_0^0 position. This influences the positions of I_1 and I_0 . That is why aging displaces both regular and diffuse maxima. Such changes are explained by formulas (16)-(28). Correspondingly, the effective values of lattice parameters calculated for regular and diffuse peaks will also differ. It is especially essential when these both maxima are observed simultaneously. The same type of effective lattice parameters dependences were observed by other authors [19,20].

CONCLUSIONS

1.   The shape of isodiffuse curves for crystals with isolated defects essentially depends on the defect force parameter b. For $b \sim 1$ defects are almost cubic and isodiffuse curves for different type defects are very similar. For non-cubic defects b differs essentially from 1 and is often negative. It enables us to determine defect type and b value from the data on diffuse scattering intensity.

2.   The presence of precipitates causes local lattice distortions due to the existence of lattice misfit between matrix and forming phase and the difference of structure amplitudes between precipitate and matrix. The type of intensity distribution depends on these mutual effects.

3.   In weakly distorted crystals with precipitates ($2M << 1$) the interference of scattering caused by the difference of structure amplitudes and distortions for elastically anisotropic crystals results in the asymmetry of isodiffuse surfaces. The intensity becomes much higher at one side of reciprocal lattice point and the latter are no longer the inversion centers for intensity distribution. The extent of asymmetry depends on value and sign of parameter C.

4.   When precipitates in weakly distorted crystals are surrounded with depletion layers isodiffuse surfaces do not pass through reciprocal lattice points but surround them. The shape of isodiffuse curves becomes crescent type.

5.   In strongly distorted crystals with precipitates belonging to the region $1 \le 2M$ both regular and diffuse maxima exist simultaneously. Diffuse scattering intensity is stretched along the diffraction vector. Positions of regular and diffuse maxima

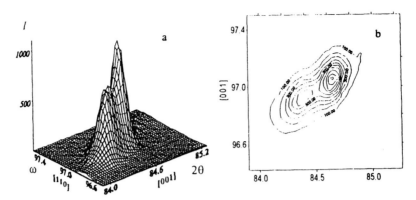

Figure 5. Experimental space intensity distributions around (222) reflection for Ni-based single crystal alloy after ageing 10 hours at 900 ^{o} C(a) and isodiffuse curves in $(1\bar{1}0)$ plane (b) in Cu K_{β} radiation

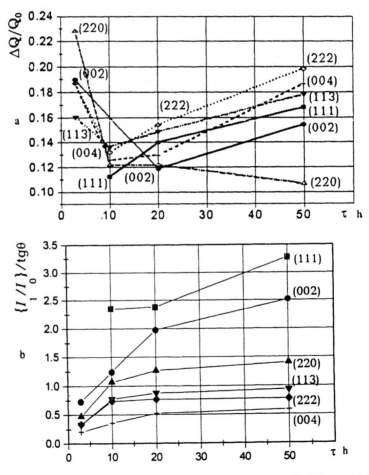

Figure 6. Relative difference between diffuse and regular maximums positions in different reflections for Ni-based single crystal depending on ageing time at 900 ^{o} C(a) and intensity ratio normalized to $tg\theta$ (b)

relatively to the ideal Bragg position depend on sign and value of volume change due to the precipitates.

6.   Experimentally reciprocal lattice space was restored around structure matrix and superstructure γ-phase points. The existence of both diffuse and regular reflections was observed. Average volume change caused by precipitates was calculated.

References

1. M.A. Krivoglaz. X-Ray and Neutron Diffraction in Imperfect Crystals., Naukova Dumka, Kiev (1983), English Transl. Springer,New York,1996.
2. P.H.Dederichs,J.Phys.F.,3:471 (1973).
3. G. Bauer, E.Seits,J.Appl.Cryst.,8:162 (1975).
4. B.C. Larson, *J.Appl.Cryst.,8:*150 (1975).
5. H. Peisl, *J.Appl.Cryst.,*8:143 (1975).
6. W. Schmats, *Treties on Material Science and Technology,*2:105 (1973).
7. H. Trinkaus, *Phys. Status Solidi,*51:307 (1972).
8. R.I. Barabash, M.A. Krivoglaz, *Metal Physics,*1:33 (1979).
9. M.A. Krivoglaz, *FMM,*9:641 (1960).
10. M.A. Krivoglaz, *FMM,*13:17 (1962).
11. V.V. Kokorin, K.V. Chuistov, *FMM,*27:1067 (1969).
12. V.V. Kokorin, K.V. Chuistov, *FMM,*27:804 (1969).
13. R.I.Barabash, M.A. Krivoglaz, *Metal Physics,*5:100 (1983).
14. A.V. Dobromislov, V.V. Glebov, N.N.Buinov, *Phys.Stst.Sol.,* A,201:87 (1973).
15. A.V. Dobromislov, V.V.Glebov, *Sov. Phys. Stat. Sol.,* 15:1245 (1973).
16. R.I. Barabash, M.A. Krivoglaz, *FMM,* 45:7(1978).
17. R.I. Barabash, M.A. Krivoglaz. *FMM,* 51:903 (1981).
18. R.I. Barabash, A.D. Rud, S.I. Sidorenko etc., *Metal physics and new technology,*18:1(1996).
19. H. Biermann, H.A. Kuhn, T. Ungar, J. Hammer, H. Mughrabi, in: *Proceedings of "9th International Conference on Strength of Metals and Alloys"*, London, England,421 (1991).
20. M.V. Nathal, R.A. Mackay, R.G. Garlick, *Material Science and Engineering*, 75:195(1985).

SHORT-RANGE DISORDER AND LONG-RANGE ORDER: IMPLICATIONS OF THE "RIGID UNIT MODE" PICTURE

Martin T. Dove,[1] Volker Heine,[2] Kenton D. Hammonds,[1,2] Manoj Gambhir,[1,2] and Alexandra K. A. Pryde[1,2]

[1] Department of Earth Sciences, University of Cambridge, Downing Street, Cambridge CB2 3EQ, UK
[2] Cavendish Laboratory, University of Cambridge, Madingley Road, Cambridge CB3 0HE, UK

INTRODUCTION

Our purpose in this contribution is to describe a set of ideas we have been developing in Cambridge which give a number of insights into the structures and stabilities of silicate phases, and the phase transitions between different structures. The basic theory is called the "Rigid Unit Mode" (RUM) model, and the key papers can be found in references 1–8. The RUM model has been used to understand issues of silicate structures such as the existence of displacive phase transitions,[1–11] the origin of the transition temperature,[1,2,4,7,8] the reason why these phase transitions are well-described by Landau theory,[1,7,8] incommensurate phase transitions,[9,10] the role of critical fluctuations,[7,8,11] the nature of high-temperature phases,[4–8,12–16] diffuse scattering in silica polymorphs,[3,5,7,17] negative thermal expansion,[18–21] zeolite catalysis,[22,23] and the excitation spectra of glasses.[24]

The RUM model is very simple in its basic ideas. The atomic structures of crystalline silicates consist of SiO_4 and AlO_4 tetrahedral structural units which are linked together at corners. In many of the important silicates, such as quartz, the feldspar family of minerals, and the zeolites, these tetrahedra are linked together to form an infinite three-dimensional network. The energies associated with distortions of the tetrahedra, as indicated by their vibrational frequencies, are much stiffer than other forces in the crystal structure (e.g. the forces that come into play when two tetrahedra swing about a common vertex). To an apparently 'crude' first approximation we can model the force constants in the crystal as having either of two values, namely a very large value for the force constants within the individual tetrahedra, and zero for all other force constants. Any harmonic phonon in the crystal will then have a frequency that reflects the extent to which its eigenvector involves infinitesimal deformations of the tetrahedra, and a phonon that can propagate without any tetrahedra having to distort will have zero frequency. These zero-frequency phonons are the "rigid unit modes" whose existence underlies the relevance of the RUM model.

Local Structure from Diffraction
Edited by S.J.L Billinge and M.F. Thorpe, Plenum Press, New York, 1998

The relevance of the RUM model for the present workshop is that any large amplitude deformations of a crystal structure, whether as static or phonon deformations, will be associated with low-energy modes of distortions: from harmonic phonon theory,[25] the square of the amplitude of an atomic displacement is proportional to the square of the harmonic frequency:

$$\left\langle |u|^2 \right\rangle = \frac{k_B T}{\omega^2} \tag{1}$$

These deformations are those that give rise to the strongest diffuse scattering, and as noted above the RUM model has been used to explain the strong diffuse scattering observed in the silica polymorphs cristobalite[26,27] and tridymite[17,28] in terms of the specific RUM phonons in these systems. Conversely, the RUM model can also give insights into the possible structural states where diffuse scattering indicates some considerable degree of structural disorder.

In this article we will review the basic ideas of the RUM model, and give a number of examples that illustrate the application to the structural disorder and diffuse scattering.

BASICS OF THE RUM MODEL

The existence of RUMs in crystalline silicates

The possibility of having RUM deformations in crystalline silicates has been recognised for some time.[29-31] For example, the structural instabilities in quartz[29] and feldspars[30] were analysed in terms of RUMs (although in terms of static deformations rather than phonon modes) some time ago, and the possible existence of RUM phonons was subsequently recognised for specific wave vectors in quartz[31] and cristobalite[26,27]. The RUM model was first suggested to us by an analysis of the incommensurate instability in quartz by Vallade and co-workers,[32-36] who demonstrated that RUMs could exist for some definite ranges of wave vectors. Our own contribution was to generalise these ideas, but before we describe this we need to discuss some general issues associated with the possible existence of RUMs.

The RUM model has been developed independently in the context of the excitations in glasses, where the equivalent zero-frequency modes are called "Floppy Modes".[37-39] In this context the basic idea is to consider any chemical bond, rather than a structural unit such as a tetrahedron, to be a rigid entity.[39] Each atom has three degrees of freedom, and each bond enforces a single constraint. In the absence of any other forces, the number of zero-frequency modes of motion in a glass or crystal is given classically by the difference between the total number of degrees of freedom in the system, F, and the total number of constraints, C, given by the chemical bonds, ensuring that proper account is taken of bonds that give identical constraints. For silica, an SiO_4 tetrahedron contains 10 bonds (Si–O and O–O), but only 9 are required to define the rigidity of the tetrahedron. Thus for each SiO_2 unit there are 9 constraints and 9 degrees of freedom, and so there will be no zero frequency modes. In our approach we count in a slightly different way,[3,7] but the end result is the same. We associate 6 degrees of freedom with each tetrahedron (rotational and translational), and 3 constraints with each shared vertex. Thus each tetrahedron has 6 constraints, and we obtain the same result that $F - C = 0$.

This then leads to an apparent contradiction: the simple counting of constraints suggests that there should be no RUMs in any crystalline silicate with full three-dimensional connectivity, but we have noted above that the existence of *some* RUMs in silica polymorphs had already been demonstrated. It is clear that the contradiction will have

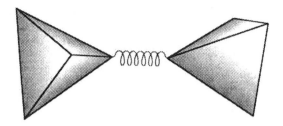

Figure 1. The split-atom representation of a pair of linked tetrahedra. An oxygen atom that is shared by two tetrahedra is represented as a pair of atoms with an equilibrium separation of zero. The spring represents the force required to separate the split atoms, and is analogous to the forces acting when two linked tetrahedra move in a manner that requires the distortions of the tetrahedra.

to be understood as a result of some of the constraints not being independent, and we have demonstrated elsewhere that symmetry plays an important role in this context.[3,7] For a system containing N atoms, and hence $3N$ normal modes, we have found that in general the number of RUMs will be of order $N^{1/3}$ or $N^{2/3}$, which is a small number compared with $3N$, and probably broadly in line with one's initial expectations.[5] However, more recently we have applied the RUM approach to zeolites, and in some of these cases we find that the number of RUMs is of order N,[22,23] which is certainly counter-intuitive.

The calculation of RUMs

Before we discuss the implications of the RUM model, we should outline the methods we use to determine the number of RUMs for any given system. Any approach must properly account for the effect of symmetry on the number of independent constraints, and for crystalline materials it would be useful to be able to determine the wave vectors of the RUMs. Vallade proposed an analytic method for quartz,[36] but we felt that this was not easily programmed in a general form. Our approach has been to use a molecular lattice dynamics method[25,40] with perfectly rigid tetrahedra. In order to develop the lattice dynamics formalism we have introduced the "split-atom" method,[3,7] which is illustrated in Figure 1. The basic idea is to split the atoms that are shared by two tetrahedra into two separate atoms, one for each tetrahedron, with an equilibrium separation of zero. A pair of split atoms are held together by inventing a harmonic force between them, and the force constant is related to the actual stiffness of the tetrahedra since the separation caused by relative displacements and rotations of two linked tetrahedra is proportional the distortions of the tetrahedra that would otherwise occur. The dynamical matrix for this system will give zero frequencies for any phonon mode that does not cause distortions of the tetrahedra: these zero-frequency solutions represent the rigid unit modes for the system. These calculations can be performed for any wave vector. The results can include the eigenvectors of the RUMs, and the calculation of the neutron scattering intensity for a range of scattering vectors. The method is not restricted to the use of tetrahedra, and will work for polyhedra of any shape with any number of neighbours.

The package of programs for the RUM analysis of any material,[41] together with manuals and test data, are freely available for any academic user, and can be accessed using a WWW browser from http://www.esc.cam.ac.uk/mineral_sciences/crush/. The package includes the basic rigid unit mode program, which we have called CRUSH, in two forms: a

Table 1. Numbers of rigid unit modes for symmetry points in the Brillouin zones of some aluminosilicates, excluding the trivial acoustic modes at $k = 0$. The "—" indicates that the wave vector is not of special symmetry in the particular structure. The numbers in brackets denote the numbers of RUMs that remain in any lower-symmetry low-temperature phases. (Taken from Reference 5.)

k	Quartz $P6_222$	Cristobalite $Fd3m$	Tridymite $P6_3/mmc$	Sanidine $C2/m$	Leucite $Ia3d$	Cordierite $Cccm$
$0,0,0$	1 (0)	3 (1)	6	0	5 (0)	6
$0,0,\frac{1}{2}$	3 (1)	—	6	1	—	6
$\frac{1}{2},0,0$	2 (1)	—	3	—	—	6
$\frac{1}{3},\frac{1}{3},0$	1 (1)	—	1	—	—	6
$\frac{1}{3},\frac{1}{3},\frac{1}{2}$	1 (1)	—	2	—	—	0
$\frac{1}{2},0,\frac{1}{2}$	1 (1)	—	2	—	4 (0)	2
$0,1,0$	—	2	—	1	—	—
$\frac{1}{2},\frac{1}{2},\frac{1}{2}$	—	3 (0)	—	0	0	—
$0,1,\frac{1}{2}$	—	—	—	1	—	—
$0,0,\xi$	3 (0)	2 (0)	6	—	0	6
$0,\xi,0$	2 (0)	2 (2)	3	1	0	6
$\xi,\xi,0$	1 (1)	1 (0)	1	—	4 (0)	6
ξ,ξ,ξ	—	3 (0)	—	—	0	—
$\frac{1}{2},0,\xi$	1 (0)	—	2	—	0	2
$\xi,\xi,\frac{1}{2}$	1 (1)	—	0	—	0	0
$\frac{1}{2}-\xi,2\xi,0$	1 (1)	—	1	—	—	6
$\frac{1}{2}-\xi,2\xi,\frac{1}{2}$	1 (1)	—	0	—	—	0
$0,\xi,\frac{1}{2}$	0 (0)	—	1	1	—	—
$\xi,1,\xi$	—	1 (0)	—	—	0	—
$\xi,\zeta,0$	1 (0)	0	1	—	0	6
$\xi,0,\zeta$	0 (0)	0	2	1	0	0
$\xi,1,\zeta$	—	0	—	1	0	—
ξ,ξ,ζ	0 (0)	1 (0)	0	—	0	0

standard version that has been vectorised to allow it to operate efficiently on a vector computer, and a parallel version that uses MPI routines to allow calculations to be performed for a three-dimensional grid of wave vectors. Also included are programs to set up a CRUSH input file (including the ability to idealise the size and shapes of tetrahedra) and a number of other ancillary and analysis programs. All programs have been written in FORTRAN77 in a portable style, but some need to be linked to the NAG library.

RUMs in crystalline silicates

We have determined the number of RUMs in a number of crystalline silicates,[5] and some representative examples for some of the simpler systems we have investigated are given in Table 1. The results of our calculations are restricted to special points, lines and places of special symmetry in the Brillouin zone. For all these systems there are no RUMs for a general wave vector; examples where we have to consider the general wave vectors will be given later in our section on zeolites.

The results of Table 1 show a number of features that appear to be of general relevance. First, in all these systems there are *some* RUMs, rather than the zero number suggested by the simple constraint counting discussed earlier. In all cases there are RUMs for wave vectors along special lines in reciprocal space, and in some cases there are RUMs for wave vectors on planes in reciprocal space. Second, in cases where there is a phase transition that is easily analysed, the number of RUMs in the lower-symmetry phase is smaller than in the higher-symmetry phase. This is particularly marked for the planes of RUMs in the high-symmetry phases of quartz and cristobalite, and in leucite none of the RUMs survive the phase transition. In the lower-symmetry phases the motions that are associated with the RUMs now necessitate distortions of the tetrahedra, and therefore have a non-zero frequency. This result is consistent with our explanation that it is the crystal symmetry causing some of the constraints in the crystal to be equivalent that allows a non-zero number of RUMs in any particular case, since the lowering of the symmetry is accompanied by a decrease in the number of RUMs.

Table 1 does not give all the results for each system. To our surprise we have found that the RUMs are not restricted to have wave vectors of special symmetry. We will show an example of this in our section on tridymite. For many systems, including some of the examples given in Table 1, and specifically noting that quartz is one of these, the RUMs can exist with wave vectors on curved lines or on the surfaces of exotic shapes in reciprocal space. This result suggests that it is generally important to search for RUMs for all wave vectors in the Brillouin zone, rather than for a few representative examples, and this precludes the use of hand calculations of the form that Vallade attempted for quartz.

It is interesting to ask whether RUMs can also exist in materials containing octahedral units. For example, it is known from experiment and by intuitive reasoning that there are lines of RUMs along the edges of the Brillouin zone in the cubic phase of the perovskite structure. However, we have investigated a number of materials containing octahedra, and in general we found that unless the octahedra are aligned with the symmetry axes there is rarely enough symmetry to cause enough of the constraints to become equivalent.[42] We therefore conclude that cubic perovskite is a special case, which perhaps is not surprising since there are not many other ways to pack octahedra together in a way that allows the orientations of the octahedra to align with symmetry axes. The cases of mixed octahedra and tetrahedra are no better in this respect, unless, as in the new ceramic ZrW_2O_8 with negative thermal expansion, some of the vertices of the octahedra or tetrahedra are not shared with other units so that the topology itself lowers the overall number of constraints.[19,20]

CRISTOBALITE

Disorder in the high-temperature phase of cristobalite

The phase transition in the silica polymorph cristobalite has been studied by a number of workers in recent years. Most of the work has been motivated by the fact that "there is something wrong" about the crystal structure of the high-temperature β-phase as deduced by analysis of the Bragg reflections.[43] The crystal structures of the two phases of cristobalite are shown in Figure 2.[43] In the low-temperature α-phase the angle subtended by the Si–O–Si bond is around 149°, and the length of the Si–O bond is around 1.60 Å. These are typical values for crystalline and amorphous silicates. But in the β-phase, the average positions of the atoms give an Si–O–Si angle of 180° and an Si–O distance of 1.54 Å. Moreover, the temperature factors of the central oxygen in the Si–O–Si linkage is elongated in the directions perpendicular to the Si–Si vector.[43] Of course, it is easy to see the cause of the problem. The Bragg diffraction gives information only about *average positions*, and not about *average distances* between atoms. This latter information is found in the total scattering data, and analysis of the total scattering data for β-cristobalite shows that actually the bond angles and distances are quite different from those suggested by the average positions and more like the typical values as found in the α-phase.[15] It might be expected therefore that the crystal structure of β-cristobalite is disordered, with the Si–O bonds tilted

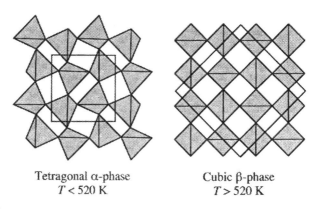

Tetragonal α-phase
T < 520 K

Cubic β-phase
T > 520 K

Figure 2. Representations of the α-phase (left) and β-phase (right) of cristobalite, showing the SiO_4 tetrahedra as solid objects. The phase transition is represented as occurring through rotations of the tetrahedra. However, in this representation of the structure of the β-phase the tetrahedra are aligned to give linear Si–O–Si bonds, which is an unfavourable alignment as discussed in the text.

away from the Si–Si vector at an angle of around 17°. Quite how this might happen has been the subject of many of the recent studies.

The RUM model has allowed us to think again about the nature of the disorder in the β-phase of cristobalite. There have been two previous propositions concerning the actual structure of this phase,[44,45] both of which assume that β-cristobalite is built from domains of a lower symmetry structure. We mention here the more intuitive model of Ghose and Hatch, namely that the structure is composed of domains of α-cristobalite.[45] This interpretation of the structure of the β-phase must now be discounted following the analysis of the total scattering data,[15] but it is interesting from one aspect, namely that it provided an *intuitive* interpretation of the disorder. It has proved to be very difficult to imagine many alternative models.

Rigid unit modes and dynamic disorder in cristobalite

The new insights into the nature of the β-phase come from the fact that there are RUMs with wave vectors lying in planes in reciprocal space. The results of our calculations are given in Table 1, but these results were predated by measurements of streaks of diffuse scattering in electron diffraction.[26,27] These streaks were shown to be fully consistent with the RUMs, and no significant features of diffuse scattering were found that could not be explained by the existence of RUMs.[3,5] The RUM at the [1,0,0] zone boundary of the Brillouin zone could provide the instability for the phase transition if it were considered from the viewpoint of the "soft mode" theory.[1–7,12,13] Another RUM at zero wave vector provides the instability that would generate the domains of the alternate domain theory of Wright and Leadbetter.[5,12,13] It is easy to imagine that if all these RUMs are operating with low frequency, they will all rotate the tetrahedra away from their average orientations with large amplitude, with the result that the crystal structure will appear to be disordered. The RUMs are dynamic phonons rather than static displacements, and therefore the disorder will be dynamic rather than static.

Experimental evidence

We then have to ask how well the RUM interpretation of the disorder of β-cristobalite stands up to further scrutiny. We have already mentioned one test, namely that the measurements of the diffuse scattering in electron diffraction are in good agreement with the

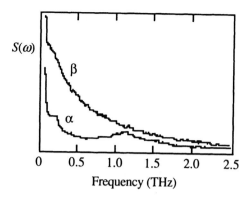

Figure 3. Inelastic neutron scattering data for the two phases of cristobalite. There is a large increase in the low-energy spectrum for the β-phase, suggesting the presence of low-frequency rigid unit modes in this phase.

predictions of the RUM model. Further verification of the existence of RUMs is given by inelastic neutron scattering measurements,[12] shown in Figure 3. The data for the inelastic signal in both phases of cristobalite show that there is a significant enhancement of the number of low-energy excitations (frequency below 1 THz) in the β-phase, which is consistent with the calculated changes in the numbers of RUMs in the two phases given in Table 1. On the other hand, it has been pointed out that the RUM model is strictly only valid within the harmonic approximation, which implies only infinitesimal RUM rotations of tetrahedra, and the model cannot assume that when whole planes of RUMs are added together simultaneously the model will retain its validity.[46] Against this objection it might be argued that the electron diffraction and inelastic neutron scattering data suggest that the RUM model remains valid even when a large number of RUMs are activated, but these data cannot rule out the possibility that a small subset of RUMs will dominate over all others. If this situation should occur, it may well map back onto the domain model.

In order to check this possibility we have performed two types of molecular dynamics simulations using samples with 4096 unit cells. (As an aside it should be noted that for this sort of analysis molecular dynamics simulations have to be carried out with large samples, since a good sampling of reciprocal space is required in order to differentiate between the situation where phonon modes of a wide range of wave vectors are activated with more-or-less equal amplitude, and the alternative situation where the behaviour is dominated by only a few phonon modes.) The first simulations were performed using realistic model interatomic potentials,[14] and these showed that the disorder was not consistent with any preferred directions of the Si–O bonds as would be suggested by a domain model. The second simulations were performed using the split-atom method described above,[16] using a fixed volume and Si–O bond length that would require tilting of the tetrahedra by about 17°. This one-parameter model was investigated in detail to determine whether all RUMs could be activated equally, or whether a small subset of RUMs would dominate behaviour. The latter situation would then lead to long correlation lengths. The results of the analysis showed quite clearly that all RUMs in β-cristobalite were activated more-or-less equally (some had higher amplitude, but not significantly so) at all temperatures. A snapshot of a (1,1,1) layer of one configuration is shown in Figure 4a. It can be seen from this figure that the distortions of the local structure differ in different regions of the crystal, showing that there is no domain pattern but that there is disorder without distortions of the tetrahedra. It is important to appreciate one point from this snapshot image. It appears as if the distortions of the structure are fairly random, but they are not. They are almost completely determined by the very small subset of normal modes of the system that are RUMs. The picture is

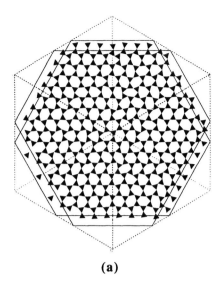

(a)

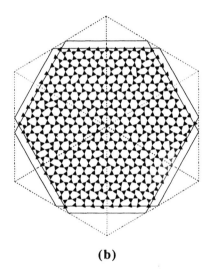

(b)

Figure 4. Snapshots of single configurations of (111) planes of SiO$_4$ tetrahedra in β-cristobalite obtained from the split-atom molecular dynamics (a) and the RMC analysis of total neutron scattering (b).

reinforced by calculations of the scattering function $S(\mathbf{q})$ in the one-parameter molecular dynamics simulation for wave vectors along [0,0,1], for which there are two RUMs. The scattering is indeed dominated by the scattering from the low-frequency RUMs, and in Figure 5 we show the results from calculations over a whole decade of temperatures (100–1000 K). It is apparent that the intensity of $S(\mathbf{q})$ does not have a significant temperature dependence, which is exactly as would be expected if the scattering comes from a disordered state with constant amplitude of disorder rather than from thermally-activated fluctuations such as harmonic phonons. Indeed, calculations of $S(\mathbf{q})$ for wave vectors with no RUMs has a normal temperature dependence.

Our final test of the RUM interpretation of the nature of β-cristobalite comes from our analysis of the total neutron scattering data.[15] In the β-phase there is a considerable amount

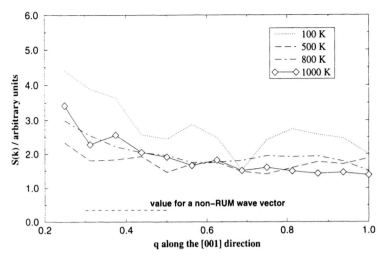

Figure 5. Calculation of the scattering function $S(\mathbf{q})$ in the one parameter model of β-cristobalite for four temperatures.

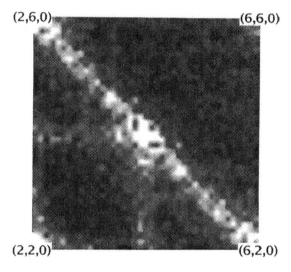

Figure 6. Calculated [001] section of diffuse scattering in β-cristobalite constructed from the RMC configurations. The streaks are consistent with the RUM predictions and electron diffraction.

of diffuse scattering from powder samples. This has been analysed using the Reverse Monte Carlo (RMC) approach with constraints consistent with the pair distribution function extracted from the data, and the results are described in more detail in the chapter by David Keen. Here we cite three essential results. The first is that the orientations of the Si–O bonds are consistent with a model that allows rotational freedom. The second is that a snapshot of the atom positions of a single configuration, Figure 4b, has a very similar appearance to that given by the split-atom simulations, with no signs of any domains. The third is from calculations of the three-dimensional diffuse scattering using the RMC results. A sample of the results is shown in Figure 6. The streaks of strong diffuse in this section of reciprocal space are fully consistent with the RUM analysis and the earlier electron diffraction results.[26,27]

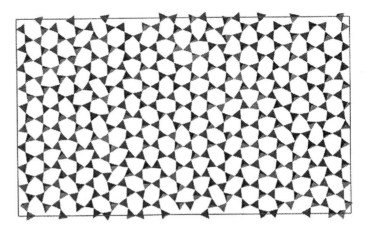

Figure 7. Snapshot of a single configuration of the (0,0,1) plan of the high-temperature phase of tridymite generated by the RMC analysis of total neutron scattering data.

TRIDYMITE

Crystal structure and disorder

The crystal structure of tridymite, another polymorph of silica, is very similar to that of cristobalite, in that the two materials are made from sheets of tetrahedra linked in a hexagonal array. In cristobalite these sheets are parallel to the cubic (1,1,1) planes, and are stacked together in an "ABCABC" manner analogous to the stacking of close-packed layers in atomic cubic-close-packed structures. In tridymite the stacking is more analogous to an "ABAB" stacking, and this yields a hexagonal crystal structure with the layers parallel to the (0,0,1) planes. Tridymite undergoes a relatively large number of phase transitions on cooling,[47] which have been studied by a large number of workers. There are problems in that some of the phase transitions are affected by the quality of the sample (tridymite invariably contains a large number of stacking faults), and different workers have identified a plethora of incommensurate phases with long-period modulations.

The average atomic positions in the high-temperature phase of tridymite as determined by the analysis of Bragg diffraction data yields the same linear Si–O–Si connectivity as we have discussed in cristobalite above, and again the RUM picture can explain the dynamic disorder that allows neighbouring tetrahedra to rotate to more favourable orientations without destroying the long-range order.[5] The RUMs in the hexagonal phase of tridymite are given in Table 1, and as in cristobalite there are planes of RUMs in reciprocal space. The short-range structure of tridymite has been analysed using total neutron scattering measurements and RMC (see the chapter by David Keen). In Figure 7 we show a snapshot of one layer of a configuration generated by the RMC analysis, and this should be compared with the snapshot of cristobalite in Figure 4 (these are the same basic layers). The snapshot of tridymite shows a similar disordered arrangement of the orientations of the tetrahedra, strinkly similar to that of cristobalite (Figure 4).

Diffuse scattering

As in cristobalite, the RUMs in tridymite can be identified in diffuse scattering using electron diffraction.[28] Indeed, the RUMs given in Table 1 have all been identified this way.[17] However, the electron diffraction data for the high-temperature phase showed a new

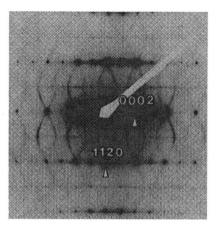

Figure 8. Diffuse scattering from tridymite seen in the electron diffraction from a single section of reciprocal space of the high-temperature phase of tridymite.

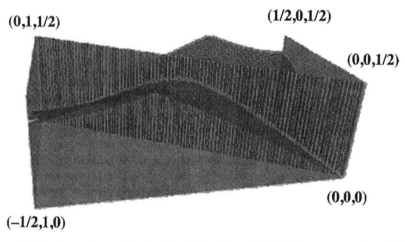

(0,1,1/2)   (1/2,0,1/2)   (0,0,1/2)   (0,0,0)   (−1/2,1,0)

Figure 9. Equi-value surfaces of the function $S(\mathbf{q})$ calculated for the high-temperature phase of tridymite.

feature. An example section of reciprocal space is shown in Figure 8. As well as containing lines of diffuse scattering that correspond to planes of RUMs intersecting this section of reciprocal space, the diffraction pattern also contains a repeated curved feature of diffuse scattering. We were able to confirm that this feature also arises from RUMs. This was the first indication that RUMs can exist with wave vectors lying on an exotic surface in reciprocal space. The full three-dimensional surface has been obtained by performing calculations with CRUSH using a fine three-dimensional grid in reciprocal space, and constructing the scattering function in which the summation is over all the phonon modes, not just the RUMs:

$$S(\mathbf{q}) = \sum_{n} \frac{1}{\omega_n^2(\mathbf{q}) + \Omega^2} \tag{2}$$

where Ω is a small number whose function is to prevent the divergence of $S(\mathbf{q})$ when sampling a wave vector containing a RUM with zero frequency. A three dimensional image can then be obtained by visualising equi-value surfaces of the function $S(\mathbf{q})$. The result for the high-temperature phase of tridymite is shown in Figure 9, and this object gives exactly the same curves in the section corresponding to the diffraction pattern in Figure 8. We have performed a similar analysis for a number of materials, including the examples given in Table 1, and we have found that curved features (whether lines or surfaces) are surprisingly

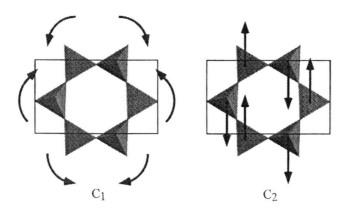

C_1 C_2

Figure 10. Representations of the RUM eigenvectors that determine the two component structures of OS-tridymite.

common. It would be interesting to try to observe some of these additional curved RUM surfaces using electron diffraction. An exotic curved surface of RUMs has also been found in the ceramic ZrW_2O_8.[19,20]

Incommensurate phase transitions

On cooling tridymite undergoes phase transitions to two other hexagonal phases, a C-centred orthorhombic phase called OC, an incommensurate phase called OS, and then to a primitive orthorhombic superlattice structure called OP.[48] There are additional phase transitions on further cooling, but the details are dependent on the quality of the samples. A RUM search was carried out on the OC-phase, and it was found that there is a line containing one RUM with wave vectors along $\mathbf{a}*$, and the eigenvector of the RUM at the incommensurate wave vector $\mathbf{q}_{IC}$ along $\mathbf{a}*$ will fix the incommensurate structure.[10]

An incommensurate structure may be described as the sum of two modulated distortions imposed on the parent structure:[48]

$$\text{Modulated structure} = \text{Parent structure} + A\left(\mathbf{C}_1 \cos\mathbf{q}_{IC}.\mathbf{r} - \mathbf{C}_2 \sin\mathbf{q}_{IC}.\mathbf{r}\right) \qquad (3)$$

where $\mathbf{C}_1$ and $\mathbf{C}_2$ are two component difference structures with a unit cell equal to that of the parent phase. Strictly speaking the difference structures involve infinitesimal displacements and so an amplitude factor A is required to give the modulation an appropriate magnitude. This description is of great use in visualising the incommensurate structure, and it also has important implications for its symmetry. The symmetries of $\mathbf{C}_1$ and $\mathbf{C}_2$ are uniquely determined by the RUM, and this limits the possible symmetries of the resulting incommensurate structure. The distortions associated with $\mathbf{C}_1$ and $\mathbf{C}_2$, which were obtained from the RUM phonon eigenvector, are shown in Figure 10.

The structure of the OS-phase of tridymite can be determined, in principle, for any incommensurate modulation by using the RUM eigenvectors for the corresponding wave vector. To test the correctness of this approach we note that calculations using $\mathbf{q}_{IC} = \mathbf{a}*/3$ should give the structure of the three-fold lock-in phase OP-tridymite.[49] The result of this construction is shown in Figure 11, where we compare the predicted and measured structures of the OP-phase. The match between the two structures is quite reasonable. However there are small differences because the structure can relax further within its space group symmetry. This has the effect firstly of squaring-up the $\pm\mathbf{C}_1$ and $\pm\mathbf{C}_2$ regions from a sinusoidal modulation. Secondly it expands the size of the $\pm\mathbf{C}_1$ region, although not at the

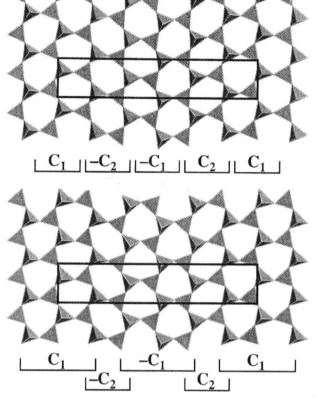

Figure 11. The construction of the OP phase of tridymite (top) and the experimental structure (bottom) showing the regions of the two component structures.

expense of the $\pm C_2$ regions but rather overlapping with them as shown in Figure 11. This overlapping of C_1 and C_2 may account for the special stability of the three-fold superlattice and hence for the discontinuous change in q_{IC} at the OS / OP phase boundary.

ZEOLITES

Crystal structures and rigid unit modes

The examples of cristobalite and tridymite have shown how RUMs can be superposed on a crystal structure dynamically, giving a dynamically disordered structure, and we have shown how this leads to strong one-phonon diffuse scattering. The other major source of diffuse scattering is from static defects, and in the case of zeolites we have shown that localised deformations of the crystal structure can arise from superpositions of static RUM distortions.[22,23]

The crystal structure of one common zeolite, LTA, is shown in Figure 12, where we highlight the way in which the tetrahedra are linked in such a way to form large cavities in the structure which may be large enough to hold small molecules. It should be stressed that the connectivity of tetrahedra is as complete as in the denser silicate phases, and there are no non-bridging Si–O or Al–O bonds. Another way to view this and other zeolite structures is by comparison with the crystal structure of sodalite. This material has a cavity that is

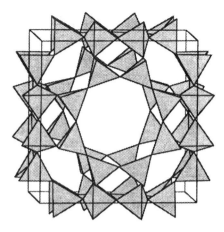

Figure 12. The crystal structure of Zeolite-LTA showing the framework of linked tetrahedra and the large internal cavity.

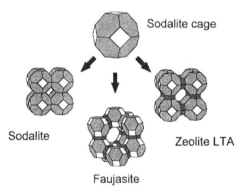

Figure 13. The formation of a number of zeolite structures from the basic sodalite cage.

incorporated into the crystal structures of some other basic zeolites, as illustrated in Figure 13. The difference between these structures lie in the topological arrangements of the basic sodalite cavities and in the channels between the cavities.

The point about highlighting the analogy with sodalite is that we initially found that the highest-symmetry version of the sodalite structure has one RUM for each wave vector,[5] or put another way, there is a complete band of RUMs in the three-dimensional reciprocal space. This suggests that the sodalite structure has an increased level of inherent flexibility. Our calculations then showed that both Zeolite-LTA and Faujasite (Zeolite-Y) have four RUMs per wave vector (i.e. four bands of RUMs), and we have found a number of other examples of zeolites that have a large number of RUMs; these are given in Table 2.[22] Since in these cases the number of RUMs is a significant fraction of the total number of normal modes, we give this fraction instead of the actual number of RUMs. In the case of zeolites the phonon density of states calculated with our CRUSH program shows a continuum of normal modes at low energies, as shown in Figure 14. The lower-energy modes cause only minimal distortions of the tetrahedra, and we have called these "Quasi-rigid unit modes" (QRUMs).[5] There is not a precise definition of a QRUM, and we simply assign a

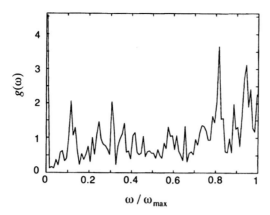

Figure 14. The density of states of Zeolite-LTA calculated using the CRUSH program. Note the continuum at low frequencies.

reasonable cut-off energy of around 1 THz (following the experimental energy scale for the RUMs in β-cristobalite as shown above). The fraction of QRUMs for a number of zeolite examples is shown in Table 2.

Table 2: Percentage of all normal modes in some selected zeolites that are either RUMs or QRUMs

Zeolite	% RUMs	% QRUMs
LTA	2.8	6
RHO (*Im3m*)	1.4	7
RHO (*I43m*)	0.7	5
Sodalite	2.8	6
Faujasite	1.4	5
UTD-1	0	8
Paulingite	0.6	6
Chabasite	0	6
Natrolite	0	5
ZSM-5	0	4

Zeolite flexibility

It is tempting to imagine that with so many RUMs and QRUMs zeolite structures will be completely floppy and flexible in an infinite number of ways, but this is not so. The possible distortions are restricted to those given by the RUM and QRUM eigenvectors, and these impose fairly strict constraints. The situation reminds us of our comment about the flexibility of the structure of β-cristobalite being restricted to that allowed by the RUMs, but of course in the zeolites there are many more RUMs that come into play.

In β-cristobalite we interpreted the dynamic disorder as arising from the superposition of dynamic RUMs as phonon modes. In the zeolites we have tackled this superposition from an alternative perspective, namely the formation of static superpositions of RUM deformations.[22,23] The RUM deformations can be included in these superpositions with arbitrary amplitude and phase. For example, for a system with one RUM per wave vector, each RUM deformation $\mathbf{R_k}$ of wave vector $\mathbf{k}$ can be assigned an amplitude A_k and phase γ_k. A spatial deformation pattern $\mathbf{L}$ that defines a set of atomic displacements can then be constructed as a Fourier transform:

Figure 15. Local deformation in the sodalite structure, showing the distortion on the central ring that does not propagate to distant neighbours

$$L = \sum_k A_k R_k \exp(i\gamma_k) \tag{4}$$

There is considerable flexibility in the choice of A_k and γ_k, but this does not allow us to overcome the restrictions imposed by the limited set of eigenvectors R_k. We call the deformation described by L a "local RUM", since A_k and γ_k can be chosen to give a localised deformation.

We have developed a FORTRAN program called LOCALRUM to allow the calculation of the local RUMs for a given zeolite structure using the eigenvectors R_k calculated by CRUSH. Since the choice of values for A_k and γ_k is arbitrary, the LOCALRUM program can work with initial random values for the amplitudes and phases of the different RUMs, or else their values can be specifically tuned to give a resultant local RUM that is close to a trial deformation.

Calculations of Local RUMs

Our first calculation of a local RUM concerned the location of the Ni site in sodalite,[22] which was thought to be a good representation of the basic atomic environment in the zeolite faujasite. In this case the Ni ion could held in place in the channels by a pinching action involving oxygen atoms moving towards the cation as the tetrahedra rotate. This is illustrated in Figure 15, which shows a section of the local RUM distortion. It is clear from this figure that the distortions of the structure about the Ni cations do not propagate very far in the structure, i.e. they are localised. These are the same deformations that will bind Ni within the faujasite structure.

A second calculation, on the new zeolite UTD-1,[23,50] which contains 14-membered rings of tetrahedra, illustrates the way in which the constraints of the RUM eigenvectors impose limitations on the degree of overall flexibility. A section of the structure is shown in Figure 16. This figure summarises the main result that only parts of the structure are flexible, because the RUM eigenvectors involve significant motions of only a subset of all tetrahedra.

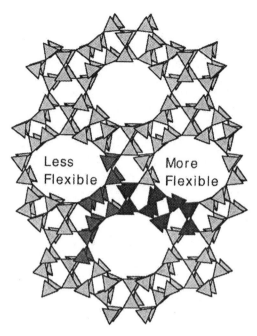

Figure 16. Section of the UTD-1 zeolite showing the flexible and rigid components of the structure.

CONCLUSIONS

In this chapter we have shown how the RUM model has enabled us to understand the local deformations in silicate crystals, whether in a dynamic sense as in the high-temperature phases of cristobalite and tridymite, or in a static sense as in the incommensurate phase transition in tridymite and the localised deformations in the zeolites. All these processes should give rise to diffuse scattering. We have noted that some of our results have been confirmed by measurements of diffuse scattering using electron diffraction. The other link to experiment is through total scattering measurements, and these have been performed on the silica polymorphs. This aspect is described in more detail in the Chapter by David Keen.

In this article we have focused more on applications and insights of the RUM model than on the details of the model. Many aspects of the RUM model have been confirmed by experiment, and we have given some of the critical results. However, in its basic sense the RUM model gives an over-simplified description of real silicates, which do not have perfectly rigid tetrahedra and completely floppy joints. For the present purposes, attempts to go beyond this simplification do not alter or add to the physics insights the RUM model has given. But in our overall model we do take explicit account of the finite stiffness of the tetrahedra (the split-atom method is formally equivalent to assigning a finite stiffness to the tetrahedra) and the forces between tetrahedra, and a number of new results follow.[1,2,4,6-8,51] The most striking of these is that the transition temperatures for displacive phase transitions in silicates are proportional to the stiffness of the tetrahedra.[4,7] The ability of the tetrahedra to distort also has important implications for the thermodynamics of phase transitions.[51] Moreover, the finite stiffness allows us to use the QRUM extension of the basic model.

Our feeling is that the RUM model provides an intuitive link between two routes to understanding diffuse scattering in silicates and the structure it suggests, namely between the use of one's imagination to visualise atomic structures, and large-scale calculations. Our imagination is always limited because disordered crystals can be complex objects. For example, Hua and co-workers[26,27] were able to visualise the RUM distortions that give rise to the diffuse scattering at a single wave vector in cristobalite, but it would have been truly heroic had they been able to deduce all the RUMs in the system by eye. It is because it is not possible to make this step that domain models of the disorder in β-cristobalite have been so popular. At the other extreme are the phonon calculations using realistic force constant models. A good model ought to be able to reproduce the one-phonon diffuse scattering reasonably accurately (and with molecular dynamics simulations it should be possible to account for multi-phonon processes also). However, simply reproducing experimental data means that you may understand the general principles (e.g. phonon theory), but it does not necessarily follow that you understand the particulars of the case in hand. The RUM model allows us to isolate the important details and make quantitative predictions, whether in interpreting the short-range rearrangements of the cristobalite structure or in deducing potential binding sites in zeolites.

ACKNOWLEDGEMENTS

This work has been supported by EPSRC and NERC through the award of research grants and research studentships. Some of our calculations were carried out using the Hitachi supercomputers in the Cambridge High Performance Computing Facility. We appreciate the help of Michele Warren in producing Figures 4 and 7. MTD acknowledges the collaboration of Dave Keen (ISIS, UK) and Ian Swainson (Chalk River, Canada) in parts of this work. He is grateful to the organisers of the meeting for the opportunity to take part in the workshop.

REFERENCES

1. M.T. Dove, A.P. Giddy, and V. Heine. *Ferroelectrics* 136:33 (1992)
2. M.T. Dove, A.P. Giddy, and V. Heine. *Trans. Am. Cryst. Assoc.* 27:65 (1993)
3. A.P. Giddy, M.T. Dove, G.S. Pawley, and V. Heine. *Acta Cryst.* A49:697 (1993)
4. M.T. Dove, V. Heine, and K.D. Hammonds. *Min. Mag.* 59:629 (1995)
5. K.D. Hammonds, M.T. Dove, A.P. Giddy, V. Heine, and B. Winkler. *Am. Min.* 81:1057 (1996)
6. M.T. Dove, M. Gambhir, K.D. Hammonds, V. Heine, and A.K.A. Pryde. *Phase Trans.* 58:121 (1996)
7. M.T. Dove. in: *Amorphous Insulators And Semiconductors*, Thorpe, M.F. and Mitkova, M.I ed. Kluwer, Amsterdam (1997)
8. M.T. Dove. *Am. Min.* 82:213 (1997)
9. F.S. Tautz, V. Heine, M.T. Dove, and X. Chen. *Phys. Chem. Min.* 18:326 (1991)
10. A.K.A. Pryde and V. Heine. *Phys. Chem. Min.* (submitted)
11. P Sollich, V. Heine, and M.T. Dove. *J. Phys.: Cond. Matt.* 6:3171 (1994)
12. I.P. Swainson and M.T. Dove. *Phys. Rev. Lett.* 71:193 (1993)
13. I.P. Swainson and M.T. Dove. *Phys. Rev. Lett.* 71:3610 (1993)
14. I.P. Swainson and M.T. Dove. *J. Phys.: Cond. Matt.* 7:1771 (1995)
15. M.T. Dove, D.A. Keen, A.C. Hannon, and I.P. Swainson. *Phys. Chem. Min.* 24:311 (1997)
16. M. Gambhir, V. Heine, and M.T. Dove. *Phase Trans.* 61:125 (1997)
17. M.T. Dove, K.D. Hammonds, V. Heine, R.L. Withers, Y. Xiao, and R.J. Kirkpatrick. *Phys. Chem. Min.* 23:55 (1996)
18. I.P. Swainson and M.T. Dove. *Phys. Chem. Min.* 22:61 (1995)
19. A.K.A. Pryde, K.D. Hammonds, M.T. Dove, V. Heine, J.D. Gale, and M.C. Warren. *J. Phys.: Cond. Matt.* 8:10973 (1996)
20. A.K.A. Pryde, K.D. Hammonds, M.T. Dove, V. Heine, J.D. Gale, and M.C. Warren. *Phase Trans.* 61:141 (1997)
21. V. Heine, P.R.L. Welche and M.T. Dove. *J. Am. Cer. Soc.* (submitted)
22. K.D. Hammonds, H. Deng, V. Heine, and M.T. Dove. *Phys. Rev. Lett.* 78:3701 (1997)
23. K.D. Hammonds, V. Heine, and M.T. Dove. *Phase Trans.* 61:155 (1997)

24. M.T. Dove, M J Harris, A C Hannon, J M Parker, I.P. Swainson, and M Gambhir. *Phys. Rev. Lett.* 78:1070 (1997)
25. M.T. Dove. *Introduction to Lattice Dynamics.* Cambridge University Press, Cambridge (1993)
26. G.L. Hua, T.R. Welberry, R.L. Withers, and J.G. Thompson. *J. Appl. Cryst.* 21:458 (1988)
27. T.R. Welberry, G.L. Hua, and R. L. Withers. *J. Appl. Cryst.* 22:87 (1989)
28. R.L. Withers, J.G. Thompson, Y. Xiao, and R.J. Kirkpatrick. *Phys. Chem. Min.* 21:421 (1995)
29. H. Grimm and B. Dorner. *Phys. Chem. Sol.* 36:407 (1975)
30. H.D. Megaw. *Crystal structures: a working approach*, W. B. Saunders, Philadelphia (1973)
31. H. Boysen, B. Dorner, F. Frey, and H. Grimm. *J. Phys. C: Sol. State Phys.* 13:6127 (1980)
32. B. Berge, J.P. Bachheimer, G. Dolino, M. Vallade, and C.M.E. Zeyen. *Ferroelectrics.* 66:73 (1986)
33. J. Bethke, G. Dolino, G. Eckold, B. Berge, M. Vallade, C.M.E. Zeyen, T. Hahn, H. Arnold, and F. Moussa. *Europhys. Lett.* 3:207 (1987)
34. G. Dolino, B. Berge, M. Vallade, and F. Moussa. *Physica*, B156:15 (1989)
35. G. Dolino, B. Berge, M. Vallade, and F. Moussa. *J. de Phys. I.* 2:1461 (1992)
36. M. Vallade, B. Berge, and G. Dolino. *Journal de Physique, I.* 2:1481 (1992)
37. H. He and M.F. Thorpe. *Phys. Rev. Lett.* 54:2107 (1985)
38. Y. Cai and M.F. Thorpe. *Phys. Rev.* B40:10535 (1989)
39. M.F. Thorpe. in: *Amorphous Insulators And Semiconductors*, Thorpe, M.F. and Mitkova, M.I ed. Kluwer, Amsterdam (1997)
40. G.S. Pawley. *Phys. Stat. Sol.* 49b:475 (1972)
41. K.D. Hammonds, M.T. Dove, A.P. Giddy and V Heine. *Am. Min.* 79:1207 (1994)
42. K.D. Hammonds, A Bosenick, M.T. Dove and V Heine. *Am. Min.* (submitted)
43. W W Schmahl, I.P. Swainson, M.T. Dove and A Graeme-Barber. *Zeit. Krist.* 201:125 (1992)
44. A.F. Wright and A.J. Leadbetter. *Phil. Mag.* 31:1391 (1975)
45. D.M. Hatch and S. Ghose. *Physics and Chemistry of Minerals*, 17:554 (1991)
46. F. Liu, S.H. Garofalini, R.D. Kingsmith, and D. Vanderbilt. *Phys. Rev. Lett.* 71:3611 (1993)
47. D. Cellai, M.A. Carpernter, R.J. Kirkpartick, E.K.H. Salje, and M. Zhang. *Phys. Chem. Min..* 22:50 (1995)
48. J.D.C. McConnell and V. Heine. *Acta Cryst.* A40:473 (1984)
49. K. Kihara. *Z. Kristallogr.* 146:185 (1977)
50. C.C. Freyhardt, M. Tsapatsis, R.F. Lobo, K.J. Balkus, and M.E. Davis. *Nature* 381:295 (1996)
51. P. Sollich, V. Heine, and M.T. Dove. *J. Phys.: Cond. Matt.* 6:3171 (1994)

VIBRATIONAL ENTROPY AND LOCAL STRUCTURES OF SOLIDS

Brent Fultz

Division of Engineering and Applied Science
mail 138-78
California Institute of Technology
Pasadena, California 91125

1. HISTORICAL INTRODUCTION

Josiah Willard Gibbs made the seminal advances in applying thermodynamics to the study of solids, from which modern materials science was born [1].* In more recent times there is some perception that the field of thermodynamics is old-fashioned. A more accurate perception, perhaps, is that the contributions of van der Waals, Boltzmann, Kirkwood, Guggenheim and Nerst have proved impressively durable. Much of our motivation today for studying the structures of solids is provided by thermodynamics. Without thermodynamics, many studies of structures of solids would amount to little more than taxonometric classifications.

To date, thermodynamics and structure have connected in two important ways. One connection is between electronic structure and local atomic structure. This effort seeks to understand relationships between the structures of solids and their internal energy, E. The second connection is between the state of order in solids and the entropy of atom configurations, S_{config}.

Without entropy there could be no thermodynamics. Entropy is, of course, defined as:

$$S \equiv k_B \ln\Omega \quad , \tag{1}$$

where Ω is the number of equivalent configurations available to a macrostate of a system. The first task in calculating S is to decide which entities are configured. Possibilities include:

configurations of atoms on sublattices	S_{config}
instantaneous positions of atoms as they vibrate about lattice sites	S_{vib}
magnetic moments of atoms	S_{mag}
electrons in different electronic states	S_{el}
magnetic moments of nuclei	S_{nucl}

---

* Physical chemists also claim Gibbs as a father figure.

Local Structure from Diffraction
Edited by S.J.L Billinge and M.F. Thorpe, Plenum Press, New York, 1998

Because we seek differences in the free energies of solid phases, the issue is not to identify the size of the entropy, but to identify differences in entropy between different macrostates or phases of a material. For example, the nuclear spin distribution is typically random for different structural phases of a solid. So although the entropy of nuclear spins is large, it does not change during a structural phase transition, and can therefore be neglected. Similarly the magnetic entropy of two paramagnetic phases well above their Curie points may be large but equal, so S_{mag} can be neglected. It is often assumed that S_{el} is negligible, since the energy widths of bands in solids are on the order of a few eV, whereas the thermal spread of electron occupancies at 1000 K is about 0.1 eV. With one outer electron per atom and a bandwidth of 10 eV, the electronic entropy is of order $S_{el} \cong k_B \ln(10.1/10) = 0.01\ k_B/atom$, which is usually negligible compared to other sources of entropy. We note that with narrower bands, however, the electronic entropy may not be negligible for the thermodynamics of solid phases.

For the thermodynamics of chemical order-disorder transitions in alloys, it has been known for some time that the configurational entropy, S_{config}, plays an important role. In a typical problem of ordering in an equiatomic binary alloy, the state of order can range from perfect, where the two species of atoms are arranged precisely onto two distinct sublattices, to random mixing on the two sublattices. Over this range the configurational entropy changes from $S_{config} = 0$ to $S_{config} = k_B \ln 2 = 0.69\ k_B/atom$. The configurational entropy is not so simple when the alloy has a state of order intermediate between these two extremes. Nevertheless, Gibbs himself nearly succeeded in developing a statistical mechanics capable of handling these problems, and Ising [2], Onsager [3], and others [4-7] made key contributions earlier this century. With the development of the cluster variation method of Kikuchi [8-10], calculations of S_{config} should now be regarded as a mature.

Even more recently, our understanding of the electronic energy of a solid has grown significantly owing to the acceptance of the local density approximation for the electron exchange and correlation energy. With electronic structure calculations for the internal energy of a solid, E, and accurate cluster approximation methods for S_{config}, two key pieces of the free energy are in hand. A number of phase diagrams have been calculated with *ab initio* free energies as:

$$F(T) = E - T\ S_{config} \ . \tag{2}$$

So then, what is the role for the vibrational entropy, S_{vib}, in solid state phase transitions? This question is not new. In the Jubelband for W. Nerst's 60th birthday, Fritz Lange published results from a cryogenic calorimetry study on white and gray tin [11]. His results, which show a difference in vibrational entropy of a very large 0.8 $k_B/atom$, were celebrated in the *Modern Theory of Solids* by Seitz [12]. Slater's book, *Introduction to Chemical Physics* has a major section on vibrational entropy [13]. Occasionally theoretical papers argued that the change in vibrational entropy could be significant [14-20], and could be comparable to the change in configurational entropy in the order-disorder transition in β-brass, for example [15,16,18]. Bakker's group performed analytical calculations for 1-D alloys, and numerical calculations for 2-D alloys that showed the possibility for a large ΔS_{vib} in order-disorder transitions [18,20]. Moraitis and Gautier performed semi-empirical calculations of ΔS_{vib} for the formation of alloys from pure elements, and argued that ΔS_{vib} was not negligible [17]. Grimvall analyzed calorimetric data to identify significant differences in vibrational entropies of various pure materials and compounds [21,22].

There has been less experimental evidence for the importance of ΔS_{vib} in solid-solid phase transitions. For example Hawkins and Hultgren reported the results of a detailed calorimetric study on the difference in vibrational entropy of ordered and disordered Cu_1Au_1, and found no effect: $\Delta S_{vib} = 0.0 \pm 0.03\ k_B/atom$ [23]. In the 1950's a curious controversy arose concerning the role of vibrational entropy in the classic disorder - $L1_2$ order transition in Cu_3Au. From rather noisy electrical resistivity data taken at cryogenic temperatures, Bowen [24] concluded that the ordering transition in Cu_3Au involved a change in vibrational entropy of 0.355 k_B/atom. This large value of ΔS_{vib} is comparable to the entire entropy of the order-disorder transition in Cu_3Au, which is about 0.40 k_B/atom [25]. Bowen's result was checked in a much more sophisticated experiment by Flinn, McManus and Rayne [26] who measured elastic constants for ordered and disordered Cu_3Au, and obtained Debye temperatures of 283.8 K for the ordered alloy, and 281.6 K for the disordered alloy. This is equivalent to a change in vibrational entropy of a mere 0.023 k_B/atom. What is amusing about this early controversy, which was evidently settled against the importance of vibrational entropy, is that both electrical resistivity data and elastic constants are poor indicators of differences in vibrational entropy. The arbiters of this controversy must be heat capacity data or measurements of the full phonon density of states. Curiously, these methods have shown recently that Bowen was closer to the truth, although probably for the wrong reasons [27].

Studies of changes in vibrational entropies of phase transitions began with seriousness in the 1990's. The present paper focuses primarily on recent studies from our group at Caltech [27-34]. In addition to the experimental results from Caltech, the Barcelona group of Planes, Ortin, et al., performed calorimetric measurements of the ΔS_{vib} in martensitic transitions and found ΔS_{vib} to be on the order of 0.3 k_B/atom [35,36]. Results from a number of theoretical studies have been reported recently. Moruzzi, et al., suggested the use of the Debye-Grüneisen approximation for calculating thermodynamic properties from first principles interatomic potentials [37]. Sanchez, et al., and Mohri, et al., used the Debye-Grüneisen approximation in alloy phase diagram calculations and found critical temperatures to be changed by typically 20 % [38,39]. Cahn argued for the importance of vibrational entropy in martensitic transitions [40], where the shear processes are deterministic and do not generate configurational entropy in the alloy. Calculations oriented towards the vibrational spectra of solids have been performed by Clark and Ackland [41] and Garbulsky and Ceder [42,43]. Very recently, other groups [44,45] have begun to calculate ΔS_{vib} for alloys approximated with embedded atom potentials.

Sections 2 - 4 in this chapter explain the physical meaning of vibrational entropy. The explanations range from the classical to the quantum statistical mechanical, as the vibrational entropy is related first to the heat capacity of the solid, and later to the phonon density of states·(DOS) of a material. Sections 5 and 6 describe the phonon formalism and how phonons are measured by inelastic neutron scattering. Sections 7 - 9 present experimental measurements of vibrational entropy and phonon DOS curves. Examples are provided to show how the vibrational entropy originates with specific features of the phonon DOS. Harmonic and anharmonic effects are both shown to be important.

2. CLASSICAL THERMODYNAMICS

Heat is, of course, a form of energy, stored in internal coordinates of a material. The change in entropy of a solid, dS, is related to the heat input, dQ, in a reversible process as $dQ = TdS$. The heat input to the material per degree per mole is the heat capacity at constant pressure, $C_p(T)$, or constant volume, $C_V(T)$. Measuring the heat capacity as a function of temperature is the direct way to obtain the change in entropy over a range in temperature. What is important for phase changes, however, is the difference in entropy between two states of a material. The direct way to determine differences in entropy at constant pressure is to measure differences in heat capacity of two states of a material over a range of temperatures, $\Delta C_p(T) \equiv C_p^\beta - C_p^\alpha$. This can be done by differential calorimetry, and then calculating:

275

$$\Delta S^{\beta-\alpha}(T) = \int_0^T \frac{\Delta C_p(T')}{T'} dT' \quad . \tag{3}$$

Classical thermodynamics says nothing about the nature of the heat (i.e., is the heat a change in chemical bonds, phonons, electrons, or magnetic moments?). It is often possible to set up experiments to study one type of internal coordinate, however. If the structural and magnetic coordinates of a material remain constant during the measurements, calorimetric methods using Eq. 3 may be a good way to obtain $\Delta S_{vib}^{\beta-\alpha}(T)$. In particular, the integrand $\Delta C_p(T)/T$ is often largest at low temperatures where there is no significant atomic diffusion, so there are often no changes in atomic configurations during the measurements.

3. CONCEPT OF VIBRATIONAL ENTROPY

Although classical thermodynamics tells us how to measure the differences in vibrational entropy of two alloy phases, it does little to answer the question, "What is vibrational entropy?" This section provides a conceptual answer, which is developed in more detail in the next section. This section also shows that in the classical limit, the difference in vibrational entropies of two solid phases depends only on their phonon spectra.

Assume that two solid phases have the same types and numbers, N, of atoms, each of mass m_j. Our 3-dimensional solid will have 3N normal modes of vibration. What is important for determining Ω in Eq. 1 is the total volume in phase space explored by the two solid phases at temperature T.[*] The phase that explores the larger volume in this phase space will have more distinguishable configurations and a larger vibrational entropy. The number of distinguishable configurations of a solid phase is directly proportional to the volume of this phase space, V:

$$\Omega = \frac{V}{V_0} \quad . \tag{4}$$

Suppose the volumes of the phase spaces for the two solid phases, α and β, are V_α and V_β. For comparing the difference in vibrational entropies of these two solid phases, defined as $\Delta S_{vib}^{\beta-\alpha} \equiv S^\beta - S^\alpha$, the constant V_0 is irrelevant:

$$\Delta S_{vib}^{\beta-\alpha} = k_B \ln\left(\frac{V_\beta}{V_0}\right) - k_B \ln\left(\frac{V_\alpha}{V_0}\right) = k_B \ln\left(\frac{V_\beta}{V_\alpha}\right) \quad . \tag{5}$$

The phase space is spanned by momentum coordinates, $\{p\}$, and position coordinates, $\{x\}$. Volumes in phase space are products of ranges of excursion in these coordinates as the atoms vibrate, Δp and Δx. Specifically:

$$V_\alpha = \prod_{j=0}^{3N} \Delta p_j \prod_{j=0}^{3N} \Delta x_j \quad . \tag{6}$$

In vibrating systems there will be a transfer of energy back and forth from kinetic to potential, providing the ranges of Δp_j and Δx_j. The range of Δp_j is easy to calculate from

---

[*] I apologize for the use of the word "phase" in the two different contexts of "phase space" and "solid phase." In the term "phase space", however, the word "phase" is an adjective.

the kinetic energy. At a temperature T, the maximum of Δp, denoted Δp_{max}, is obtained from:

$$k_B T = \frac{\Delta p_{max}^2}{2m} \quad , \text{ so} \tag{7}$$

$$\Delta p_{max} = \sqrt{2mk_B T} \quad . \tag{8}$$

To obtain the range Δx, we invoke the harmonic approximation. A feature of the harmonic approximation is that the vibrations can be resolved into independent normal coordinates, Q. In the harmonic approximation the maximum of a spatial normal coordinate, Q_{max}, is obtained from the potential energy of a harmonic oscillator of frequency ω:

$$k_B T = \frac{1}{2} m \omega^2 Q_{max}^2 \quad , \text{ so} \tag{9}$$

$$Q_{max} = \frac{1}{\omega} \sqrt{\frac{2k_B T}{m}} \quad . \tag{10}$$

Figure 1 illustrates our results so far (Eqs. 8 and 10). The important point is that the momentum coordinates have the same range for different solid phases, whereas the spatial coordinates scale as ω^{-1}. If two solid phases have the same types of atoms, then their volumes in phase space cannot differ because of differences in their momentum coordinates. The difference in phase space volume arises from the difference in their position coordinates because of the factor ω^{-1} for each normal mode of vibration. The physical interpretation is simple. If a solid phase has stiffer springs, it will have vibrations of higher frequency. For the same thermal energy, the atoms in a stiffer solid will not move so far from their equilibrium positions, so fewer positional configurations are possible. This stiff phase has a low vibrational entropy. We can calculate this difference in vibrational entropy with Eqs. 5, 6, 8, and 10:

$$\Delta S_{vib}^{\beta-\alpha} = k_B \ln \left(\frac{\prod\limits_{i=0}^{3N} \sqrt{2m_i k_B T} \; \prod\limits_{i=0}^{3N} \frac{1}{\omega_i} \sqrt{\frac{2k_B T}{m_i}}}{\prod\limits_{j=0}^{3N} \sqrt{2m_j k_B T} \; \prod\limits_{j=0}^{3N} \frac{1}{\omega_j} \sqrt{\frac{2k_B T}{m_j}}} \right) \quad , \tag{11}$$

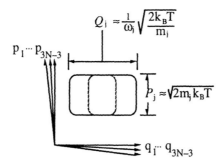

Figure 1. Ranges of coordinates in the phase space occupied by solids at temperature T.

where the subscripts i refer the β-phase and j to the α-phase. Since the masses of the atoms are the same in the two phases:

$$\Delta S_{vib}^{\beta-\alpha} = k_B \ln \left(\frac{\prod_{i=0}^{3N} \frac{1}{\omega_i}}{\prod_{j=0}^{3N} \frac{1}{\omega_j}} \right) \quad , \tag{12}$$

and the logarithm of a product is the sum of the logarithms:

$$\Delta S_{vib}^{\beta-\alpha} = k_B \left(\sum_{i=0}^{3N} \ln(\omega_j) - \sum_{j=0}^{3N} \ln(\omega_i) \right) \quad . \tag{13}$$

Equation 13 is in fact an exact result in the classical limit and the harmonic approximation. With 3N phonons in a solid, it is often more practical to work with a phonon spectrum, or density of states (DOS), $g(\varepsilon)$. The product $g(\varepsilon)d\varepsilon$ is the fraction of vibrational modes in the energy interval $d\varepsilon$, and $\varepsilon = \hbar\omega$. Transforming Eq. 13 to an integral over a continuum of states, the difference in classical vibrational entropy (per atom) of two phases, $S^\beta - S^\alpha$, depends in a straightforward way on the difference in the phonon DOS of the two phases:

$$S^\beta - S^\alpha = 3k_B \int_0^\infty \left[g^\alpha(\varepsilon) - g^\beta(\varepsilon) \right] \ln\varepsilon \, d\varepsilon \quad , \tag{14}$$

where the difference of the two normalized $g(\varepsilon)$ avoids problems with the dimensionality of the argument of the logarithm. In the harmonic approximation, the phonon density of states is all we need to calculate the vibrational entropy. To understand details of the phonons, however, we may need to know their polarizations and dispersions. (These topics are discussed in section 5.) A more detailed understanding of individual phonons may allow us to identify correlations between the atomic structure and the vibrational entropy.

4. VIBRATIONAL ENTROPY IN QUANTUM STATISTICAL MECHANICS

This section derives the phonon part of the heat capacity at constant pressure, $C_p(T)$, from elementary quantum statistical mechanics. The phonon energy in the lattice, E_{ph}, is obtained by summing the energies of all phonons, each of energy ε. This total energy, E_{ph}, includes all products of ε with the phonon DOS, $g(\varepsilon)$, and the Bose-Einstein phonon occupancy factor, $n(\varepsilon)$:

$$E_{ph}(T_1) = 3 \int_0^\infty g(\varepsilon) \, \varepsilon \, n(\varepsilon, T_1) \, d\varepsilon \quad , \tag{15}$$

where the phonon occupancy factor (including the half-occupancy from the zero point vibration) is:

$$n(\varepsilon, T) = \frac{1}{2} + \frac{1}{e^{\varepsilon/kT} - 1} \quad . \tag{16}$$

278

At a slightly higher temperature, $T_2 = T_1 + \Delta T$, the phonon energy is $E_{ph}(T_2)$:

$$E_{ph}(T_2) = 3 \int_0^\infty g(\epsilon) \left(\epsilon(T_1) + \frac{d\epsilon}{dT} \Delta T \right) \left(n(\epsilon, T_1) + \frac{\partial n}{\partial \epsilon} \frac{d\epsilon}{dT} \Delta T + \frac{\partial n}{\partial T} \Delta T \right) d\epsilon \ . \tag{17}$$

The change in phonon energy between T_1 and T_2, $dE_{ph}/d\Delta T$, gives the phonon contribution to the heat capacity at constant pressure, C_p^{ph} :

$$C_p^{ph}(T_1) = C_V(T_1) + C_{anl}^{ph}(T_1) \quad , \tag{18}$$

where we have separated the harmonic part:

$$C_V(T_1) = 3 \int_0^\infty g(\epsilon) \epsilon(T_1) \frac{\partial n}{\partial T} d\epsilon \ , \tag{19}$$

and an anharmonic part:

$$C_{anl}^{ph}(T_1) = 3 \int_0^\infty g(\epsilon) \frac{\partial \epsilon}{\partial T} n(\epsilon, T_1) \, d\epsilon + 3 \int_0^\infty g(\epsilon) \epsilon \frac{\partial \epsilon}{\partial T} \frac{\partial n}{\partial \epsilon} \, d\epsilon \ , \tag{20}$$

and have ignored small terms in ΔT. The integral in Eq. 19 does not involve any changes with temperature of phonon energies — it includes only the change in phonon occupancy with temperature. This is as expected for harmonic vibrations, where vibrational frequencies do not change with the amplitudes of thermal vibration. It is assumed that the only reason why phonon energies undergo a change with temperature is that the volume of the solid increases a bit, and the interatomic potential is anharmonic. With no change in volume, no change in phonon energies occurs with increasing temperature. The part of the heat capacity of Eq. 19, which involves no change in phonon energies, is therefore $C_V(T)$, the heat capacity at constant volume.

The anharmonic contribution of Eq. 20 is proportional to the temperature dependence of the phonon energy. In the classical limit, where $n(\epsilon, T) = k_B T / \epsilon$, it is not difficult to show that $C_{anl}^{ph}(T) = 0$. The classical picture is that as the phonon energies decrease with temperature, their occupancy factors increase inversely, leading to no net change in the heat capacity. Actually, even at high temperatures there remains a change in the zero point energy when $\partial \epsilon / \partial T \neq 0$, so a small contribution to the heat capacity still remains. This $C_{anl}^{ph}(T)$ is a more significant fraction of the heat capacity at low temperatures, however, and is positive when $\partial \epsilon / \partial T < 0$, as is typically the case.

Our treatment of anharmonic effects is not yet complete, however. There is an additional penalty to be paid in internal elastic energy as the crystal expands against its bulk modulus, B. If the crystal expands linearly with temperature, the elastic energy increases quadratically:

$$E_{el}(T) = \frac{1}{2} B v \left(\int_0^T 3\alpha(T') dT' \right)^2 \ , \tag{21}$$

where $\alpha(T)$ is the linear coefficient of thermal expansion (one-third the volume coefficient of thermal expansion), and the specific volume is v. Since $E_{el}(T)$ is positive, and the anharmonic contribution $C_{anl}^{ph}(T)$ is also positive, thermal expansion is energetically unfavorable. We require another physical phenomenon to motivate thermal expansion. This missing ingredient turns out to be the phonon entropy. As the material expands and the phonon frequencies soften ($\partial\varepsilon/\partial T < 0$), the vibrational entropy of the solid becomes larger (cf. Eq. 14 for the high temperature limit). The contribution, $-TS_{vib}$, to the free energy of the expanded solid with phonons overcomes the penalty in elastic energy. We can calculate the equilibrium value of the coefficient for thermal expansion of a solid as it is heated from temperature T_1 to $T_2 = T_1 + \Delta T$ by first calculating a free energy with elastic energy and phonon entropy:

$$F(T_1+\Delta T) - F(T_1) = E_{el}(T_1 + \Delta T) + E_{el}(T_1) - T[S_{vib}(T_1+\Delta T) - S_{vib}(T_1)] \quad . \qquad (22)$$

Using Eqs. 14 and 21, minimizing Eq. 22 with respect to α, and then taking its derivative with respect to T, we obtain the following proportionality between the coefficient of thermal expansion, α, and the heat capacity at constant volume, $C_V(T)$:

$$\alpha(T) = \frac{\gamma\, C_V(T)}{3Bv} \quad . \qquad (23)$$

In setting up our expression for $\partial\varepsilon/\partial T$, we used a constant of proportionality, γ, known as the average Grüneisen constant. It is the average (weighted by heat capacity) of the fractional change in energy of the phonon modes with the change in volume of the crystal. For an individual phonon mode, i, we often define a mode Grüneisen parameter:

$$\gamma_i = -\frac{v}{\varepsilon_i}\frac{\partial\varepsilon_i}{\partial v} \quad . \qquad (24)$$

In practice, the Grüneisen constant is often assumed to be the same for all phonon modes, but this assumption is unlikely to survive detailed scrutiny.

In summary, to account for all anharmonic contributions to the heat capacity we have to consider: 1) the $C_{anl}^{ph}(T)$ of Eq. 20, which is positive when $\partial\varepsilon/\partial T$ is negative, 2) the internal elastic energy needed to expand the crystal, 3) the increase in vibrational entropy as the crystal expands. Although items 1 and 3 can be obtained from measurements of the phonon DOS at different temperatures, item 2 cannot. While it is possible in principle to obtain information about the bulk modulus from the low energy tail of the phonon DOS, in practice this is not so accurate. It is usually more reliable to use the classical relationship between $C_p(T)$ and $C_V(T)$ to obtain the anharmonic contribution to the entropy:

$$C_p(T) - C_V(T) = 9\, B v\, \alpha^2\, T \quad . \qquad (25)$$

Nevertheless, by measuring the phonon DOS at different temperatures it is possible to obtain the $\partial\varepsilon/\partial T$ needed for the evaluation of $C_{anl}^{ph}(T)$ of Eq. 20. Furthermore, when α is a constant, it can be shown with Eq. 22 that in equilibrium at high temperatures:

$$\Delta T\, [C_p(T) - C_V(T)] = T\, \Delta S_{vib} \quad , \qquad (26)$$

and with Eq. 14 the phonon DOS can provide the change in vibrational entropy, ΔS_{vib}, over the range in temperature ΔT. One type of Grüneisen constant, γ, can be measured directly from the shifts of the phonon using the relation: $\partial\varepsilon/\partial T = -3\gamma\alpha\varepsilon$. With more detail, however, phonon DOS curves do not provide $\partial\varepsilon/\partial T$ directly, but rather provide $\partial g/\partial T$ and $\partial g/\partial\varepsilon$. Figure 2 illustrates the relationship between $\partial\varepsilon/\partial T$, $\partial g/\partial T$ and $\partial g/\partial\varepsilon$. Two normalized DOS curves are shown, one for temperature T, and the other shows some phonon softening at the higher temperature T+ΔT. What is important for Eq. 20 is how phonons within a particular energy range reduce their energy with increasing temperature. An example is illustrated by the arrow at the bottom of Fig. 2. The slope of the phonon DOS curve, $\partial g/\partial\varepsilon$, is the negative of the horizontal shift, $\partial\varepsilon/\partial T$, divided by the vertical shift, $\partial g/\partial T$:

$$\frac{\partial\varepsilon}{\partial T} = \frac{-\dfrac{\partial g}{\partial T}}{\dfrac{\partial g}{\partial\varepsilon}} \tag{27}$$

(In practice, Eq. 27 becomes difficult to use when $\partial g/\partial\varepsilon$ is small and ΔT is not. In this case it may be helpful to scale $g(\varepsilon)$ by its breadth when evaluating $\partial g/\partial T$.)

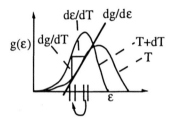

Figure 2. The relationship between the slope of the phonon DOS curve, $\partial g/\partial\varepsilon$, its temperature dependence, $\partial g/\partial T$, and the energy shift with temperature, $\partial\varepsilon/\partial T$.

Substituting Eq. 27 into Eq. 20 gives:

$$C_{anl}^{ph}(T_1) = 3 \int_0^\infty g(\varepsilon) \frac{-\dfrac{\partial g}{\partial T}}{\dfrac{\partial g}{\partial\varepsilon}}\left(n(\varepsilon, T_1) + \varepsilon\frac{\partial n}{\partial\varepsilon}\right) d\varepsilon , \tag{28}$$

Equation 28 can be used with experimental data to identify the effect of phonon softening on the heat capacity of the solid. It is a part of the anharmonic contribution to the heat capacity, but it does not account for the energy required to expand the crystal against its bulk modulus.

In summary, the phonon DOS readily provides the heat capacity at constant volume, $C_V(T)$. In practice, most of the vibrational entropy of a solid phase can be obtained from $C_V(T)$. The change in the heat capacity with temperature, owing to the softening of the phonon DOS, can also be obtained from measurements of the phonon DOS at different temperatures. The phonon DOS is less useful for determining the bulk modulus of the material, and hence the thermal expansion (Eq. 23) and the full anharmonic contribution to the heat capacity (Eq. 25). With independent measurements of the density and bulk modulus, however, the phonon DOS can be used to provide all thermodynamic information on the vibrational entropy.

5. BORN – VON KÁRMÁN MODEL OF LATTICE DYNAMICS

The most important step in understanding the vibrational entropy is knowing the phonon DOS. The Born von - Kármán model provides a formalism to calculate and model the phonon DOS of a periodic solid. This model allows for the calculation of the vibrational eigenmodes of masses (atomic nuclei) connected together by linear springs to various neighboring masses. This section presents some of the key results and equations of the Born von - Kármán formalism, but many results are presented without derivation. The equations of motion of the masses are:

$$m_k \ddot{u}_\alpha\binom{l}{k} = -\sum_{l' k' \beta} \phi_{\alpha\beta}\binom{l\,l'}{k\,k'} u_\beta\binom{l'}{k'} \quad ,\tag{29}$$

where $u_\alpha\binom{l}{k}$ is the displacement vector of the αth Cartesian component (x, y, z) for the atom k in the lth unit cell. $\phi_{\alpha\beta}\binom{l\,l'}{k\,k'}$ is the force constant involving the displacement of the atom k and the neighboring atom k'. The coupled set of linear equations of Eq. 29 has harmonic solutions, known as "phonon solutions" for $u_\alpha\binom{l}{k}$:

$$u_\alpha\binom{l}{k} = \frac{1}{\sqrt{m_k}} U_\alpha(k|q) \exp\left(i(q \cdot x(l) - \omega(q)t)\right) \quad ,\tag{30}$$

where $x(l)$ is the shift of $u_\alpha\binom{l}{k}$ from its equilibrium position in a perfect lattice, and q is the wavevector of the vibrational mode. The frequency for this particular vibrational mode is $\omega(q)$, related to the phonon energy as $\varepsilon(q) = \hbar\omega(q)$. Substituting Eq. 30 in 29, and rearranging the equations of motion we have:

$$\omega^2(q) U_\alpha(k|q) = \sum_{k'\beta} D_{\alpha\beta}\binom{q}{kk'} U_\beta(k'|q) \quad , \text{where}\tag{31}$$

$$D_{\alpha\beta}\binom{q}{kk'} = \frac{1}{\sqrt{m_k m_{k'}}}\sum_l \phi_{\alpha\beta}\binom{0\,l}{i\,j} \exp\left(i(q \cdot x(l))\right).\tag{32}$$

We have defined the dynamical matrix D in Eq. 32 in terms of the Fourier transform of the force constant matrix. Equation 31 provides an eigenvalue equation for the ω^2, and these frequencies are obtained by diagonalizing the dynamical matrix for a particular choice of phonon wavevector, q:

$$\left\|D(q) - \omega^2(q)\, \delta_{ij}\right\| = 0\tag{33}$$

To obtain the vibrational spectrum of the solid, we must diagonalize D for all wavevectors within the first Brillouin zone of the crystal, and collect a histogram of vibrational frequencies. This is the phonon DOS, $g(\varepsilon)$. The eigenvectors of the dynamical matrix, $U(k|q)$, and the phonons themselves, $\mathbf{u}(k|\mathbf{q})$, have $3 \times k$ dimensions, where k is the number of atoms in the unit cell. For example, suppose the crystal has 4 atoms per unit cell. The dynamical matrix will be 12×12, with a structure as follows:

$$
\mathbf{D}\binom{q}{kk'} =
\begin{bmatrix}
\mathbf{D}\binom{q}{11} & \mathbf{D}\binom{q}{12} & \mathbf{D}\binom{q}{13} & \mathbf{D}\binom{q}{14} \\[2ex]
\mathbf{D}\binom{q}{21} & \mathbf{D}\binom{q}{22} & \mathbf{D}\binom{q}{23} & \mathbf{D}\binom{q}{24} \\[2ex]
\mathbf{D}\binom{q}{31} & \mathbf{D}\binom{q}{32} & \mathbf{D}\binom{q}{33} & \mathbf{D}\binom{q}{34} \\[2ex]
\mathbf{D}\binom{q}{41} & \mathbf{D}\binom{q}{42} & \mathbf{D}\binom{q}{43} & \mathbf{D}\binom{q}{44}
\end{bmatrix}
, \qquad (34)
$$

where the individual matrices in Eq. 34 are themselves 3×3, for example:

$$
\mathbf{D}\binom{q}{21} =
\begin{bmatrix}
D_{xx}\binom{q}{21} & D_{xy}\binom{q}{21} & D_{xz}\binom{q}{21} \\[2ex]
D_{yx}\binom{q}{21} & D_{yy}\binom{q}{21} & D_{yz}\binom{q}{21} \\[2ex]
D_{zx}\binom{q}{21} & D_{zy}\binom{q}{21} & D_{zz}\binom{q}{21}
\end{bmatrix}
, \qquad (35)
$$

and the eigenvectors (transposed as 1×12) are of the form:

$$
U(k|\mathbf{q}) = \begin{bmatrix} U_x(1|\mathbf{q}) & U_y(1|\mathbf{q}) & U_z(1|\mathbf{q}) & U_x(2|\mathbf{q}) & U_y(2|\mathbf{q}) & U_z(2|\mathbf{q}) \end{bmatrix} \qquad (36)
$$

$$
U_x(3|\mathbf{q}) \quad U_y(3|\mathbf{q}) \quad U_z(3|\mathbf{q}) \quad U_x(4|\mathbf{q}) \quad U_y(4|\mathbf{q}) \quad U_z(4|\mathbf{q}) \,] \quad .
$$

From section 4 we know that the spectrum of eigenvalues, i.e., the phonon DOS, is all the information about phonons that is used when calculating thermodynamic properties. Knowledge of the eigenvectors of the dynamical matrix, $U(k|\mathbf{q})$, is often important for two reasons, however. First, the experimental method of coherent inelastic scattering requires knowledge of the $U(k|\mathbf{q})$ of the different phonons. The strength of inelastic scattering depends, for example, on how the motion of atom 1 in the unit cell is oriented with respect to the momentum transfer in the experiment (i.e., how the vector $[U_x(1|\mathbf{q}) \quad U_y(1|\mathbf{q})$ $U_z(1|\mathbf{q})]$ is oriented with respect to $\mathbf{Q}$). In some cases it is possible to interpret data from incoherent inelastic scattering without knowledge of $U(k|\mathbf{q})$. This is not generally true, however, when the unit cell comprises different species of atoms, and when the symmetry of the crystal is not cubic. The second reason why the eigenvectors of the dynamical matrix, $U(k|\mathbf{q})$, are important is that they may provide a better understanding of the relationships between vibrational entropy and the local atomic structure. Phonon DOS curves for different alloy phases, such as shown in sections 7 - 9 below, are often quite different. It is rarely possible to attribute these differences to a simple shift in energy of all phonons, and knowledge of the atom motions in the individual phonons, their $U(k|\mathbf{q})$, is usually required in order to understand why the local structure of the solid affects the lattice dynamics, the phonon DOS, and the vibrational entropy.

6. INELASTIC NEUTRON SCATTERING

Most of the articles in this book are concerned with the phenomenon of diffraction, which is a special case of cooperative scattering by a group of atoms. Diffraction requires a type of scattering called "coherent scattering", characterized by a precise phase relationship between the incident and scattered waves. In diffraction, the phase relationships between scatterings from individual atoms are set by the wavepath across inter-atomic distances. Diffraction is particularly useful for measuring features of atomic arrangements in materials. The scattering of a wave by an atom need not involve a precise phase relationship between the incident and scattered wave, however. This is the case of "incoherent scattering."

Besides being "coherent" or "incoherent", scattering processes are "elastic" or "inelastic" when there is, or is not, an energy change of the wave after scattering. For electron and x-ray scattering, the usual processes of interest are coherent elastic scattering, which is useful for diffraction, and incoherent inelastic scattering, which is useful for spectroscopy. Two other combinations of the word pairs {coherent , incoherent} and {elastic , inelastic} can be formed, however, and both are important for neutron scattering. "Coherent inelastic" scattering occurs when a material undergoes an internal excitation with a precise energy-wavevector relationship, such as the excitation of a particular phonon. In this case there is an energy loss to the material during scattering (inelastic), but the scattering amplitude depends on the phase relationship between the atom movements in the phonon, and the phases of the incident and scattered neutron wavevectors. This process is therefore coherent. Also, "incoherent elastic" scattering means that the elastic scattering from different atoms involves imprecise phase relationships between the incident and diffracted waves. This occurs for example in neutron scattering when the nuclei have different spin orientations. Each spin orientation causes a different phase shift during scattering, so the phase relationships between incident and scattered waves are not consistent. Some of the elastic scattering is then incoherent.

Inelastic scattering, especially of neutrons, is the most important experimental method for determining the phonon DOS of a solid, and for determining energy-wavevector relationships of individual phonons. Inelastic neutron scattering is a topic that quickly extends beyond the scope of the present article, so this section presents only a few results in a way to appeal to the reader's understanding of diffraction experiments. For further explanations, the reader is referred to several excellent books and book chapters on the topic [46-48]. In elastic scattering, the structure of the crystal affects the diffraction pattern through a "structure factor", $F(\mathbf{Q})$. The analogs for incoherent or coherent inelastic scattering experiments are the "incoherent dynamical structure factor intensity", $|G_{inc}(\mathbf{Q},\nu)|^2$, or the "coherent dynamical structure factor", $G_{coh}(\mathbf{Q},\mathbf{q},\nu)$. Here the phonon frequency is $\nu = E/h = \omega/2\pi$, the experimental momentum transfer is $\mathbf{Q}$, and the phonon wavevector is $\mathbf{q}$. Missing from our dynamical structure factors are Debye-Waller factors, phonon energies and occupancy factors, and the normalization of neutron flux owing to a change in neutron velocity after scattering. The important features included in the dynamical structure factors are the geometrical efficiencies of how particular phonons will be excited (or de-excited) by a neutron that is scattered with a particular incident energy and momentum transfer. The analogy in a diffraction experiment is that $F(\mathbf{Q})$ describes how the geometrical positions of the atoms in the unit cell affect the phase relationships in coherent elastic scattering.

The dynamical structure factor intensity for incoherent inelastic scattering is $|G_{inc}(\mathbf{Q},\nu)|^2$, which is measured at a particular energy loss, $E=h\nu$ and momentum transfer, $\mathbf{Q}$. The amplitude of the scattering depends on the alignment between the momentum transfer, $\mathbf{Q}$, and the directions of motion or "polarization" of each atom in the unit cell (described in Eqs. 30 and 36 as a 3k-component vector, $\mathbf{U}_{rk}^{\gamma}(\mathbf{q})$, which has x-, y-, and z-components for each of the k atoms in the unit cell). The dependence on alignment is the scalar product $\mathbf{Q} \cdot \mathbf{U}_{rk}^{\gamma}(\mathbf{q})$. The $|G_{inc}(\mathbf{Q},\nu)|^2$ receives smaller contributions from atoms with large mass, M_{rk}, since they have amplitudes of vibration that are inversely proportional to their mass. There is no dependence of $|G_{inc}(Q,\nu)|^2$ on the alignment of the phonon **q** and the momentum transfer, **Q**. The incoherent dynamical structure factor intensity is:

$$|G_{inc}(Q,\nu)|^2 = \sum_{rk} \frac{\sigma_{inc,rk}}{M_{rk}} \sum_{\gamma} \sum_{q} |Q \cdot U_{r_k}^{\gamma}(q)|^2 \, \delta(\nu - \nu_{\gamma}(q)) \, , \qquad (37)$$

The delta function assures the matching of energies of the phonon and the incident neutron. (The crystal vibration cannot grow if there is a mismatch of frequencies of the neutron and phonon waves.) Equation 37 can be evaluated for untextured polycrystals of one atom species. In this case the crystallographic average of Eq. 37 over the various orientations of **Q** is:

$$|G_{inc}(\nu,Q)|^2 = Q^2 \sum_{rk} \frac{\sigma_{inc,rk}}{3M_{rk}} \sum_{\gamma} \sum_{q} |U_{r_k}^{\gamma}(q)|^2 \, \delta(\nu - \nu_{\gamma}(q)) \, . \qquad (38)$$

Equation 38 is closely related to the phonon density of states, $g(\varepsilon)$ (included in Eq. 38 as the distribution of the delta functions $\delta(\nu - \nu_{\gamma}(q))$). It is sometimes possible to use incoherent inelastic neutron scattering to determine $g(\varepsilon)$ for a material. In doing so, the measured data for neutron counts versus energy at a particular value of Q are first corrected for background, and then divided by the thermal factor, $(n(\nu) - 1)/\nu$:

$$\frac{n(\nu)-1}{\nu} = \frac{1}{\nu[1-\exp(-h\nu/kT)]} \, , \qquad (39)$$

which is related to the Bose-Einstein phonon occupancy factor. In ideal experiments, this corrected scattering spectrum will be the phonon density of states of the material. Unfortunately, three problems may vitiate the procedure. First, different species of atoms in the unit cell may have strongly different incoherent cross sections, σ_{inc}. In this case there will be much weaker scattering from phonons that emphasize motions of the atoms with small σ_{inc}. In general, correcting for this effect requires *a priori* knowledge of the $U_{r_k}^{\gamma}(q)$, which is usually equivalent to knowing the phonon DOS itself. Second, at high temperatures it may be possible for two or more phonons to be excited by one neutron. This "multiphonon scattering" causes multiple energy losses (gains, too), and distorts the measured neutron energy loss spectrum. It is often possible to correct for this multiphonon scattering, however, and multiphonon scattering is often negligible for small Q (< 5 Å^{-1}) at temperatures of 300 K and below. Third, there may be a significant contribution from coherent inelastic scattering that must be treated separately.

The dynamical structure factor for coherent inelastic scattering, $G_{coh}(Q,q,\nu)$, is:

$$G_{coh}(Q,q,\nu) = \sum_{rk} \frac{1}{\sqrt{M_{rk}}} \sum_{\gamma} \sum_{g} b_{r_k} Q \cdot U_{r_k}^{\gamma}(q) \; e^{iQ \cdot r_k} \, \delta(\nu - \nu_{\gamma}(q)) \, \delta(Q-q-g) \, . \qquad (40)$$

Here b_{rk} is the coherent scattering length of the nucleus at r_k , which is related to the coherent cross section: $\sigma_{coh} = 4\pi|b_{rk}|^2$. Notice the delta function in Eq. 40 that does not appear in Eq. 37. In addition to the requirement that the frequency of the phonon matches the frequency of the neutron, $\delta(\nu-\nu_{\gamma}(q))$, coherent inelastic scattering also requires matching of the neutron and phonon wavevectors (modulo a reciprocal lattice vector, **g**), so there is the additional factor $\delta(Q-q-g)$. The idea is that phases of oscillation of the neutron wave must match the vibrational motions of all atoms along a phonon. This is a coherent scattering process, so the scattering amplitudes of the atoms and their motions must cooperate to scatter the neutron. The orientation of the momentum transfer along the vibrational direction of an atom is important (as in the previous case of incoherent scattering), hence the product $\mathbf{Q} \cdot \mathbf{U}_{rk}^{\gamma}(\mathbf{q})$. For coherent scattering it is also important that $\mathbf{Q}$ be in phase with atom motions over many unit cells. Hence the additional factor of $e^{i\mathbf{Q} \cdot \mathbf{r}_k}$ in Eq. 40.

Coherent inelastic scattering experiments are most effective when single crystals are available for study. The orientation of the crystalline axes with respect to $\mathbf{Q}$ gives control over the particular phonon that is excited. The experiment is performed by aligning $\mathbf{Q}$ along specific directions of the crystal (usually directions of high symmetry such as (111), (110), (100)). The frequency of the phonon can be measured by scanning the energy loss while maintaining a constant $\mathbf{Q}$. Individual energy spectra are fit to resonance peaks characteristic of (damped) harmonic oscillators, and the energy of the oscillator is extracted as a point on the phonon dispersion curve of ε, $\mathbf{q}$. Coherent inelastic scattering has been the most important method for measuring phonon dispersion curves, which are plots of $q_\gamma(\nu)$ for the different branches, γ, of phonons in a crystalline solid. For polycrystalline solids, the coherent inelastic scattering is averaged over all orientations of the crystallites, and hence over all orientations of $\mathbf{q}$. With a good understanding of the lattice dynamics of single crystals, it is possible to calculate the coherent inelastic scattering from polycrystals. Unfortunately, the inverse problem of going from coherent inelastic scattering spectra of polycrystals to phonon dispersion curves (or even to $g(\varepsilon)$), is usually impossible.

Coherent inelastic scattering experiments on single crystals with well-defined unit cells give rigorous phonon dispersion curves. Alloys with chemical disorder have no unique unit cell with translational periodicity, however, so here the usefulness of dispersion curve measurements is less clear. In measurements of energy spectra at constant $\mathbf{Q}$, the phonon energies are broadened in a disordered alloy. It is often possible to extract an average phonon energy, and dispersion curves can be obtained for alloys with chemical disorder. The Born - von Kármán model is often used for interpreting these average vibrational energies, but it is necessary to select an average unit cell in order to make these interpretations. Consider the case of chemically disordered Cu_3Au. Interpretations of the individual energy spectra are performed by assuming an fcc unit cell with all atoms having the same average mass. This analysis, which assumes an fcc lattice dynamics for the disordered alloy, is known as a "virtual crystal approximation." Since the virtual crystal approximation ignores the details of local chemical environments (which may be Cu-rich or locally ordered, for example) the phonon DOS obtained from the virtual crystal lattice dynamics may not be rigorous.

7. VIBRATIONAL ENTROPY AND PHONON DOS OF TRANSITION METAL ALUMINIDES

Transition metal aluminides, such as Ni_3Al, Fe_3Al, and NiAl, have been the subjects of numerous studies in metallurgy and metals physics over the past decade or so. Their mechanical properties and corrosion resistance at high temperature have motivated engineering applications as components for jet aircraft engines and as piping and valves for the processing of fossil fuels. There has also been a significant effort to understand the phase diagrams of transition metal aluminides. An understanding of the reasons for stability of the cubic phases of the nickel and iron aluminides could perhaps be used to engineer titanium aluminides with crystal structures of cubic symmetry and acceptable ductility at low temperatures.

In thermodynamic equilibrium, the alloy Fe_3Al assumes an equilibrium state of DO_3 chemical order. With cooling rates from the melt exceeding 10^5 K/s, however, it is possible to retain a high degree of chemical disorder in quenched alloys. Alloys of Fe_3Al subjected to piston-anvil quenching, for example, are nearly disordered solid solutions [49,50]. Van Dijk [51] and Robertson [52,53] have used inelastic neutron scattering to measure phonon dispersion curves along high symmetry directions in single crystals of ordered and disordered Fe_3Al. (Their disordered alloy was obtained with an Fe-rich composition.) Using Robertson's force constants up to fifth neighbors (columns 1 and 3 of Table 4 in [53]), we used the Born - von Kármán model of section 5 to calculate the phonon dispersion curves and the phonon DOS for disordered and $D0_3$-ordered Fe_3Al. In these calculations the disordered state was represented as a bcc lattice with a basis of 1 atom having an average mass of the Fe and Al atoms, and the bcc-based $D0_3$-ordered structure was represented as an fcc lattice with a 4-atom basis. The phonon DOS curves are shown in Fig. 3. The most prominent change upon ordering in the phonon DOS curves of Fe_3Al are the gap around 9 THz for the ordered alloy, and the optical modes at 10 - 11 THz.

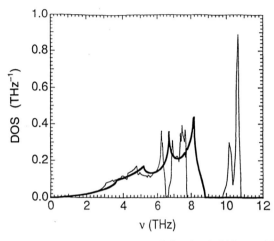

Figure 3. Phonon DOS curves for ordered (thin curve) and disordered (thick curve) Fe_3Al, calculated using force constants from Robertson [53].

Using the two phonon DOS curves of Fe_3Al shown in Fig. 3, the heat capacity at constant volume, $C_V(T)$, was calculated for the disordered and disordered phases using Eq. 19. The difference in these two heat capacity curves, $\Delta C_V(T) \equiv C_V^{dis} - C_V^{ord}$, is shown in Fig. 4. The agreement with the experimental heat capacity, $\Delta C_p(T)$, measured by differential scanning calorimetry, is qualitatively reasonable. The phonon DOS curves and the harmonic approximation with Eq. 19 are more than able to account for the experimental heat capacity. The large change in the optical modes at 10-11 THz (which are absent in the disordered alloy in the virtual crystal approximation) accounts for nearly all of this difference in C_V. The heat capacity from the phonon DOS curves overestimates the heat capacity by nearly a factor of two, however. We attribute this discrepancy to the use of the virtual crystal approximation for the lattice dynamics of disordered Fe_3Al. A realistic model of a disordered alloy would include some local environments that resemble the ordered structure on a local scale. These environments should contribute high energy features to the vibrational spectrum of the solid. This high energy may be scattered by the lack of translational symmetry of the disordered alloy, or may be transferred between the different regions of local chemical order. Although these high energy vibrations are not appropriately described as phonons with Eq. 30, these vibrations do indeed account for thermal energy and must be considered when calculating the heat capacity. If the change in the vibrational intensities at 10-11 THz is not so large as predicted with the virtual crystal approximation, the changes in heat capacity and vibrational entropy will also be smaller.

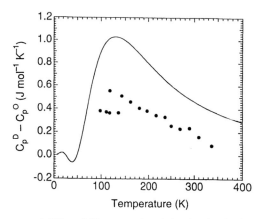

Figure 4. Points: measured differential heat capacity of disordered and ordered samples of Fe₃Al, $\Delta C_p \equiv C_p^D - C_p^O$, as a function of T. Curves were calculated using Eq. 19.

To check for the presence of high frequency vibrational modes in disordered transition metal aluminides, we prepared powdered samples of Ni₃Al by high energy ball milling [32]. These samples had a high degree of chemical disorder. Inelastic neutron scattering from Ni₃Al is dominated by coherent scattering from Ni atoms. Equation 40 for the coherent dynamical structure factor intensity, $|G_{coh}(\mathbf{Q},\mathbf{q},\nu)|^2$, reflects this fact since the ratio $b_{Ni}^2 M_{Al}/b_{Al}^2 M_{Ni} = 4.1$ for coherent scattering intensity. Although the incoherent scattering is weaker, the ratio of $\sigma_{inc,Ni} M_{Al}/\sigma_{inc,Al} M_{Ni}$ is even larger. Neutron inelastic scattering spectra were acquired at five values of Q. Since the lattice dynamics of L1₂-ordered Ni₃Al were known from previous experiments on single crystals, calculations of the crystallographic average of Eq. 40 could be performed the different Q. With these calculated spectra, we could choose how to average the experimental spectra to give a reasonable representation of the phonon DOS. The energy spectra were corrected for the thermal factor of Eq. 39, and were then averaged appropriately to give the phonon DOS curve of Fig. 5. It was not obvious how to correct the data from the disordered sample to account for differences in the coherent and incoherent scattering of Ni and Al. Two independent analyses seemed plausible. If the high frequency "optical modes" were dominated by the motions of

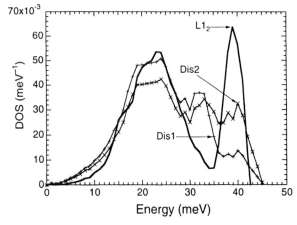

Figure 5. Phonon DOS of Ni₃Al obtained from inelastic neutron scattering spectra at several values of Q [32]. The curve "L1₂" was obtained from the annealed powder with the L1₂ structure. The two curves, "Dis1" and "Dis2", were obtained from the as-milled powder, with two assumptions about the weighting of the observed intensity at around 40 meV (see text).

Al atoms in the disordered alloy, then the weight factor used for the ordered alloy would be appropriate. On the other hand, if the Ni and Al atoms were to both contribute to the high energy modes in the same way as they do for low energy modes, no correction factor would be required for the high energy modes. Both cases are presented in Fig. 5 as Dis2 and Dis1, respectively. Although there are quantitative differences between the results from the two methods of data analysis, the two methods have an important qualitative similarity. High energy modes are present in the disordered alloy, modes that are not predicted by the virtual crystal approximation.

By examining the eigenvectors of the dynamical matrix, $U(k|q)$ in Eqs. 30 and 36, for the $L1_2$-ordered Ni_3Al and for the $D0_3$-ordered Fe_3Al, we found that the high energy vibrations were dominated by the motions of the Al atoms. Interestingly, the low energy part of the phonon DOS in these alloys shows little change upon ordering. Furthermore, the elastic constants of Fe_3Al as a function of composition indicate that ordered Fe_3Al has weaker elastic constants than disordered Fe-Al alloys. This would indicate that the origin for the larger vibrational entropy of the disordered alloy is not in the low frequency vibrational modes. Changes in the high frequency optical modes of the ordered alloy are capable of explaining the difference in the vibrational entropies of disordered and ordered Ni_3Al and Fe_3Al. It therefore appears that changes in the local motions of Al atoms are responsible for the change in vibrational entropy of transition metal aluminides upon chemical ordering.

The vibrational frequencies of the larger Al atoms may be more sensitive to local atomic arrangements than the smaller Ni atoms. Experimental evidence for this sensitivity of Al atoms to disorder in the alloy was provided by extended electron energy loss fine structure (EXELFS) [29], which showed that the mean-squared relative displacements (MSRD) of the Al atoms were more sensitive to the state of chemical disorder than were the MSRD of the Ni atoms. A similar trend was found for Fe_3Al, where the MSRD of Al atoms was shown to be more sensitive to chemical disorder than that of Ni atoms. Other evidence of how disorder in Ni-Al alloys affects vibrational modes was provided by measurements of Ni and Al thermal factors in a diffraction study by Georgopoulos and Cohen [54]. The mean-squared displacements of Ni atoms were larger than those of Al atoms, in contrast to a simple mass effect expected on the basis of Eq. 10, for example. Why should the Al atoms have such limited vibrational amplitudes, and why should they be affected more strongly by chemical ordering than the transition metal (TM) atoms in TM_3Al alloys? A contributing factor may be that the metallic radius of Al is about 1.43 Å, whereas the Ni atom radius is 1.25 Å and the Fe radius is 1.27 Å [55]. The stiffness of effective springs to the larger atoms is perhaps affected more strongly by changes in local atomic structure than is the stiffness for the smaller atom species.

8. VIBRATIONAL ENTROPY AND PHONON DOS OF ORDERED AND DISORDERED Cu_3Au

The fcc - $L1_2$ transition in the alloy Cu_3Au has been an archetype for metallurgical studies of order-disorder transitions. The free energy and the phase diagram of Au-Cu have also been topics of numerous theoretical calculations. Some of the earlier work on the difference in vibrational entropy of ordered and disordered Cu_3Au was reviewed in section 1. Following coherent inelastic neutron scattering measurements by Hallman [56], phonon dispersion curves along high symmetry directions were measured for single crystals of both disordered and ordered Cu_3Au by Katano, Iizumi and Noda [57]. They fit their results to calculations with a Born–von Kármán model, and published sets of force constants for ordered and disordered Cu_3Au.

With differential scanning calorimetry, we measured the difference in heat capacity of chemically disordered and $L1_2$-ordered Cu_3Au from 70 K - 300 K [27]. Results of these measurements are presented in Fig. 6. Figure 6 also compares these measured results of $\Delta C_p(T)$ to a differential heat capacity $\Delta C_V(T)$ calculated with the Born–von Kármán model using the force constants of Katano, et al. [57]. Agreement is qualitatively correct, although the $\Delta C_V(T)$ from the phonon DOS curves lies above the calorimetric data. We again attribute this discrepancy to deficiencies in how the virtual crystal approximation accounts for the vibrational spectrum of the disordered alloy. The virtual crystal approximation overestimates the change in the phonon DOS with disorder in alloy.

Examining the force constants of Katano, et al., for $L1_2$-ordered Cu_3Au [57], we noted that dominant forces between Au-Cu 1nn pairs are radial. We suggested an intuitive picture of the lattice dynamics involving the metallic radii of stiff spheres [27]. Upon ordering, the larger Au atoms serve to separate the more abundant, but smaller Cu atoms. The 1nn force constants involving Cu and Au atoms, $C^{Au\text{-}Cu}_{1xx}$ and $C^{Au\text{-}Cu}_{1xy}$, are indeed large and consistent with stiff spheres in contact. The highest frequency modes in the vibrational spectrum of ordered Cu_3Au involve opposing movements of 1nn Au-Cu pairs, which are controlled by these large force constants. On the other hand, the 1nn Cu-Cu pairs are not in rigid contact, being spread apart by the larger Au atoms. It is therefore not surprising that the 1nn Cu-Cu force constants are weaker. The opposing movements of 1nn Cu-Cu pairs are highly constrained by the surrounding arrangement of large Au atoms. We know little about the vibrational polarizations of the individual atoms in the disordered alloy at high frequencies, but we suggest that on the average they are more isotropic than in the ordered alloy. We suggest that with the development of $L1_2$ order, the directions of movement of both Cu and Au atoms are constrained and generally increased in frequency, owing to the stiff sphere contact between Au and Cu neighbors.

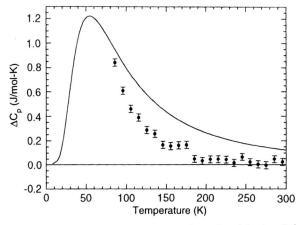

Figure 6. Difference in heat capacity of disordered and ordered samples of Cu_3Au. Points: experimental data with standard deviations between data of different sets. Solid curves were calculated as described in the text.

9. ANHARMONIC VIBRATIONS IN Co_3V

The classical relationship of Eq. 25 is commonly used to impress students of thermodynamics with the power of formal thermodynamic manipulations. It is therefore surprising that this relationship has been given so little attention in the analysis of alloy phase diagrams. The bulk modulus, linear coefficient of thermal expansion, and specific volume are widely known, or at least are not difficult to measure. Consider a typical set of numbers with $B = 3\times10^{11}$ N m^{-2}, $v = 10^{-5}$ m^3 mol^{-1}, and $\alpha = 2\times10^{-5}$ K^{-1}. With Eq. 25 we obtain an anharmonic contribution of $C_{anh}(T) = C_p(T) - C_V(T) = T\times10^{-2}$ [J (mol K)$^{-1}$]. Over a temperature interval of 100 K, with Eq. 3 this corresponds to a change in vibrational entropy of 1 J (mol K)$^{-1}$, or 1/8 k_B/atom. Especially when the thermal expansion of one phase is greater than another, the anharmonic contribution to the entropy of the phase change can be significant.

290

These anharmonic contributions to the entropy should be reflected in changes in the phonon DOS, which affects $C_V(T)$ through Eq. 19, and $C_p(T)$ through Eqs. 20 and 26 or Eqs. 23 and 25. Figure 7 shows experimental phonon DOS curves from Co_3V at elevated temperatures, obtained from inelastic neutron scattering measurements. The interpretations of the data were straightforward, since the scattering is largely incoherent. There was, however, a modest correction for multiphonon scattering that required an iterative procedure described in [34]. The phonon DOS curves in Fig. 7 show a softening with temperature, which although not large is consistent and is well beyond the 0.01 meV energy accuracy of the triple axis spectrometer used for the measurements. Some of this softening can be associated directly with changes in alloy phase with temperature. The data obtained at 1098 K and 1223 K were from the same hP24 phase, however.

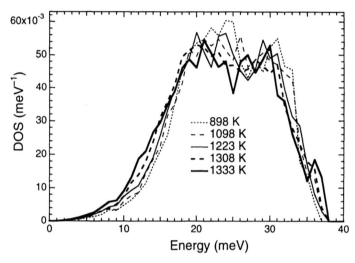

Figure 7. Phonon DOS curves of Co_3V obtained from inelastic neutron scattering spectra measured at various temperatures.

Although the vibrational entropy of a solid phase increases significantly with temperature, what is important for phase stability is the entropies of the different solid phases at the same temperature. The phonon DOS curves of Fig. 7 were therefore used with Eq. 14 to obtain a difference in harmonic vibrational entropy in the high temperature limit. The results of this calculation are presented in Fig. 8. There are two interesting aspects of these data. First, the differences in vibrational entropy of the different phases are provided in the harmonic approximation. For example, the difference in vibrational entropy between the hP24 phase at 1223 K and the fcc phase at 1333 K is about 0.09 k_B/atom. This is probably a smaller contribution to the entropy of the hP24 - fcc phase transition than the configurational entropy of chemical disordering.

The second interesting aspect of these measurements on Co_3V is the anharmonic behavior of the hP24 phase. From Eq. 24 we can predict the temperature dependence of the phonon DOS, and differences between the experimental DOS curves of Fig. 7. In the simple Grüneisen approximation where all phonons have the same mode Grüneisen constant, this difference should be proportional to: $g(\varepsilon) + \varepsilon \, dg/d\varepsilon$. Figure 9 shows the average of the phonon DOS curves obtained from the hP24 phase at 1098 and 1223 K. The calculated difference is presented immediately below it, as is the experimental difference of the two phonon DOS curves from the hP24 phase. The experimental difference curve has a qualitatively poor agreement with the calculation from the Grüneisen approximation. Although the calculated curve was scaled to approximately match the experimental difference at energies from 10 - 20 meV, the two curves agree poorly from 30 - 35 meV. The

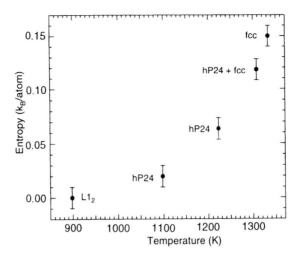

Figure 8. Differences in vibrational entropies in the high temperature limit (Eq. 14) for phases of Co_3V using phonon DOS curves of Fig. 7.

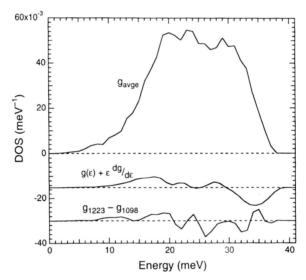

Figure 9. Difference in the phonon DOS curves of the hP24 phase measured at 1098 K and 1223 K. Smooth curve was calculated based on the usual Grüneisen approximation that all phonon modes change with volume by the same Grüneisen constant.

temperature dependence of the phonon DOS of the hP24 phase does not follow a simple Grüneisen model where the same Grüneisen constant applies to all vibrational modes.

From measured values of α, v, and an estimate of B, the anharmonic contribution to the vibrational entropy calculated with Eq. 26 is 0.11 k_B/atom over a temperature range of 200 K around 1000 K. This is larger than the 0.07 k_B/atom over this temperature range predicted with the data of Fig. 8. This difference could be consistent with the error bars shown in the figure, however, and better measurements are required to test the detailed relationship between the temperature-dependence of the phonon DOS and the anharmonic effects in Co_3V.

10. SUMMARY

There is now widespread evidence that vibrational entropy must be included in an understanding of solid state phase transitions. While the basic principles are known, the individual phenomena that contribute to differences in vibrational entropy of solid phases require much more investigation.

We do not expect all of the vibrational modes in a solid to have the same dependence on local atomic structure. In ordered and disordered transition metal aluminides, for example, there is a change in the high energy optical modes upon ordering that can account for most of the difference in vibrational entropy. From other studies performed so far, it seems that vibrational entropy is sensitive to the packing of atoms of different size.

Although much of the vibrational entropy depends on harmonic effects that can be calculated readily with the phonon DOS, anharmonic effects also make important contributions to the entropies of solid phases. Phonon DOS measurements at different temperatures will provide information about these anharmonic contributions, which have been shown to be important for the hP24 phase of Co_3V.

The hope is that there will emerge, at least for specific classes of materials, systematic trends showing how differences in vibrational entropy depend on the local atomic structure in a material.

ACKNOWLEDGMENTS

It is a pleasure to acknowledge the help from many persons involved in this work. Former students Dr. J. K. Okamoto, Prof. L. Anthony, Prof. L. J. Nagel, and present students H. Frase and M. Manley completed much of the experimental work, and set many directions, as did scientists at Caltech, Dr. C. C. Ahn, Dr. G. Le Caër, and Prof. R. Ravelo at U. Texas, El Paso. The ongoing neutron scattering work has involved collaborations with a number of scientists at national laboratories, in particular Drs. R. McQueeney, G. Kwei, R. Robinson and formerly R. M. Nicklow. I am especially grateful to my long-standing collaborator, Dr. J. L. Robertson of the Oak Ridge National Laboratory, for his gracious enthusiasm and many ideas. This work was supported by the U. S. Department of Energy under contract DE–FG03–96ER45572.

REFERENCES

1. J. W. Gibbs, *Trans. Conn. Acad.* 3:108 (1878).
2. E. Ising, *Z. Physik* 31:253 (1925).
3. L. Onsager, *Phys. Rev.* 65:117 (1944).
4. H. A. Bethe, *Proc. Roy. Soc. London A* 150:552 (1935).
5. J. G. Kirkwood, *J. Chem. Phys.* 6:70 (1936).
6. L. D. Landau, *Sov. Phys.* 11:26 (1937). *ibid.* 11:545 (1937).
7. E. M. Lifshitz, *Fiz. Zh.* 7:61 (1942). *ibid.* 7:251 (1942).
8. R. Kikuchi, *Phys. Rev.* 81:988 (1951).
9. D. de Fontaine, *Solid State Phys.* 34:73 (1979).
10. J. M. Sanchez and D. de Fontaine, *Phys. Rev. B* 21:216 (1980).
11. F. Lange, *Z. Physik. Chem.* 110A:360 (1924).
12. F. Seitz. *Modern Theory of Solids.* McGraw-Hill, New York (1940) p. 483.
13. J. C. Slater. *Introduction to Chemical Physics.* McGraw-Hill, New York (1939) p. 215.
14. J. S. Rowlinson, *Proc. Roy. Soc. A* 214:192 (1952).
15. C. Booth and J. S. Rowlinson, *Trans. Faraday Soc.* 51:463 (1955).

16. P. J. Wojtowciz and J. G. Kirkwood, *J. Chem. Phys.* 33:1299 (1960).
17. G. Moraitis and F. Gautier, *J. Phys. F:Metal Phys.* 7:1421 (1977).
18. H. Bakker, *Philos. Mag. A* 45:213 (1982).
19. J. A. D. Matthew, R. E. Jones and V. M. Dwyer, *J. Phys. F: Metal Phys.* 13:581 (1983).
20. H. Bakker and C. Tuijn, *J. Phys. C* 19:5585 (1986).
21. J. Rosen and G. Grimvall, *Phys. Rev. B* 27:7199 (1983).
22. A. F. Guillermet and G. Grimvall, *J. Phys. Chem. Solids* 53:105 (1992).
23. D. T. Hawkins and R. Hultgren, *J. Chem. Thermodynamics* 3:175 (1971).
24. D. B. Bowen, *Acta Metall.* 2:573 (1954).
25. O. Kubaschewski and J. A. Catterall. *Thermochemical Data of Alloys.* Pergamon, London (1956) p. 63.
26. P. A. Flinn, G. M. McManus and J. A. Rayne, *J. Phys. Chem. Solids* 15:189 (1960).
27. L. J. Nagel, L. Anthony and B. Fultz, *Philos. Mag. Lett.* 72:421 (1995).
28. J. Okamoto, C. C. Ahn and B. Fultz, *Proceedings of the XIIth International Congress for Electron Microscopy.* L. D. Peachey and D. B. Williams, eds. San Francisco Press (1990) p. 50.
29. L. Anthony, J. K. Okamoto and B. Fultz, *Phys. Rev. Lett.* 70:1128 (1993).
30. L. Anthony, L. J. Nagel, J. K. Okamoto and B. Fultz, *Phys. Rev. Lett.* 73:3034 (1994).
31. B. Fultz, L. Anthony, J. L. Roberston, R. M. Nicklow, S. Spooner and M. Mostoller, *Phys. Rev. B* 52:3280 (1995).
32. B. Fultz, L. Anthony, L. J. Nagel, R. M. Nicklow and S. Spooner, *Phys. Rev. B* 52:3315 (1995).
33. L. J. Nagel, B. Fultz, J. L. Robertson and S. Spooner, *Phys. Rev. B* 55:2903 (1997).
34. L. J. Nagel, B. Fultz and J. L. Robertson, *Philos. Mag. B* 75:681 (1997).
35. A. Planes, L. Manosa, D. Rios-Jara and J. Ortin, *Phys. Rev. B* 14:7633 (1992).
36. L. Manosa, A. Planes, J. Ortin and B. Martinez, *Phys. Rev. B* 48:3611 (1993).
37. V. L. Morruzzi, J. F. Janak and K. Schwarz, *Phys. Rev. B* 37:790 (1988).
38. J. M. Sanchez, J. P. Stark and V. L. Moruzzi, *Phys. Rev. B* 44:5411 (1991).
39. T. Mohri, S. Takizawa and K. Terakura, *J. Phys. Condens. Matter* 5:1473 (1993).
40. J. W. Cahn, *Prog. Mater. Sci.* 36:149 (1992).
41. S. J. Clark and G. J. Ackland, *Phys. Rev. B* 48:10899 (1993).
42. G. D. Garbulsky and G. Ceder, *Phys. Rev. B* 49:6327 (1994).
43. G. D. Garbulsky and G. Ceder, *Phys. Rev. B* 53:8993 (1996).
44. R. Ravelo, J. Aguilar, M. Baskes, J. E. Angelo, B. Fultz and B. L. Holian, "Free energy and vibrational entropy difference between ordered and disordered Ni$_3$Al", submitted to *Phys. Rev. B.*
45. J. Althoff, D. Morgan, D. de Fontaine, M. Asta, S. M. Foiles and D. D. Johnson, "Vibrational spectra in ordered and disordered Ni$_3$Al", submitted to *Phys. Rev. B.*
46. W. Marshall and S. W. Lovesey. *Theory of Thermal Neutron Scattering.* Oxford, London (1971).
47. G. Kostorz and S. W. Lovesey, *Treatise on Materials Science and Technology Vol. 15 Neutron Scattering.* G. Kostorz, ed. Academic Press, New York (1979) p. 1.
48. G. L. Squires, *Introduction to the Theory of Thermal Neutron Scattering.* Dover, New York, (1978).
49. Z. Q. Gao and B. Fultz, *Philos. Mag. B* 67:787 (1993).
50. B. Fultz and Z. Q. Gao, *Nucl. Instr. and Methods in Phys. Res. B* 76:115 (1993).
51. C. Van Dijk, *Phys. Lett. A* 34:255 (1970).
52. I. M. Robertson, *Solid State Commun.* 53:901 (1985).
53. I. M. Robertson, *J. Phys.:Condens. Matter* 3:8181 (1991).
54. P. Georgopoulos and J. B. Cohen, *Scripta Metall.* 11:147 (1977).
55. E. T. Teatum, K. A. Gschneider, Jr. and J. T. Waber, *Compilation of Calculated Data Useful in Predicting Metallurgical Behavior of the Elements in Binary Alloy Systems*, Los Alamos Laboratory Report LA-4003 (1968).
56. E. D. Hallman, *Can. J. Phys.* 52:2235 (1974).
57. S. Katano, M. Iizumi and Y. Noda, *J. Phys. F: Met. Phys.* 18:2195 (1988).

DIFFUSE SCATTERING BY DOMAIN STRUCTURES

Friedrich Frey

Institut für Kristallographie und Angewandte Mineralogie
Ludwig-Maximilians-Universität München
80333 München, Germany

WHAT IS A DOMAIN STRUCTURE ?

There is no common use of the term "domain structure" in the literature and therefore it is broadly used to specify "coarsened" crystalline structures. Domain formation in a crystal describes a spatial arrangement of different regions which are related to one another in some regular way and which are separated by walls. Common crystals made up by "mosaic" blocks which are separated irregularly by unspecified defects such as dislocations, small angle boundaries or irregular micro-strain fields (and other defects too) are not considered to be domain-structured. Along the same line, a specific defect arrangement in a crystalline structure together with its surrounding strain field, sometimes called "micro-cluster" or "microdomain", does not form a domain structure irrespective of whether or not these clusters show a tendency to order.

The term "domain" usually implies either a spatially varying structure forming separate blocks such as occurring in twin-domain structures (Figure 1), or a spatial variation of a physical property which may be visualized by different configurations, such as the orientation of an electric dipole moment in ferroelectric domains. Whereas an electric dipole moment can again be related to a structural origin viz. the separation of the centre of gravities of negatively and positively charged atoms or molecules, a magnetic domain structure refers to ordering of magnetic moments. The array of these moments may exhibit a domain pattern irrespective of the underlying chemical structure. Structural domains demand a somewhat more detailed consideration under the chemical aspect: different domains may be chemically homogeneous, as in the case of twin domain structures, or chemically heterogeneous, which occurs, for example, in feldspar structures with their complicated Ca/Na- or Al/Si- rich/poor domains. This aspect can be visualized more easily by considering a binary alloy consisting of two chemical species A, B. Depending on the interaction energy between A and B, dissimilar atoms tend to be neighbours or not. Either we have a tendency for A-B ordering in domains which are embedded in an otherwise disordered A/B matrix or a tendency to form A-A-A.. or B-B-B...-type domains (or clusters) in the disordered matrix. Under the action of a driving force, e.g. by variation of the temperature, the disordered "matrix" structure my become thermodynamically unstable. In the former case this process may give rise to "out-of-step" domains (see below) or to completely segregated parts with - somewhat hypothetical- only one residual "domain" boundary between an A-type and a B-type crystallite. (In this context the fascinating fields of domain structure formation in polymers or in porous materials should be mentioned. For example, cylindrically shaped aggregates of block-copolymers with internal "amorphous" structure, organize themselves to an (ordered, e.g. hcp-) domain-like structure. In zeolite-type structures or in other host-guest structures with open tunnels, guest molecules may order in domains which may be related to interesting physical properties. A further discussion is, however, beyond the scope of this paper).

The different aspects which are considered to be essential for the definition of a domain structure, can be summarized by symmetry arguments. (The order parameter concept covers basically the same aspects). Individual domains in a domain structured crystal may be derived from a - possibly hypothetical - higher symmetric "aristophase" obeying the concept of symmetry groups, and may be related by a symmetry operation which is either a point group element or a translational element or a "black and white" symmetry element. The symmetry operation is an element of the point group or the Laue class or a translational vector of the aristophase, which is not an element of the internal structure of the daughter (=domain) phase. Black and white (or colour) symmetry has to be used for magnetic domain structures and may also be used for chemical domain ordering. While preserving the same lattice, the disordered (usually high temperature) phase which is specified by "grey color", decomposes into black and white domains, possibly embedded in the grey matrix (Figure 2). This symmetry approach takes into account the symmetry of statistically defined structures (mixed crystals, alloys).

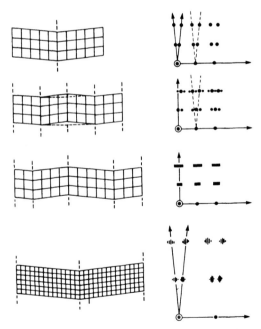

Figure 1. Twin domain structures and corresponding reciprocal lattices. From above: twin pair, periodic twins, non-periodic twinning, periodic twinning with long superperiod.

Domains can exhibit a new order by themselves (superorder), thus creating new symmetries of the superstructure. It depends on this superorder whether or not an averaged structure can be defined meaningfully. If one considers incommensurate structures we have also to include symmetry elements of the superspace group and if one intends to include domain-like ordering in quasicrystalline structures, one has to include also non-crystallographic symmetries.

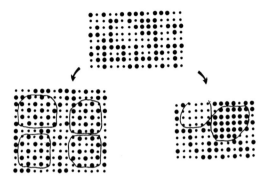

Figure 2. Chemical domains: ordered domains (left), coherent precipitations (right)

Apparently the boundaries between the different domains are essential and must be, in principle, clearly definable. Their existence may even be used for the definition of a domain. This does not mean that the boundaries or domain-walls are simple atomic planes rather than extended intermediate structural states which mellow the transition from one domain to the next one. Occasionally it might be helpful to define an extended domain wall as a new domain with a "gradient" structure which accomodates those of the neighboured domains (Figure 3). On the average, this gradient structure matches the structure of the aristophase. Domain walls "carry" the symmetry change from one to another domain. The strict symmetry relation between the domains may be violated as a consequence of intergrowth: misfits at the planes of coincidence may produce straining effects or single dislocations destroying an exact symmetry relation. There is a stepwise transition from fully coherent domains to fully incoherent crystal parts (Figure 4).

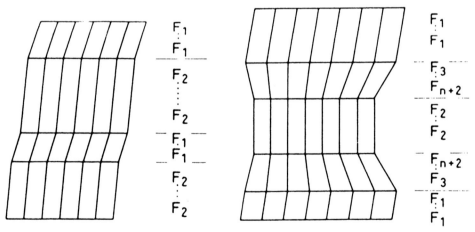

Figure 3. Lamellar domains with structures F_1, F_2 and congruent (left) and incongruent (right) plane of intergrowth. In latter case the walls are modelled by a gradient structure.

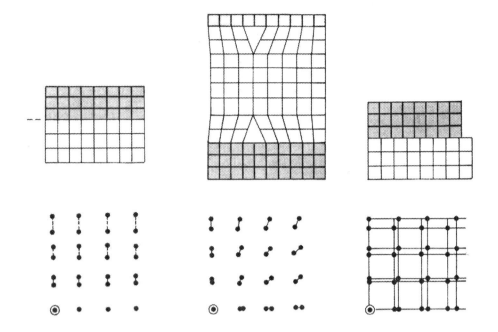

Figure 4. Coherent, semicoherent, and incoherent intergrowth of domains and corresponding reciprocal lattices. Note: The term "domain" looses its sense in the latter case.

Possibly best known are twin-domain and antiphase-domain structures. A twin is a rational intergrowth of congruent or enantiomorphic individuals of the same crystal species in two or more well-defined orientation states. Twin elements are rotational axes, mirror planes or centres of inversion. The contact interface (domain boundary) is in most cases a low-energy boundary with good structural fit. However, no criterion is available to decide what is a "low-energy" boundary in more complicated structures. Twin-domain structures are formed during crystal growth, transformation from a higher symmetry phase (aristophase) or under mechanical pressure. In many examples typical morphological features occur, such as lamellar and polysynthetic twinning (polytypic structures, feldspars, aragonite), others show no distinct features (e.g. switchable domains in ferroic crystals or quartz). If the twin element is not a symmetry element of the lattice symmetry, there is a superposition of differently oriented reciprocal lattices. It depends on the (ir)regularity of the twin-domain structure whether or not (part of) the individual reciprocal lattices coalesce into more or less diffuse spots (Figure 1). If the twin element is a (pseudo) symmetry element of the lattice symmetry, but not of the point group symmetry, the reciprocal lattices coincide (merohedral or pseudomerohedral twins). If we have an equal volume fraction of the individuals, the Bragg reflection pattern exhibits a higher symmetry. If the intergrown twin domains are small, the term domain structure should then perhaps be replaced by the term "disordered structure" or "super-structure" depending on the degree of a possible domain ordering.

Anti-phase domains are related to a lost translation of type $1/2t_n$ where t_n is a translational vector of the aristophase. An example is given by a binary alloy which undergoes a disorder - order transformation by cooling. Accordingly, diffracted waves suffer a phase shift of π for certain directions. More general types of "out-of-phase" domains are due to other fractional translational vectors. A particularly interesting type of out-of-phase domains are responsible for the formation of the so-called shear structures (Figure 5).

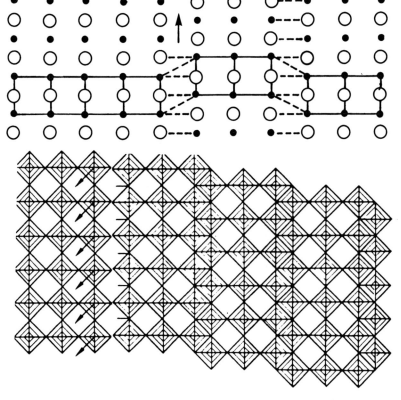

Figure 5. Out-of-phase domains: principle and formation of shear-domain structures

The example of chemical domain ordering gives us a good picture of the occurrence of real domain structures: the length scale may range from a few unit cells to mm size, i.e. macroscopic lengths. Therefore, there is no clear cut separation between a domain structure and a new "structure" on the one side and intergrown crystallites an the other side. With other words, there is a continuous transition between the terms " (disordered) structure" - "domain-structure" - (segregated) "polycrystal" or "microstructure". The size may be affected by many factors such as the origin of the formation (e.g. coarse domains if growth determined), actual temperature (fluctuating sizes close to a phase transformation temperature), electrostatic influence in polar crystals (minimization of large electrostatic energy of an untwinned crystal by creation of domain boundaries), energy of the domain boundaries, stress, internal strain distribution, impurities, and some other factors too. The long-range interaction of strain-fields are, for example, particularly important because they may govern the formation of domain structures in ferroelastic materials (Marais et al.[1]). It is not intended to discuss the various origins of domain pattern formation further in this article.

Depending on the size scale of the domains, there are various methods to recognize and investigate domain structures. Apart from macroscopic methods (morphology, etching, pyro- and piezoelectric probing), optical methods (polarisation microscopy, optical activity), or topography, transmission electron microscopy (TEM) investigations are particularly valuable. Sometimes TEM is the best tool to study local features of domain structure, i.e. domain walls. The method is, however, largely two-dimensional and particularly well suited to study surface domain patterns (which are out of the scope of this article). On the other hand, the average and the distribution of domains are not easy to obtain by TEM investigations, in particular if long-range non-periodic or periodic domain structures and accompanying strain fields are present. Diffraction methods are somewhat complementary in as far as the bulk is accessible and the whole diffracting specimen gives an average information about the domain structure in the bulk as well as an insight into long-range correlations. Domain structures superimposed on the crystallographic structure give rise to non-conventional diffuse scattering or more or less diffuse "reflections". It should be noted that also for this type of investigations the "gap" between EM and diffraction investigations becomes closer with the availability of powerful synchrotron sources (and also high flux reactors): high-intensity and high-resolution instruments allow for a study of smaller samples, and weaker and finely structured diffuse phenomena. Particularly neutron methods are helpful to decide between a static from a dynamic domain behaviour (e.g. domain wall excitations, critical size fluctuations,..). The aspects of dynamics or kinetics, however, are not covered in this article. Only some specific experimental X-ray and neutron diffraction methods are outlined to study static domain structures.

There is no textbook exclusively related to domain scattering. This is partly due to lack of a general theory. Various aspects are summarized, e.g., by Cowley[2], Jagodzinski[3], Jagodzinski and Frey[4], Boysen[5], and Frey[6]. For simple reasons lamellar domains are frequently observed in real domain structured crystals about which an extensive literature exists (see below).

The plan of this article is as follows: First a general outline of the mathematical background is given to describe analytically domain structures and their diffraction patterns. Then different domain models with increasing complexity are outlined to learn about the general characteristics of domain-related diffuse scattering. A special section is devoted to the situation of lamellar domains. This is followed by a short selection of examples taken from work carried out in the home institute of the author. There is a final section about some experimental developments to record diffuse X-ray and neutron intensities.

GENERAL DIFFRACTION THEORY

A study of a domain structure means to solve the individual average structures, to determine the orientation relation between the (twin- and out-of-phase) domains, the size und shape of a domain, the domain size distribution, the distribution of domains and the correlation lengths when domains tend to superorder. An even more complicated and largely unsolved problem concerns the internal "structure" of domain walls, their local orientations, and their extension. This task is particularly complicated when a domain boundary has a different chemical composition (e.g. due to enrichment of impurity atoms), when the congruence between the accompanying domains is lost and when straining and relaxations are associated with the wall formation.

Diffraction phenomena by domain structures can be described in an analytical way by use of Fourier transforms. Vectors in real and reciprocal (=Fourier) space are denoted by $r = (x,y,z)$ and $H = (\xi,\eta,\zeta)$, respectively. The coordinates (x,y,z) and (ξ,η,ζ) refer to arbitrary values in units of basis vectors in respective space.

$$A(H)= \int a(r)\exp\{2\pi iHr\}\ dr \quad \text{and} \quad a(r)= \int A(H)\exp\{-2\pi iHr\}\ dr \qquad (1a,b)$$

shortly written as $a(r) \leftrightarrow A(H)$

We will use the symbols $\rho(r)$ and $F(H)$ for structure (of a unit cell) and corresponding structure factor, respectively. For the purpose of this article some useful laws of Fourier transformation are:

a(-r) ↔ A(-H) which means, if a(r)=a(-r) then A(H)=A(-H)

a$^+$(r) ↔ A$^+$(-H), a$^+$(-r) ↔ A$^+$(H), where "$^+$" denotes "complex conjugate" (c.c.)

Complex functions have to be considered if anomalous X-ray scatterers are present in a structure.

a(r)+b(r) ↔ A(H)+B(H)

s a(r) ↔ s A(H) (s = scalar quantity)

a(r-r$_o$) ↔ exp{2πiHr$_o$}A(H) and A(H-H$_o$) ↔ exp{-2πiH$_o$r}a(r)

a(r)•b(r) ↔ A(H)*B(H)

a(r)*b(r) ↔ A(H)•B(H)

a(r)*a$^+$(-r) ↔ |A(H)|^2 (Note: |A(-H)|^2 ≠ |A(H)|^2 if a(r) is complex)

where the symbol * denotes the operation of convolution:

$$a(r)*b(r) = \int a(r')•b(r-r')\ dr' \ \text{and}\ A(H)*B(H) = \int A(H')B(H-H')\ dH' \qquad (2a,b)$$

Whereas the distributive law remains valid

$$a(r)*[b(r)+c(r)] = a(r)*b(r) + a(r)*c(r)$$

the associative law of multiplication is no longer valid if mixed products oocur:

$$[a(r)*b(r)]•c(r) ≠ a(r)*[b(r)•c(r)]$$

Particularly important functions and their Fourier transforms are:

(1) Patterson function of a structure ρ(r):

$$P(r) = ρ(r)*ρ^+(-r) \ (\text{Note: } P(r) = P(-r)\ \text{holds only if } a(r) \text{ is real !})$$

$$P(r) ↔ |A(H)|^2 \sim I(H)\ (= \text{"intensity" apart from other factors})$$

(2) δ-"functions" δ(r), δ(H):

For a definition of this functional see, e.g., Cowley[2]. We use it here in its "relaxed defini-tion" δ(r)=1 if r=0 and zero elsewhere.

$$δ(r) ↔ 1\ (\text{const}) \ \text{and}\ δ(H) ↔ 1$$

In consequence, sharp phenomena in real space correspond to broad phenomena in reciprocal (Fourier-) space and vice versa.

$$\delta(r-r_0)*\delta(r-r_0') = \delta(r-r_0-r_0')$$

$$\delta(r-r_0) \leftrightarrow \exp\{2\pi iHr_0\}$$

(3) Lattice functions:

Finite lattice (basis a_1,a_2,a_3; t_n= translational vector= $n_1a_1+n_2a_2+n_3a_3 = (n_1,n_2,n_3)$)

$$\Sigma\delta(r-t_n) \leftrightarrow \Sigma\exp\{2\pi iHt_n\} = \{\sin(\pi N_1h)/\sin(\pi h)\} \{..N_2..\} \{..N_3..\}$$

where the summation over n_1, n_2, n_3 extends to N_1, N_2, N_3, respectively.

Infinite lattice:

$$l(r) = \Sigma\delta(r-t_n) \leftrightarrow L(H) = C \Sigma\delta(H-H_0) \quad \text{(reciprocal lattice)}$$

with H_0 = reciprocal lattice vector with integer coordinates h,k,l with respect to a reciprocal basis a_1^*, a_2^*, a_3^*. The normalizing factor C will be omitted in the following.

(4) Box function:

$b(r)=1$ for $-A\leq r\leq +A$ ($A=\alpha a_1+\beta a_2+\gamma a_3$) and zero elsewhere
(α,β,γ are scalar quantities)

$$b(r) \leftrightarrow B(H) = \{\sin(\pi\alpha h)/(\pi h)\} \{..\beta..\} \{..\gamma..\}$$

$$p(r)= b(r)*b(-r) \leftrightarrow \{\sin(\pi\alpha h)/(\pi h)\}^2 \{\sin(\pi\beta k)/(\beta k)\}^2 \{\sin(\pi\gamma l)/(\pi l)\}^2$$

Quite generally a domain structure can now be decribed by (cf. Boysen[5]):

$$\rho_D = [\rho_1*l_1][\Sigma b_{1j}*d_{1j}] + [\rho_2*l_2][\Sigma_{2j}*d_{2j}] + [\rho_3*l_3][\Sigma_{3j}*d_{3j}] +... \tag{3}$$

The l_i, b_{ij}, and d_{ij} describe the individual lattices in domains of type i, the shape of domain i with size j, and the distribution of domains i with size j, respectively. Clearly the b_{ij} include a possible domain size distribution of a domain of type i. As mentioned above, domain walls may be included as an own domain type i, if feasible, and may even be split up into a sequence of "domains" in case of a complicated wall structure.

Fourier transformation gives:

$$F_D = [F_1L_1]*[\Sigma B_{1j}D_{1j}] + [F_2L_2]*[\Sigma B_{2j}D_{2j}] + [F_3L_3]*[\Sigma B_{3j}D_{3j}] +... \tag{4}$$

and the intensity is therefore given by:

$$I_D = \Sigma_i|[F_iL_i]*[\Sigma_j B_{ij}D_{ij}]^2+2\Sigma_{i\neq k}[F_iL_i]*[\Sigma_j B_{ij}D_{ij}][F_k^+L_k^+]*[\Sigma_j B_{kj}^+D_{kj}^+] \tag{5}$$

A further mathematical treatment of this very general expression seems to be not feasible, if no additional information is available from crystal chemistry or from experimental observations. Only some qualitative conclusions can be drawn from this general expression. The first term is usually the leading term, while the second one containing crossterms, provides, in general, only smaller additional changes. The first term corresponds to sharp or diffuse reflections which are not only determined by structure factors F_i, but also by (the Fourier transforms of) the shape, size and distribution functions. If the domain distribution is coherent with the underlying basic lattice, the D_{ij} define a set of δ-functions, (multiplied with the B_{ij}), i.e. sharp reflections (equ.(5) term 1), which may be accompanied by diffuse or satellite scattering. If the domain ordering process is not three dimensional, the "reflections" may be extended in streaks or planes which are (in another meaning) "diffuse" in reciprocal space. If the d_{ij} are incoherent, the cross terms are averaged out, but now the first term decribes diffuse phenomena (e.g. by domain size fluctuations). Interferences between domains of different kinds may be responsible for additional extinction rules of sharp and diffuse reflections. This is discussed in more detail by Jagodzinski and Frey[4].

Basic domain structures

Most helpful for an understanding of domain scattering are considerations of particular domain arrangements. In the following, basic domain problems with increasing complexity are discussed to learn about the diffraction features of a particular domain order.

(1) Two domain types with coherent lattices and random domain distribution

Type 1: $\rho_1(r)$, $l(r)*\rho_1(r)$; $F_1(H)$, $L(H) \cdot F_1(H)$
Type 2: $\rho_2(r)$, $l(r)*\rho_2(r)$; $F_2(H)$, $L(H) \cdot F_2(H)$

Both domain types occur with same a-priori probability and a random mixture is assumed

Domain structure: $\rho_D = [l(r)*\rho_1(r)]b(r) + [l(r)*\rho_2(r)](1-b(r))$ (6)

average structure: $\langle\rho\rangle = 1/2\ (\rho_1+\rho_2)$; $\langle F\rangle = 1/2\ (F_1+F_2)$;
difference structure: $\Delta\rho = 1/2\ (\rho_1-\rho_2)$; $\Delta F = 1/2(F_1+F_2)$;

With $b'(r)=2b(r)-1$ (b' symmetric; $b(r)=1/2(1+b'(r))$; $(1-b(r))=1/2\ (1-b'(r))$)
the domain structure can be written as:

$\rho = \{l(r)*(\langle\rho\rangle+\Delta\rho)\}1/2\ (1+b'(r)) + \{l(r)*(\langle\rho\rangle-\Delta\rho)\}1/2\ (1-b'(r)) =$

$\{l(r)*\langle\rho\rangle\} + \{l(r)*\Delta\rho\}\ b'(r)$ (7)

Fourier transformation gives:

$L(H)\ \langle F(H)\rangle + \{L(H)\ \Delta F(H)\}*B'(H)$; ($B'(H) =2B(H)-\delta(H)$) (8)

There are superimposed sharp and diffuse contributions at any reciprocal lattice point, where the sharp reflections are governed by the average structure factor, the diffuse ones by the difference structure, and the "diffuseness" is due to convolution with B' (more precisely: $B^2(H)$). The general appearance depends on the $\Delta F/\langle F \rangle$-ratio. If there is no average structure $\langle \rho \rangle$, $\langle F \rangle = 0$, only diffuse maxima centred at the reciprocal lattice points can be observed. This latter statement could be violated at some points in reciprocal space: If the structure factor ΔF shows, occasionally, a steep gradient at a position close to a reciprocal lattice point, the convolution product $[L\Delta F]*B$ may have its maximum not exactly at a lattice point. The diffuse maximum occurs "surprisingly" at an incommensurate position which has, however, nothing to do with an incommensurate structure. (Note: This is a quite general aspect of the position of diffuse reflections which should also be taken into account if "wrong" positions or strange profiles are observed in powder patterns).

(2) Distribution of one domain type in a matrix structure with coherent lattice

In a matrix structure $\rho_1(r)$, domains with a different structure $\rho_2(r)$, but the same coherent lattice $l(r)$ are embedded (cf. example (1)). It is assumed that the domains have approximately equal sizes described again by the shape function $b(r)$. The distribution of the domains is denoted by ($t_m = m_1a_1 + m_2a_2 + m_3a_3$ are the centres of the domains):

$$d(r) = \sum_m \delta(r - t_m)$$

The total structure can be written as:

$$\rho_{tot}(r) = l(r)*\rho_1(r) - [l(r)\ b(r)]*\rho_1(r)*d(r) + [l(r)\ b(r)]*\rho_2(r)*d(r)$$

$$= l(r)*\rho_1(r) - [l(r)\ b(r)]*[\rho_2(r) - \rho_1(r)]*d(r) \qquad (9)$$

shortly: $\rho_1(r) = \rho(r)$; $\rho_2(r) - \rho_1(r) = \Delta\rho(r)$, and $F(H)$, $\Delta F(H)$, correspondingly.

$$F_{tot}(H) = L(H)\ F(H) + [L(H)*B(H)]\ \Delta F(H)\ D(H) \qquad (10)$$

$$I(H) \sim |F(H)|^2 = |L(H)F(H)|^2 +$$

$$+ \{[L(H)F(H)][L(H)*B(H)]\ \Delta F(H)D(H)\} + \{cc\}$$

$$+ |L(H)*B(H)|\Delta F(H)D(H)|^2 \qquad (11)$$

Term 1 in equ. (11) denotes sharp reflections governed by the basic structure, terms 2 and 3 denote also contributions to the sharp reflections (multiplication with $L(H)$!), whereas term 4 describes diffuse scattering. Backtransformation of term 4 gives with $l(r) = l(-r)$ and under the assumption of centrosymmetric domains, $b(r) = b(-r)$, which is realistic in most cases:

$$[l(r)b(r)]*[l(r)b(r)]*\Delta\rho(r)*\Delta\rho(-r)*d(r)*d(-r) \qquad (12)$$

The first convolution product in (12) gives simply $l(r)p(r)$ ($p(r) = b(r)*b(r)$). The difference structure $\Delta\rho(r)$ is a non-periodic function. We ignore $\Delta\rho$ in the following to study the influence of $d(r)$ only. With $D(r) = d(r)*d(-r)$ term 4 is written:

$$l(r) [p(r)*D(r)] = l(r)D'(r)$$

The diffuse intensity distribution is governed by the Patterson function of the distribution.

$$|L(H)*D'(H)|^2$$

It depends now on the specific distribution $d(r)$ to analyse the problem further. Limiting cases may be discussed under the assumptions that the domain sizes are more or less well defined (small fluctuations) and that the perfect lattice extends far beyond the correlation length of the distribution $d(r)$. Moreover, we know that $d(r)$ cannot be completely random because the minimum distance between two domains is given by the domain size.

(a) First we assume a periodic domain ordering with a symmetric modulation function (Figure 6a). This is exemplified for the one-dimensional (1D) case, i.e. there is no modulation with respect to y,z, and, in consequence, there arise $\delta(k),\delta(l)$ functions in reciprocal space. The period of $l(x) = a_1$, width of a domain $= 2A$ (in units of a_1) and period of $d(r) = 4A$. $D(r)$ is a pyramid function with period $4A$ which can be written as the convolution product of $p(x)*D(x)$.

$p(x)=1-x/(2A)$ for $|x|\leq 2A$ and zero elsewhere

$D(x)=\Sigma_M\delta(x-M\bullet 4A)$

Fourier transformation gives $(M=\pm 1,\pm 2,\pm 3,...)$

$$P(h)\bullet D(H) = \sin^2(2\pi Ah)/(\pi h)^2\ 1/4A\ \Sigma\delta(h-M/(4A)) \tag{13a}$$

This product must be convoluted with $L(h)=\Sigma_n\delta(h-n)$ (n in units of a_1^*). There are satellite reflections around each Bragg reflection. Because the function $P(h)$ has zero points at positions $h=M/2A$, only satellites of odd order occur (Figure 6a). The distribution function shows its characteristic intensity features close to the Bragg points.

(b) If we assume a periodic, but asymmetric distribution function, e.g. a domain width 1A and a distance of 3A to the next one (period 4A, Figure 6b), we have basically the same qualitative behaviour. There are, however, more satellites of even and odd order at positions $h= M\bullet 1/(4A)$ due to the asymmetry reflected by $p(x)= 1-x/A$ and the corresponding $P(h)$ which has zero points at $h= M\bullet 1/A$. If the asymmetry is described by a general periodic distribution function (1D case)

$d(x) = \Sigma_M\delta(x-M\bullet 2A)$

$D(h) = \Sigma_M\delta(h-M/2A)$ (neglecting a normalizing factor)

we can replace $b'(x)$ and $B'(h)$ from example (1) by

$b''(x)= 2b(x)*d(x)-1$ and $B''(h)=2B(h)D(h)-\delta(h)$

and the diffuse part writes

$\Delta F(h)\{L(h)*B''(h)\} = \Delta F(h)\ C\int\Sigma\delta(h-n-h')\ 2\{\sin\pi h'A/(\pi h')\} - \delta(h')\ dh'$

which gives the general appearance of the satellites governed by the behaviour of the Fourier transform of b(r) and weighted by ΔF:

$$\sum_M 2(\sin\pi M)/(\pi M) \sum_M \delta(h-M/2A) - \sum_n \delta(h-n) \tag{13b}$$

Clearly the diffuse intensity of case (1) is now dissolved into satellites. (The subtraction of a δ-like contribution does not play a role). For the case of periodic domain patterns there exist theories (Korekawa[7], Böhm[8,9]) which provide analytical intensity expressions for the main and satellite reflections. From their intensity characteristics the internal domain structure can be derived such as displacive or density modulations or combinations of them. The theories may also be extended to more complicated shape functions b(r), such as triangular or saw-tooth-like functions.

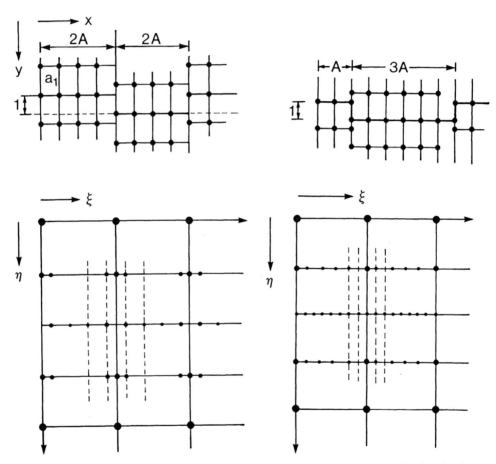

Figure 6. Periodic domain distribution with symmetric (a) and asymmetric distribution function (b).

(c) If the periodic domain distribution only exists over small distances, we have to multiply (in direct space) with a box function corresponding to the correlation length of the domain ordering which means, in Fourier space, another convolution which concerns only the satellite reflections which become diffuse. In total, we have a modulated diffuse intensity distribution close to the (sharp) Bragg reflections.

(3) Domains with same structures and out-of-phase lattices

Assuming a fault vector T, this situation can be described by a modified expression for the domain structure of example (1) ($\rho_1 = \rho_2$)

$$\{l(r)*\rho(r)\}1/2 \ (1+b'(r)) + \{l(r-T)*\rho(r)\} \ 1/2 \ (1-b'(r)) =$$

$$[l(r)+l(r-T)]*\rho/2 + \{[l(r)-l(r-T)]*\rho/2\} \ b'(r) \tag{14}$$

Fourier transformation gives:

$$[L(H)(1+\exp\{2\pi iTH\})]F(H)/2 + [L(H)(1-\exp\{2\pi iTH\})]F(H)/2 * B'(H) \tag{15}$$

We have again sharp and diffuse peaks. An explicite example for this type is given below.

(4) Strained domains

If straining is present in a domain or a domain wall (which may also be treated as separate domain), the lattice inside such a domain can be written as (Boysen[5])

$$l_s(r) = \Sigma_n\delta(r-t_n-s(n))$$

where $s(n)$ describes the atomic displacements in the n.th cell. Fourier transformation gives

$$L_s(H) = \Sigma_n\exp\{2\pi it_nH\} \ \exp\{2\pi is(n)H\}$$

For particular cases, such as a harmonic strain modulation $s(n)$, the theory of modulated crystal structures[7,8,9] can again be applied. This aspect will not be discussed further in detail. It should only be emphasized here that, in contrast to a domain size effect, the higher order satellites become important with increasing value of H, or, in the case of superimposed disorder, the diffuse intensity increases at cost of the accompanying reflection. Here "disorder" means fluctuations of phase and amplitude of an s-wave, or, in the case of more complicated strain modulations, the consideration of a Fourier series and corresponding superimposed sets of satellites. For long wavelengths of the strain modulations, the particular phenomena are observed close to or even within the tails of the Bragg reflections. An adequate treatment demands then a detailed line profile analysis up to high H-values. Measurement of the widths of diffuse phenomena along different directions give an estimate of the anisotropic size of the domains. In practice, however, one has to take care of other diffuse intensity contributions (e.g. TDS) which may obscure the profiles.

(5) Domains with non-coherent lattices

If there are matching problems at domain boundaries or domain size fluctuations, the perfect lattice function exists only over short distances. The centres of the domains are no more defined by one lattice function, i.e. we have a basically non-periodic $d(r)$ function and a more continuous Patterson function $D(r)$. Fourier transformation yields then a major contribtion only to the peak $I(0)$, i.e. to the small-angle scattering regime which is, however, out of the scope in this article.
Most of the real domain structures are somewhere in between these limiting cases.

LAMELLAR DOMAINS

A special paragraph is devoted to lamellar, i.e. one-dimensional, domain ordering because this phenomenon and related stacking fault problems are frequently observed. Planar domains occur in numerous metallic structures which are made up by stacking of closed packed atomic layers (e.g. in fcc and hcp metals). Particularly interesting are polytypic structures with a large variety of stacking periods. Lamellar domain ordering is also observed in layered structures with a high anisotropy of chemical bonding: stacking faults may act as nuclei and may also act as boundaries between sequences of differently stacked layer sequences (lamellar domains). For these problems theories were developed which use fault probabilities such as a-priori probabilities for the occurrence of certain fault layers, and a-posteriori probabilities which describe the occurrence of a certain layer if the preceding one is of a certain type. More probabilities of that kind take into account next-but-one-interactions and so on. These theories use either the difference equations method (Wilson[10], Jagodzinski[11,12]) or the equivalent matrix method (Kakinoki and Komura[13]) or the direct calculation of the Patterson function (Cowley[14]) and allow an identification of the specific type of fault as well as a quantitative determination of the probabilities (frequency of faults) from the diffuse intensity distribution.

To give an impression of the complex formulae in this type of treatment, the case of two types of lamellae stacked upon another along a mean a_3-direction is shortly outlined here[15]. The domains have structures ρ_1, ρ_2, lattice constants $a_3 = \langle a_3 \rangle \pm \Delta(a_1, a_2, a_3)$ and a coherent plane of intergrowth (cf. Figure 3). The stacking is determined by the (complex) probablities $p_{ik}(q)$ which describe the chance to find a layer of type k (k=1,2) q layers apart from a reference layer i and includes a phase vector Φ_q. p_i is the a-priori probablity to find a layer (i=1,2), α and β describe the probabilities that a layer of type i or k, respectively, is continued. The intensity distribution for the most simple case of only next nearest neighbours interactions is given by

$$I(\zeta) = \sum_m \left[A_m (1-|\lambda_m|^2)/\{1-2|\lambda_m|\cos(2\pi(\zeta+\Phi_m))+|\lambda_m|^2\} - \right.$$

$$\left. -2B_m|\lambda_m|\cdot\sin(2\pi(\zeta+\Phi_m))/\{1-2|\lambda_m|\cos(2\pi(\zeta+\Phi_m))+|\lambda_m|^2\} \right] \quad (m=1,2) \qquad (16)$$

λ_m and Φ_m are determined by $\lambda_m = |\lambda_m| \exp(2\pi i\Phi_m)$ where λ_m are the solutions of a quadratic equation $\lambda^2 - \lambda(\alpha \exp(2\pi iH\cdot\Delta) + \beta \exp(-2\pi iH\cdot\Delta)) - 1 + \alpha + \beta = 0$. The constants A_m and B_m can be calculated from the structure factors $F_i, \alpha, \beta, \Phi_q$ in a cumbersome way. Each λ_m describes a reflection: $\lambda_m = 1$ and 0 describe the limiting situation of a sharp reflection and a diffuse streak. Depending on the values of α, β, Δ, different contributions of sharp and superimposed diffuse reflections may be evaluated. The situation of incoherent planes of intergrowth (Figure 3) is much more complicated.

Cowleys theory[14] is also developed for the general case of different kinds of layers. Different translational vectors and arbitrary fault vectors may be combined with a change of scattering density thus defining a planar domain boundary. The general expression for the total diffracted intensity is (also) rather lengthy and is not reproduced here. It was successfully applied to Wadsley-type shear domain structures (cf. Figure 5) and close packed structures.

Another approach is to treat the faulted layer sequences as lamellar domain structure. This approach is more concerned with the observation of planar boundaries and related domain structures in mixed crystals and inorganic compounds especially in low-symmetric systems[16]. For various reasons superstructures are formed in mixed crystals: the supercell is made up from subcells which are structurally very similar. An impurity atom or some other small defect (vacancy, interstitial) may therefore change the correct sequence of the subcells. If one kind of defects dominates, there is a chance of a slightly "favourite" subcell close to this defect. Even in case of small defect concentrations, e.g. a small amount of impurities, there will be a competition between the correct sequence of the subcells and a minimum energy of the defect subcell. Depending on the specific structure there is a chance that these particular "wrong" subcells tend to avoid one another (another case would be the tendency to form a defect-cluster which is not of interest here). For reason of a minimum free energy, the wrong subcells prefer to cumulate in planes thus minimizing the surface energy of the wrong subcell. In consequence, there is a tendency to form planar boundaries. Moreover, due to the tendency to avoid one another, the "faults" arrange in more or less equidistantly spaced planes, or, in other words, there is often even a tendency towards an average periodicity of the lamellar domain stacking which determines the mean thickness of the domains. If one kind of faulting dominates we have equal domain boundaries with the same type of fault vector.

Basic diffraction features of such a lamellar type of domain ordering can be understood by the following simple consideration[17]. Let us define the basic structure by a cell with lattice a_1, a_2, a_3 and a structure ρ and structure factor F. The cell is now subdivided into subcells with $a_i' = a_i/m_i$ where m_i are integers. For the sake of simplicity, the fault vector (out-of-step vector) T is assumed to have two components, one parallel to the (a_1, a_2)-plane and one perpendicular it, i.e. parallel to a_3 (cf. Figure 7):

$$T = s_1 \cdot a_1' + s_3 \cdot a_3' = s_1/m_1\, a_1 + s_3/m_3\, a_3 \quad (s_i \text{ integers}: 0 \leq s_i \leq m_i - 1)$$

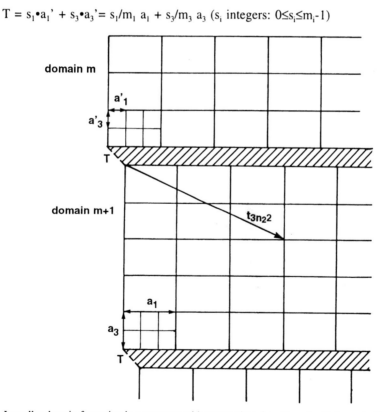

Figure 7. Lamellar domain formation in a structure with superorder where the subcells have similar structures.

The average domain thickness (measured along a_3) is given by a comparably small number N_3 as compared to N_1 and N_2, the number of cells parallel to a_1 and a_2, which go to infinity. The crystal consists of P domains where $(P \cdot N_3)$ is of the same order as N_1 and N_2, i.e. P is assumed to be large. The origin of a domain p (p=1,2,...P) can be chosen at

$$pT + pN_3 a_3 = p\{s_1/a_1 + (N_3 + s_3/m_3)a_3\}.$$

With $t_n = n_1 a_1 + n_2 a_2 + n_3 a_3$ the scattered amplitude A(H) can now be written as

$$A(H) = F(H) \sum\sum\sum\sum \exp\{2\pi i(T+t_n)H\}$$

where the sums over n_1, n_2, p, n_3 run from 0 to N_1-1, N_2-1, P-1, N_3-1, respectively.

$$A(H) = F(H) \sum\exp\{2\pi i n_1 a_1 H\} \sum\exp\{2\pi i n_2 a_2 H\}$$

$$\sum\exp\{2\pi i p[(s_1/m_1)a_1 + (N_3+s_3/m_3)a_3]H\} \sum\exp\{2\pi i n_3 a_3 H\} =$$

$$= G(\xi) \, G(\eta) \, G(\xi,\zeta) \, L_4(\zeta) \qquad (17)$$

Whereas the first three sums give three sets of δ-functions, the last one gives the Laue-function of a finite lattice. These 4 conditions have to be obeyed in reciprocal space:

$$G(\xi) = \sum\delta(\xi - h); \; G(\eta) = \sum\delta(\eta - k) \text{ (h,k=integer)}; \; L_4(\zeta) = \{\sin(\pi N_3 \zeta)/\sin(\pi\zeta)\}$$

$$G(\xi,\zeta) = \sum\delta(\{s_1/m_1 \cdot \xi + (N_3+s_3/m_3) \cdot \zeta\}p) \; \text{(p=integer)}$$

Apart from the influence of the structure factor we have strong and weak reflections, depending on the variation of $L_4(\zeta)$. The weak ones are asymmetrically located around the positions of the reciprocal lattice belonging to the structure of the domain (Figure 8a). The special case of a an out-of-step vector with a component only perpendicular to the domain boundary is shown in Figure 8b : a fault-vector $a_3' = 1/2a_3$ was chosen which describes a (periodic) anti-phase domain structure. If the aristophase has a cubic structure a lamellar type ordering of this type would lead to a superposition of three variants shown in Figure 8c, and if we have to obey, for example a fcc lattice, a final schematic pattern would look like that of Figure 8d. A famous example is the lamellar antiphase-domain structure in Cu_3Au perpendicular to a_1 and with fault vectors $(a_2+a_3)/2$. The diffraction pattern is due a superposition of three variants and shows a characteristic grouping of weak reflections. In fact, due to domain size fluctuations and restricted correlation lengths, the weak reflections become broad and may coalesce in one diffuse maximum, while the integrated intensity remains unaffected (cf., e.g. Warren[18]). More complex centred lattices and corresponding extinction rules are treated by Smith[19] in the context of feldspar structures.

By the same simple treatment the qualitative diffraction phenomena of <u>disordered</u> lamellar domain structures (parallel to a_1, a_2) with coplanar fault-vectors can be derived[17]. For the sake of simplicity we assume a disorder with respect to only two positions $\pm s a_1$ (s denotes a fractional value of a_1) or $\alpha \cdot s \cdot a_1$ ($\alpha = \pm 1$). There is a chance of twin domains,

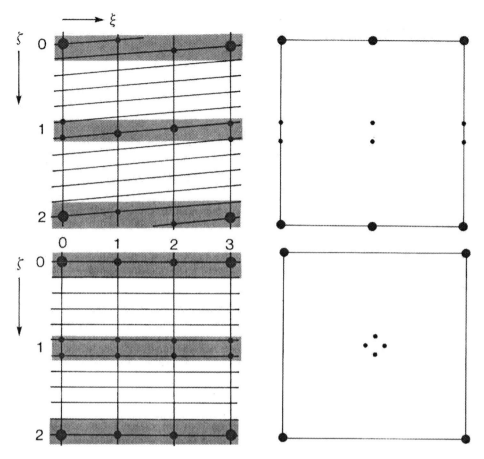

Figure 8. Schematic diffraction patterns of periodic lamellar domains with general fault vector (a), fault vector perpendicular to the domain stacking (b), cubic cell, unique stacking direction (c), and superposition of symmetric variants with fcc lattice (d).

or polytypic domains, or any other type of disordered domains. The origin of a domain which is p periods away from a reference layer is given by

$$t = s \cdot \Sigma_j \alpha_j \cdot a_1 + n_1 a_1 + n_2 a_2 + p a_3; \quad \text{(j sums over the layers)}$$

If there is no further conditional probability, the values of α fluctuate randomly from one layer to the next. The sum $\Sigma \alpha_j$ is even or odd, if p is even or odd, respectively, which can be expressed by

$$\Sigma \alpha_j = p + 2 z_p \quad \text{with } -p \leq z_p \leq 0.$$

The scattered amplitude of the domain crystal consisting of P domains can be written as:

$$A(H) = F(H) \, \Sigma\Sigma exp\{2\pi i(n_1 a_1 + n_2 a_2)H\} \, \Sigma exp\{2\pi i[s(p+2z_p)a_1 + pa_3]H\}$$

$$A(H) = F(H) \, \Sigma\Sigma exp\{2\pi i(n_1 \xi + n_2 \eta)\} \, \Sigma exp\{2\pi i[2sz_p \iota + p(s\xi + \zeta)]\} \tag{18}$$

The first two sums refer again to sums over δ-functions $\delta(\xi-h)$, $\delta(\eta-k)$, h,k being integers. The last sum depends on the value of s. Assuming $s=1/4$ which is observed in several compounds, we have

for $h=4n$ $(n=0,\pm1\pm,2,..)$: $2sz_ph=$ integer, $sh=$integer

$\Sigma\exp\{2\pi ip\zeta\}$ shows peaks for $\zeta=$integer

for $h=4n\pm2$: $2sz_ph=$integer, $sh=$ integer/2 $(=m/2)$

$\Sigma\exp\{2\pi ip(n/2+\zeta)\}$ shows peaks for $\zeta=m/2$.

for h=odd: no general conclusion is possible because z_p is unknown. However, the sum will vanish only at particlar ζ-values and diffuse streaking parallel to a_3^* through h=odd occurs most likely.

An analogous consideration of disordered lamellar domain formation by stacking hexagonal closed packed planes along a_3 (usual axes a_1,a_2,a_3 in a hexagonal system) with erratic or more cooperative faulting by vectors $\pm(1/3a_1+2/3a_2)$ gives also sharp reflections for $(h+k)=3n$ (n integer), $\zeta=$integer and more or less diffuse ones (along ζ) at $(h+k) = 3-n\pm1$. An example (cobalt) is given in the next section.

A general theory of diffraction by crystals with planar domains was developed by Adlhart[20]. Instead of using an increasing number of probabilities to describe increasing domain sizes, domain size distributions are used which can be described by different functions. A domain of type k has a structure ρ_k and a lattice defined by a_{1k},a_2,a_3. Different domains of same type k may have different sizes $a_{1k}\cdot A_k$. These sizes have a distribution $w_k(A)$. Different domain size distributions are assumed to be statistically independent. There are coherent planes of intergrowth parallel to a_2,a_3 (assumed to be equal in all domains). The diffracted intensity occurs therefore on rods perpendicular to these lamellar planes. The crystal is made up of a cyclic sequence of K different types of domains. Each group is counted by l, and the total number of domain groups is L_D (not to be confused with the symbol for a lattice function). The origin of the lattice G_{kl} of a particular domain D_{kl} (of type k in group l) is displaced by an arbitrary fault vector T_k. Then the domain structure can be written as:

$\rho_D = \Sigma_k\Sigma_l\ G_{kl}(r-r_{kl})$ where the r_{kl} are the vectors to the origins of each domain

$G_{kl}(r) = \Sigma_v\delta(r-va_{1k})$

where the sum extends from 0 to A_{kl}, the width of domain k,l. The diffracted intensity is given by Fourier transformation of the Patterson function of this domain structure. After some lengthy calculations and averaging processes, the author arrives at an exact result for the scattering of a domain group:

$I(H) = P_s(H) + 2\ Re\{1/(1-R(H))\} \bullet P_d(H)\}$ (Re = real part of $\{...\}$) (19)

P_s is the Fourier transform of the average Patterson function of uncorrelated domains:

$P_s(H) = \Sigma_k\ \langle G_k(H)G_k^+(H)\rangle F_kF_k^+$

which governs the sharp contribution,

$$P_d(H)=\Sigma_k\Sigma_{k'}\langle G_{k'}(H)\rangle\langle G_k(H)\rangle F_{k'}F_k^*B_{k+1}(H)\bullet..\bullet B_{k'-1}(H)\bullet\exp\{2\pi iH(T_k+..+T_{k'-1})\}$$

where the B_k denote average domain sizes: $B_k(r)=\langle\delta(r-a_{1k}\bullet A_{kl})\rangle$, and

$$R(H) = B_1(H)B_2(H)\bullet...\bullet B_k(H) \; \exp\{2\pi iH(T_1+T_2+...+T_k)\}$$

Bragg scattering occurs if $R(H)=1$. An example is given below.

EXAMPLES

In this section some examples are presented which are taken either from work of the author's group or of colleagues in the home institute of the author.

(1) Lamellar stacking in closed packed structures: cobalt, zincblende

Cobalt is one of the first examples where X-ray diffuse scattering was studied quantitatively (cf. Wilson[10] and references therein). One of the main results was the observation that only reciprocal lattice rows parallel to the stacking direction of the closed packed planes, i.e. parallel to $(111)_{fcc}$ or $(001)_{hcp}$, through h,k with h+k=3n±1, exhibit diffuse reflections. This is quite understandable from the qualitative arguments given above. A quantitative study was carried out by means of neutron diffuse scattering[21]. Neutron scattering was used to get rid of surface effects because the near-surface structure might be quite different from that of the bulk. (It was known from earlier work that small grains of Co show a quite differing structural behaviour, not to be discussed here further). A quantitative analysis with four terms of the kind given by equ.(16) was carried out to take care of the coexistence of fcc and hcp domains with different amounts of (stacking) disorder. This procedure works quite well for temperatures T below T_m ($\approx$ 700K) which indicates the phase transformation temperature from the low T hcp into the fcc phase. Fault probabilities could be extracted quantitatively. Close to and above T_m the fitting with this commonly used formula was not satisfactory. The reason was that in the vicinity of the transformation temperature very small hcp domains occur which could not be treated in the usual way. Adding terms of the form[11]

$$2K_m\{2\lambda_m^2-\lambda_m(1+\lambda_m^2)\cos(2\pi\zeta) + \lambda_m^{W+1} \cos(2\pi\zeta(W+1)) -$$

$$2\lambda_m^{W+2} \cos(2\pi\zeta W) + \lambda_m^{W+1}\cos(2\pi\zeta(W-1))\}\bullet(1-2\lambda_m\cos(2\pi\zeta)+\lambda_m^2)^{-2} \qquad (20)$$

a reasonable fit was achieved (W is the thickness of the hcp-type slabs, K_m are constants) (Figure 9), and the widths of the hcp domains were found to be between 25 and 40 closed packed layers. In other words, close to T_m there are embryonic hcp-domains which have to be considered in the context of the phase transformation. It is worthwhile to note that the fcc-phase shows a completely different behaviour.

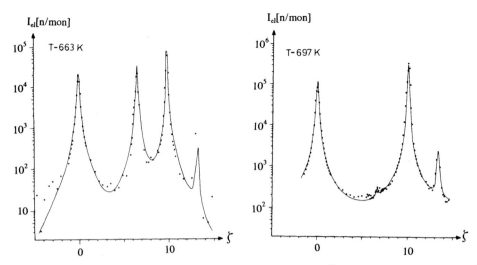

Figure 9. Purely elastic neutron diffraction of pure cobalt taken from [21]: Diffuse scattering along (10.ζ) measured at 2 temperature points close to the martensitic hcp-fcc transformation (~700K) indicate intergrown hcp (peaks at ζ=0 and 1) and fcc (ζ=2/3 and 4/3) lamellar domains. Fitting with the theory given by equs. (16) and (20). Note the logarithmic scale.

An even more complex behaviour was found in the case of zincblende[22], ZnS, which is formally related to the closed packed structures by filling one half of the tetrahedral voids. There is again a cubic (blende) to hexagonal (wurtzite) phase transformation (at ~ 1300K) which is, however, absent in almost fault-free zincblende single crystals. The quantitative analysis of the diffuse streaking from different specimens within the framework of the same theory (no terms equ.(20)), revealed puzzling results: In crystals with a large amount of stacking faults there are heavily disordered cubic domains coexisting with relatively well ordered hexagonal domains. Approaching the transformation temperature wurtzite domains vanish and recover again above 1400K. Other specimens showed a complete different behaviour and, in addition, polytypic sequences, other than the basic fcc or hcp domains, occur. An example of the diffuse rods which have to be analysed is shown in Figure 10. Obviously, the complexity of the domain ordering behaviour is governed by thermal statistics, and polytypic variants are, most likely, "dead ends" of structural fluctuations. Coexisting fcc, hcp and polytypic domains may affect the phase transformation behaviour and my also affect the properties of ZnS material.

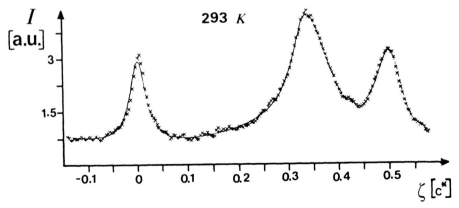

Figure 10. Intensity profile (10.ζ) of a heavily faulted ZnS crystal fitted with the theory given by equ. (16) (4 terms). Figure taken from ref.[22]

314

(2) Domain ordering during phase transformation: enstatite

The pyroxenes are chain-silicates and form an important group of rockforming minerals. Enstatite, $MgSiO_3$, occurs in three basic ambient-pressure polymorphs with reconstructive phase transformations between them. Orthorhombic protoenstatite (PE) is stable above 1300 K, while both orthorhombic orthoenstatite (OE) and monoclinic clinoenstatite (CE) exist at lower temperatures. On cooling from high temperatures (PE phase) a disordered domain structure of intergrown CE and OE domains develops[23]. The structures may be visualized as being composed of two structural layers parallel to the a_2-a_3 plane. The layers consist of SiO_4-chains and bands of MgO_6 octahedra which differ mainly in the orientation of the so-called M1 octahedra which will denoted by "+" and "-" symbols. The different polymorphs differ by different stackings of the layers parallel to the b-c plane with PE: +-+-+-, CE ++++ or ---- (twin domains) and OE ++--++-- (Figure 11). OE may be understood formally as microtwinned CE on a unit cell scale. Due to slight differences of equally oriented layers in CE and OE the unit cell of CE contains two such layers: $2a_{CE}sin\beta$ $\approx a_{OE}$ (β = monoclinic angle). In the course of the PE - OE/CE transformation OE domains separated by CE lamellae, i.e. out-of-step domains, are formed with fault vectors $T=\Delta_1$-$a_{1/OE}+\Delta_3a_{3/OE}$. This gives rise to diffuse streaking and anisotropic broadening of reflections[23,5] (Figure 12). By means of the theory outlined above (equ.(15)) the sharp and diffuse contributions of the reflections can be written as

$$I_{s/d}\sim(1 \pm cos(2\pi(\Delta_1\xi+\Delta_3\zeta))) \quad \text{(subscripts s and d relate to + and - sign, respectively)}$$

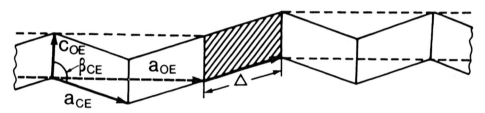

Figure 11. Out-of-step domains in orthoenstatite. The domain wall (shaded area) are characterized by a general (irrational) fault vector (see text). Figure taken from ref.[23].

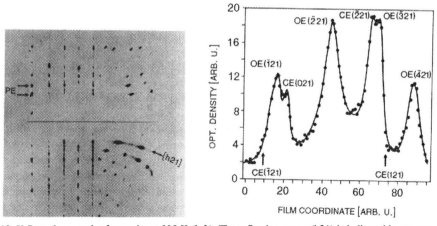

Figure 12. X-Ray photograph of enstatite at 985 K (left). The reflection group (h21) is indicated by an arrow along which an optical densitometer record was taken (right). The profile was fitted by intergrown OE and twinned CE domains. Figure taken from ref.[23]

It was found in this example that reflections (h21) with h=even have a "sharp" contribution of only 7%, whereas those with h=odd have a "sharp" contribution of 93%. Measured profiles were compared with calculated ones (Figure 12b) using $\Delta_1=1/2$ and $\Delta_3=a_{1/OE}/a_{3/OE}$ tang($\beta-\pi/2$). In other words, the analysis of the diffuse scattering allowed for a determination of an irrational fault vector. A detailed discussion is given in ref.[23].

(3) Lamellar exsolutions: pyroxenes

The case of two types of (lamellar) domains fairly often occurs in the course of exsolution processes. For this case equ.(19) is rearranged

$$I(H) = \langle G_1(H)G_1^+(H)\rangle|F_1|^2 + \langle G_2(H)G_2^+(H)\rangle|F_2|^2 + 2Re\{1/(1-R(H)\cdot$$

$$\{\langle G_2(H)\rangle\langle G_1(H)\rangle F_2 F_1^+\cdot exp(2\pi iHT_1)+\langle G_1(H)\rangle\langle G_2(H)\rangle F_1 F_2^+\cdot exp(2\pi iHT_2)$$

$$+\langle G_1(H)\rangle\langle G_1(H)\rangle|F_1|^2\cdot exp(2\pi iH(T_1+T_2))\cdot B_2(H) +$$

$$+\langle G_2(H)\rangle\langle G_2(H)\rangle|F_2|^2\cdot exp(2\pi iH(T_1+T_2))\cdot B_1(H)\}\} \qquad (21)$$

In the simple case with $a_{11}=a_{12}=T_1=T_2 = (a,0,0)$, there are two domains with the same lattice, but different structures and fault vectors T equal to the lattice constant. Then an average superstructure may be defined and the diffraction pattern shows (diffuse) satellites where the degree of diffuseness depends on the average size of the domains and the domain distribution functions $w_k(A)=\Sigma_n v_k(A)\cdot\delta(A-n)$ ($v_k(A)$ is an arbitrary distribution function). Figure 13 shows the calculated diffuse scattering for two types of domains with constant average superstructure (period M = $10\cdot a_{11}$) and various distribution functions $v_k(A)$ (equal for both types of domains). Note that the positions of the diffuse satellites are generally not at commensurate positions in reciprocal space which is due to the actual distribution of the different periods. The satellite positions do not correspond to $1/\langle M\rangle$ but $\langle 1/M\rangle$ which is usually not the same. Note that the domains are fully commensurate on a local scale and the underlying lattices are also commensurate.

This theory was applied by Adlhart [20] to a lunar pyroxene which shows exsolution domains of Ca-rich (augite "A") and Ca-poor (pigeonite "P") lamellae as observed by X-ray photographs (Figure 14). The related A - P splitting of characteristic reflection (004) could be explained by a symmetric distribution function. Additional splittings into (A1/A2/A3) and (P1/P2/P3) groups along other crystallographic directions indicate other sets of super-imposed domains which could (partly) be analysed in terms of asymmetric distribution functions. The parameters of the distributions and the structure factors could be extracted by a comparison of the experimental pattern with a calculated one.

DIFFUSE X-RAY AND NEUTRON DIFFRACTION: TECHNIQUES[25]

There are the well known properties of X-rays and slow neutrons which make either an X-ray or a neutron diffraction experiment more convenient, and there are many problems where only the combination of both diffraction methods give access to the solution of the problem. X-rays probe the charge distribution (electrons) which may differ from the distribution of the nuclei which interact with neutrons (magnetic interactions are not considered here). This is the basis of the so-called X-N technique which could also be exploited for a study of ferroelectric domains not carried out so far (?). Due to the form factor fall off in X-ray scattering and a likewise constant scattering length of neutrons, the neutrons have their advantage when a complete interpretation of diffuse scattering up to

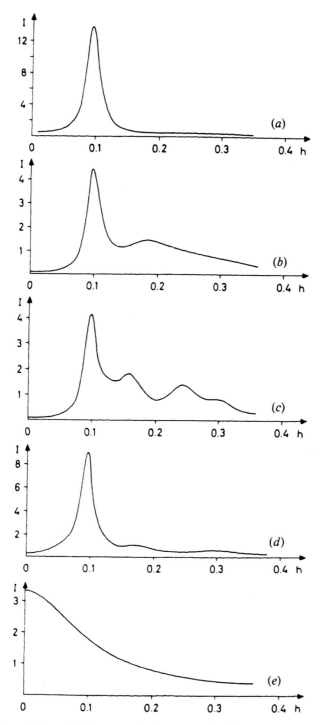

Figure 13. Model calculations of diffuse scattering from a crystal with two types of domains and constant superstructure of $10a_1$ with various distribution functions. (a) Two equal Gaussians, (b) one Gaussian and one domain with constant size, (c) two different box functions, (d) two different Gaussians, (e) two exponentials. Figure taken from ref.[20] with kind permission of the author (for details see there).

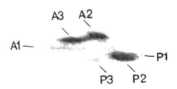

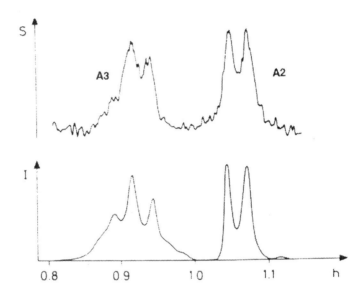

Figure 14. X-ray photograph and densitometer scanning (cf.arrow) of the (004) reflection group of a lunar pyroxene. The diffraction pattern can be understood by intergrown augite (A) and pigeonite (P) domains and further complex intergrowth of A and B-types domains related to the prior thermal history of the moon. The calculated profile (bottom) is based on two boxfunctions (A2 and A3 lamellae) with mean sizes $24 \cdot a_{11}$ and $11 \cdot a_{12}$, and size distributions of ± 5 and ± 2 units, respectively. Figure taken from ref.[20] and also from unpublished results of Jagodzinski and Peterat[24] with kind permission of the authors.

high Q-values ($Q = 2\pi \sin\theta/\lambda$) is required. This is generally the case when relatively small atomic displacement vectors in the domain structure formation are involved or when the difference structure must be determined from diffuse data only. It is, however, fair to say that no dedicated neutron diffractometer for diffuse scattering at high Q exists because

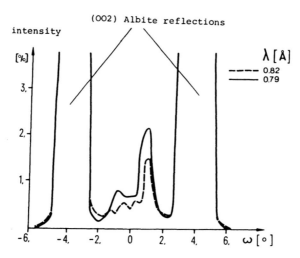

Figure 15. Interfacial diffuse scattering in an albite feldspar with twin domains: contrasting the relative atomic contributions by use of anomalous scattering (reproduced from ref.[25] with kind permission of the authors).

measurements at high Q always suffer from a limited Q-resolution. The scattering power of X-rays increases with Z^2 (Z = number of electrons). This makes it more difficult to detect light elements (e.g. H,O) in the presence of heavy ones with X-rays rather than with neutrons which are scattered by the atomic nuclei in a somewhat erratical way. The analogous argument holds if trying to discriminate between neighbouring elements (e.g. Al/Si). The method of anomalous X-ray dispersion allows, however, to contrast and identify certain elements. Even small concentrations of impurity atoms as low as 10^{-6} can be determined by this method if the impurity atoms are located at specific sites, e.g.in domain boundaries, or if certain other defect structures exist with characteristic diffuse scattering in reciprocal space. This (even weak) diffuse scattering can then be contrasted by tuning the wavelength across a particular absorption edge of the interesting species. This was demonstrated in case of diffuse streaks due to boundaries between lamellar domains in albite-feldspar[26] (Figure 15).

A basic question in all diffuse scattering investigations is, whether the order problem is of static or dynamic nature. Typical energies of dynamic fluctuations, if any, are of the same order as those of thermal neutrons, e.g. 35 meV for neutrons with wavelength of 1.5 Å, which is comparable to CuK_α radiation (which has an energy of 8keV!). This means that an energy analysis of scattered neutrons allows for a study of the dynamics. For a decision between static or dynamic origin, the integral diffuse intensity, as measured with a conventional diffractometer, must be compared with the purely elastic component which can be recorded by placing an (energy-) analysing crystal in the diffracted beam which is set to zero energy transfer. It should be noted that with the advent of the X-ray synchrotron sources it is nowadays possible to measure energy changes down to ~10 meV with moderate resolution. However, no X-ray experiment of that type is known in context with disorder diffuse scattering.

True absorption is generally much weaker in the neutron case. Large samples can be used which seems to be important for diffraction experiments of bulky materials. Moreover, the question of sample environment is less serious than in the X-ray case (see below).

General requirements for (optimized) measurements of diffuse scattering, in particular at a synchrotron source, and necessary considerations of resolution and other data corrections, are summarized elsewhere [5,6,27,28]. Only some aspects will be mentioned, where significant progress has been made recently. Diffuse intensities being spread out in reciprocal space are usually weaker by several orders of magnitude as compared to Bragg intensities. High brilliance synchrotron sources promise dramatic progress, but, in the case of slowly varying diffuse scattering phenomena, the high resolution is often unnecessary, if not to say disadvantageous! In these cases an experimental set-up at a laboratory rotating anode is often superior. To obtain as much information as possible, large regions of reciprocal space need to be explored. To reduce measuring times this requires, besides high intensities, rapid data collection. This is most easily obtained by linear or area detectors.

Perhaps most important is a very low background. Sample incoherent background scattering like fluorescence and Compton scattering for X-rays or incoherent scattering for neutrons deserve special attention and should be reduced as far as possible. Inelastic scattering from phonons is always present and has to be separated. It varies approximately proportionally with temperature and this fact may be used for its discrimination. For X-rays this is the only experimental possibility, whereas with neutrons analysers, time-of-flight recording or, for very high resolution, backscattering or spin-echo methods can be applied to reduce the inelastic components. The problem is particularly serious close to Bragg peaks where the inelastic scattering (TDS) has a peak too. One has to be very careful therefore when investigating diffuse phenomena in this region. Other sources of contaminating spurious background scattering are beam defining elements, sample support and environment, and air scattering of the primary beam. Under this aspect experimental improvements are particularly designed collimators such as He-flooded beam pathways, small vacuum chambers around the sample which can be placed in a large Eulerian cradle, or, as an optimum, a diffractometer entirely enclosed in a large vacuum chamber. Domain ordering phenomena may depend on parameters such as temperature, pressure, electric field,...That means, a variation of these parameters can help to separate different contributions and to interpret more or less complex diffuse diffraction patterns. Then the sample environment, e.g. heating and cooling devices, must be particularly carefully designed to avoid additional spurious scattering as far as possible.

The classical 2D area detector is the photographic film. Perhaps the most important progress in recording diffuse X-ray data is made by the availability of multiwire detectors, imaging plates and CCD cameras. For 2D position-sensitive (proportional) counters problems may arise from inhomogeneities of the wire array as well as the limited dynamic range when a Bragg reflection is accidentally recorded. Image plate (IP) systems have the major advantage of a larger dynamic range, 10^5-10^6, compared to 10^2-10^3 of an X-ray film. IP's allow data collection either in plane geometry or in Weissenberg geometry, both with low and high temperature devices. Clearly CCD detectors are well suited for diffuse scattering too, a basic prerequisite being a low intrinsic noise which can be achieved by cooling with liquid nitrogen. With these new detectors extended diffuse data sets can be collected. The standard technique is the so-called "monochromatic Laue" or "NOROMOSIC" (NOn ROtating MOnochromatic SIngle CRystal) technique, where the crystal is rotated in small angular steps and an image is taken at each step. Plane or cylindrical geometry may be applied. The evaluation of the large amount of data and the reconstruction of the reciprocal space are quite demanding[29].

Neutron diffuse scattering may be recorded at a few dedicated instruments such as D7 at the ILL/Grenoble, DNS at FZ/Jülich or SXD/ISIS-RAL which are equipped with a bank of detectors. The single crystal diffractometer D19/ILL, equipped with a multi-wire area detector is also used for collecting diffuse data. The flat-cone machine E2/HMI-Berlin is equipped with a linear PSD, has an option to record also higher order layers and can also be operated in an "elastic" mode with multi-crystal analyser. The instrument D10/ILL has only a single detector, but can be operated as a low background two-axes or three-axes diffractometer. Neutron sensitive image plates are available, and efforts are currently made towards an applicability for diffuse neutron work[30].

ACKNOWLEDGEMENTS

The author is grateful to Dr. Hans Boysen for many fruitful and stimulating discussions and to Prof. Waltraud Kriven for careful reading of the manuscript.

REFERENCES

1. S.C.Marais, V.Heine, C.M.M.Nex, and E.K.H.Salje. Phys.Rev.Lett. 66:2480 (1991)
2. J.M.Cowley. Diffraction Physics, North-Holland, Amsterdam (1981)
3. H.Jagodzinski. Progr. Crystal Growth and Charact. 114:47 (1987)
4. H.Jagodzinski and F.Frey, in: Int.Tables Vol.B, Ch.4.2, U.Shmueli, ed., Kluwer, Dordrecht (1993)
5. H.Boysen. Phase Trans. 55:1 (1995)
6. F.Frey. Z.Kristallogr. 212:257 (1997)
7. M.Korekawa. Habilitation Thesis, Univ. München, 1967
8. H.Böhm. Acta Crystallogr. A31:622 (1975)
9. H.Böhm. Z.Kristallogr. 143:56 (1976)
10. A.J.C.Wilson. X-Ray Optics, Methuen, London (1962)
11. H.Jagodzinski. Acta Crystallogr. 2:201,208,298 (1949)
12. H.Jagodzinski. Acta Crystallogr. 7:17 (1954)
13. J.Kakinoki and Y.Komura. J.Phys.Soc.Japan 7:30 (1952), 9:169,177 (1954)
14. J.Cowley. Acta Crystallogr. A32:83 (1976)
15. H.Jagodzinski and M.Korekawa. Suppl.3, Geochimica et Cosmochimica Acta 1:555 (1972)
16. P.E.Champness and G.W.Lorimer, in: Electron Microscopy in Mineralogy, H.R.Wenk, ed., Springer, Berlin (1975)
17. K.-H.Jost. Röntgenbeugung an Kristallen, Heyden, Rheine (1975)
18. B.E.Warren. X-Ray Diffraction, Addison-Wesley, Reading Mass., 1969
19. J.V.Smith. Feldspar Minerals, Vol.I, Springer, Berlin (1974)
20. W.Adlhart. Acta Crystallogr. A37:794 (1981)
21. F.Frey and H.Boysen. Acta Crystallogr. A37:819 (1981)
22. F.Frey, H.Jagodzinski, and G.Steger. Bull.Minéral.Cristallogr. 109:117 (1986)
23. H.Boysen, F.Frey, H.Schrader, and G.Eckold. Phys.Chem.Minerals 17:629 (1991)
24. H.Jagodzinski and M.Peterat. Unpublished results
25. H.Boysen and F.Frey. Phase Transitions (in press)
26. H.Berthold and H.Jagodzinski. Z.Kristallogr.193:85 (1990)
27. F.Frey. Acta Crystallogr. B51:592 (1995)
28. T.R.Welberry and B.D.Butler. Chem. Rev. 95:2369 (1995)
29. M.Estermann and W.Steurer, in: Quasicrystals, C.Janot and R.Mosseri, eds., World Scientific, Singapore (1995)
30. Workshop on the Use of Neutron Image Plates, Brookhaven Nat.Lab., Febr. 22-23 1996, Proceedings to appear in J.Neutron Res. (http://bnlstb.bio.bnl.gov/-www_root/webdocs/workshop/nip_main.htmlx)

RECENT "LOCAL" STRUCTURAL STUDIES: METALLIC ALLOYS, SUPERCONDUCTORS AND PROTEINS

George H. Kwei,[1] Despina Louca,[1] Simon J.L. Billinge[2] and H.D. Rosenfeld[3]

[1]Los Alamos National Laboratory, Los Alamos, NM 87545-1663
[2]Department of Physics and Astronomy and Center for Fundamental Materials Research, Michigan State University, East Lansing, MI 48824-1116
[3]Dupont Central Research and Development, Experimental Station, Wilmington, DE 19880-0328

INTRODUCTION

Rietveld structural refinement with the Bragg scattering from a powder diffraction pattern provides structural information that is *averaged* over several hundred angstroms or more. While this information has been extremely useful in our understanding of materials, there are systems whose true *local* structure differs substantially. In these cases, pair distribution function (PDF) analysis of the same powder data can provide a much more accurate description of important structural changes associated with changes in their properties.[1,2] Examples include systems that undergo electronic changes that result in local structural instabilities,[3-6] systems which involve transitions between structures of different crystallographic symmetry but whose *local* structure remain more or less invariant,[7-9] systems which can actually be described as superpositions of simpler structural variants,[10-12] or systems where the local order in the structure changes.[13-18] Our work on the electronic oxides, which includes the high transition temperature cuprate superconductors, the perovskite-like ferroelectrics and the colossal magnetoresistance manganites, has been published and will not be discussed further here.

Instead we will emphasize our progress to date on the study of metallic alloys, and somewhat older but ongoing work on the exploration of the use of *differential* pair-distribution-function (DPDF) analysis of more complex structures to greatly reduce the number of pair correlations that are observed and to simplify the interpretation of the data. So far, metallic alloys have not been studied much using PDF techniques; but the techniques are ideally suited for the study of phenomena such as disorder and phase instability and we hope that our work will represent a small start in what will become an active area. We will describe our use of isotopic-substitution neutron diffraction for DPDFs about the Cu-sites in $YBa_2Cu_3O_{6+\delta}$ (YBCO) and near-resonant x-ray diffraction to measure the local structure about a central heavy atom in systems of biological interest. The work on YBCO is part of a search for changes in the axial Cu-O bond through T_c, where small changes in this bond may be obscured by contributions from the more plentiful, nearly equidistant Y-O bonds. The work on biological systems is motivated by a desire to better understand their structure and changes in the structure with biological function near a heavy metal site. At very high x-ray energies, the lighter atoms will only contribute a Compton component, and this technique will readily yield interatomic distances for multiple metal sites. Near resonance, difference PDFs will give the radial distribution of atoms about that site, much as in x-ray absorption fine structure (XAFS) measurements or high-resolution nuclear magnetic resonance (NMR) experiments, but extending to larger radial distances.

Although we are developing DPDF techniques out of the necessity to solve these interesting problems, the use of the contrast provided by the different scattering lengths from different isotopes in neutron scattering or different x-ray scattering factors near the absorption edges in x-ray scattering will certainly find more common use, especially as the systems we study become more complex. The use of neutron scattering with isotopically labeled YBCO and the use of near-resonant x-ray diffraction to study biological systems represent only the first step towards exploiting the versatility of the DPDF technique.

CRYSTAL AND LOCAL STRUCTURE OF α-PLUTONIUM

Plutonium sits at a boundary in the actinide series that separates $5f$ electron localization and itinerancy. To its left in the periodic table, the f-electrons are delocalized and contribute to the bonding; to its right, the electrons are localized and do not. One manifestation of this is the eight valence states plutonium exhibits in its chemistry. These multiple valence states often lead to the coexistence of different valence states in metallic plutonium so that even pure plutonium may be thought of as a "self-intermetallic" alloy. Another manifestation is the extremely rich structural polymorphism and instability which is exemplified by the six allotropic phases below its melting point at 914 K as it is heated at ambient pressure (as shown in Figure 1).[19] All this suggests that the local structure may differ from the average crystallographic structure. We have embarked on a study of the local structure of the various phases of plutonium in order to provide a better understanding of bonding in this system and to gain insight into the forces that drive the many phase transitions.

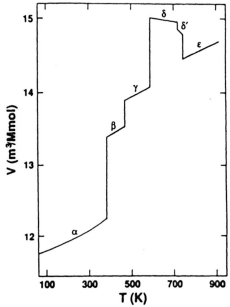

Figure 1. Atomic volumes for the various structural phases of plutonium at ambient pressure (Reference 19, courtesy of the University of California Press). The successive phases (with atoms per unit cell and space group) are α (16, monoclinic $P2_1/m$), β (14, body-centered monoclinic $I2/m$), γ (8, face-centered orthorhombic $Fddd$), δ (4, face-centered cubic $Fm\bar{3}m$), δ' (2, body-centered tetragonal $I4/mmm$) and ε (2, body-centered cubic $Im\bar{3}m$).

We begin our studies with α-plutonium because it is stable from 0 to 395 K and is the most easily accessible from an experimental point of view. It is also the most accessible from a theoretical point of view because the $5f$ electrons are delocalized and, unlike the other phases, band structure calculations have been reasonably successful in predicting its properties.[20] The complex crystal structure has been determined by Zachariasen and Ellinger[21]: α-plutonium is monoclinic (space group $P2_1/m$) with eight atom types all located at either (x,1/4,z) or (-x,3/4,-z) positions, for a total of 16 atoms per unit cell. It has many unusual properties. It is a hard and brittle metal but it is extremely soft vibrationally, with a Debye temperature of ≈200 K,[22] and it has one of the largest coefficients of thermal expansion of any metal. The resistivity exhibits the most anomalous behavior of all nonmagnetic materials: it is pretty anisotropic and shows a maximum ca. 80 K along the <100> direction.[23] The temperature dependence of the magnetic susceptibility is flat, typical of a Pauli paramagnet. Hall effect measurements show that the Hall coefficient changes sign a number of times.[24]

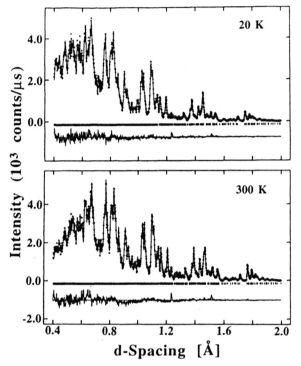

Figure 2. Part of the neutron powder diffraction data for α-Pu at 20 and 300 K are shown in panels (a) and (b), respectively. The points shown by plus (+) marks represent data collected on the +153° detector bank of HIPD. The continuous line through the data is the calculated profile from Rietveld refinement, assuming the same isotropic atomic displacement parameter for all atoms. The tick marks below the data indicate the positions of the allowed reflections for α-Pu in the monoclinic $P2_1/m$ space group. The lower curve in each panel represents the difference between the observed and calculated profiles. Small unindexed Bragg peaks at d-spacings of 1.23 and 1.52Å are from the vanadium sample tube.

The sample used was enriched to 99.9% ^{242}Pu in order to minimize the absorption from the broad neutron resonance from the ^{239}Pu isotope at 0.48 eV. Approximately 12.2 gm of sample in a granular form, was sealed in a double-walled vanadium sample tube with an Ar/He atmosphere. The sample tube was mounted on the tip of a closed-cycle He refrigerator. Neutron powder diffraction data were collected using the High Intensity Powder Diffractometer (HIPD) at the Manuel Lujan Jr. Neutron Scattering Center at Los Alamos. Data were collected to give good statistics at each of the five different temperatures (20, 80, 140, 220 and 300 K). Data were also collected for background and empty container runs, as well as incident spectrum measurements using a vanadium rod, all to be used in the PDF analysis. The crystallographic structure analysis was carried out with standard GSAS Rietveld refinement package.[25] The PDF analysis and real-space refinements were carried out with in-house transformation and refinement packages.

The diffraction data collected at sample temperatures of 20 and 300 K are shown in Figure 2. There are a number of notable features in these data: first, the diffraction pattern is extremely complex consistent with a monoclinic structure; second, the intensity at small d-spacings drops off quickly with temperature, suggesting a steep dependece of the atomic displacement parameters on temperature; and finally, there appears to be a fair amount of diffuse scattering at higher temperatures that most likely results from thermal diffuse scattering.

Table 1. Structural parameters for α-Pu. The space group at all temperatures is $P2_1/m$, with all atoms at (x,1/4,z) sites.[a]

Temperature (K)	20	80	140	220	300
Lattice Parameters:					
a	6.11369(21)	6.12139(19)	6.13280(19)	6.15111(17)	6.17477(17)
b	4.76500(15)	4.77264(14)	4.78237(14)	4.79788(13)	4.81758(13)
c	10.8731(4)	10.8832(4)	10.8965(4)	10.9176(3)	10.9440(3)
β	101.837(3)	101.829(3)	101.816(3)	101.799(2)	101.776(2)
Cell Volume:	310.016(12)	311.202(12)	312.815(12)	315.396(11)	318.703(11)

Occupancies (f), Atomic Positions (x,1/4,z) and Thermal Parameters (U_{iso}):

		20	80	140	220	300
Pu(1)						
	x	0.3372(4)	0.3377(4)	0.3366(4)	0.3373(4)	0.3369(5)
	z	0.15977(26)	0.15948(26)	1.5889(27)	0.15823(28)	0.1583(3)
	U_{iso}	0.004(11)	0.070(11)	0.149(12)	0.250(14)	0.384(16)
Pu(2)						
	x	0.7659(5)	0.7670(5)	0.7680(6)	0.7701(6)	0.7725(7)
	z	0.16995(27)	0.17001(27)	0.16980(28)	0.1704(3)	0.1708(3)
Pu(3)						
	x	0.1292(6)	0.1304(5)	0.1317(6)	0.1342(6)	0.1368(6)
	z	0.33845(28)	0.33852(27)	0.33894(29)	0.3396(3)	0.3399(4)
Pu(4)						
	x	0.6576(6)	0.6567(6)	0.6567(6)	0.6571(6)	0.6577(7)
	z	0.4579(3)	0.4575(3)	0.4573(3)	0.4563(3)	0.4554(4)
Pu(5)						
	x	0.0269(4)	0.0267(4)	0.0272(4)	0.0279(4)	0.0285(5)
	z	0.61983(29)	0.62019(28)	0.6203(3)	0.6207(3)	0.6214(4)
Pu(6)						
	x	0.4657(4)	0.4664(4)	0.4661(5)	0.4657(5)	0.4649(5)
	z	0.64853(27)	0.64881(26)	0.64905(28)	0.64925(28)	0.6503(3)
Pu(7)						
	x	0.3220(5)	0.3229(5)	0.3239(5)	0.3252(5)	0.3291(5)
	z	0.92709(28)	0.92736(27)	0.92697(29)	0.9267(3)	0.9264(3)
Pu(8)						
	x	0.8712(4)	0.8708(4)	0.8708(4)	0.8706(5)	0.8705(5)
	z	0.89437(26)	0.89439(25)	0.89486(27)	0.89529(28)	0.8949(3)
R_{wp}/R_{exp}(%):		1.90/1.22	1.78/1.14	1.80/1.15	1.74/1.09	1.77/1.09
χ^2_{red}:		2.665	2.926	2.890	3.358	3.601

[a]Units for lattice parameters are Å and °, units for cell volume are Å³. Numbers in parentheses following refined parameters represent one standard deviation in the last digit(s). Units for thermal parameters U_{iso} (constrained to be the same for all sites) are 100 Å².

The structures at each of the temperatures can be refined for lattice constants, atomic positions, and a single isotropic atomic displacement parameter (it is possible to refine separate displacement parameters for each atom, but the correlation between parameters becomes large and the values obtained become less reliable). The results of these refinements are shown in Table 1. Many of the observed features are already known.[22] The cell volumes increase rapidly in going from 20 to 300 K, with all the lattice parameters increasing by about 1% (a little more for *b* and a little less for *c*), and with the monoclinic angle decreasing by about 0.06%. As pointed out earlier, all atoms lie on sheets *b*/2 apart at y = 1/4 and 3/4. This leads to many nearest-neighbor pairs separated by approximately 2.4 Å, much smaller than the nearest-neighbor distance of 3.3 Å in the fcc δ-phase. The isotropic displacement parameters also increase sharply as has been observed in a previous determination of these parameters from neutron powder diffraction work.[22] The positional parameters change with temperature generally by about 1% or less. It is interesting to note that the value of χ^2_{red} for the refinements increases with temperature: this is almost entirely due to the increase in diffuse scattering.

The pair distribution function for α-Pu at 20 K shown in Figure 3 is obtained from the neutron powder scattering data by the transformation

$$\rho(r) = \rho_o + \frac{1}{2\pi^2 r} \int_o^{Q_{max}} Q[S(Q)-1] sin(Qr) dQ$$

after correcting the data for scattering by the sample tube and the background, and dividing by the incident neutron spectrum as measured from the scattering from a vanadium rod. For this PDF, as well as all the others for α-Pu, a Q_{max} of 30 Å^{-1} was used. This value appears to be optimal for reducing truncation errors while not losing any resolution in the pair correlations. Also shown in Figure 3 is a model PDF calculated from the crystallographic structural parameters listed in Table 1 for 20 K. It is immediately apparent that the measured and model PDFs agree quite well, especially at larger *r*. The correlation peaks remain at sharp over the entire range and this implies that the structure remains well ordered to large distance. Thus the kind of disorder found by Cox *et al.*[26] from XAFS experiments for the near-neighbor environment in 3.3 at.%Ga doped Pu alloy in the fcc δ-phase (see below) is not present in pure Pu in the α-phase. Deviations in the measured peak heights in the small *r* region may be interesting. These arise from correlated motion, and contain information about the forces that join the near neighbors.

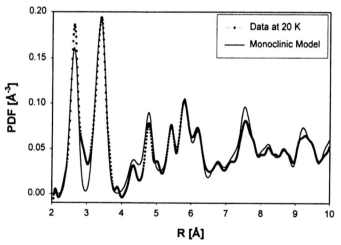

Figure 3. Comparison of experimental and model pair distribution functions for α-Pu at 20K. The model PDF is calculated from the crystal structure reported in Table 1.

The temperature dependence of the PDFs are shown in Figure 4 where only PDFs for 20, 140 and 300 K are shown. All peaks decrease in intensity as the temperature is increased as would be expected from the increased atomic vibrational amplitudes at higher temperatures. Nevertheless, all peaks overlap and this suggests that the increased diffuse scattering at higher temperatures must be thermal in origin rather than displacive.

More qualitative and quantitative information can be extracted from these data by modeling the real space structure. However, the complexity of the monoclinic structure results in a distribution of bonds, each of which may have a different temperature dependence, contributing to each correlation peak. Under these conditions, a real space refinement, without the symmetry constraints imposed on a crystallographic refinement, becomes difficult. The approach we need to take is to make plausible changes in the model until better agreement is obtained, but it may be difficult to obtain a unique model in this way. We hope to start this in the near future.

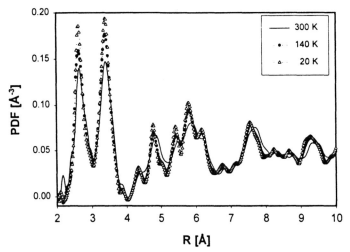

Figure 4. Comparison of experimental PDFs for α-Pu at various temperatures (20, 140 and 300K). PDFs for 80 and 220K are omitted for the sake of clarity.

Little is known about the local structure of the other phases of plutonium. Cox et al.[26] have reported an x-ray absorption fine structure (XAFS) study of the structure of 3.3 at.%Ga doped plutonium alloy. Gallium enters the lattice substitutionally (up to about 10 at.%) and at higher concentrations it is known to stabilize the high temperature fcc δ-phase of plutonium at lower temperatures. At 3.3 at.% doping, the δ-phase is stable at room temperature, but occasionally converts to the α-phase at lower temperatures. The XAFS studies at the Ga K-edge showed that its nearest neighbors are well ordered (with a sharp peak), as are atoms in the second and third coordination shells. The Pu L_{III}-edge data showed that the plutonium environment is substantially less well ordered, with the peaks from the second and third coordination shells completely absent. This finding is surprising for a substitutional alloy and it would be interesting to explore this further with neutron PDF techniques to examine the nature of this disorder.

Since so little is known about the bonding in plutonium, additional information about the interatomic forces will be useful and may be available. Since the volume changes with different phases are so large, the bonding and the participation of the directional $5f$ electrons in that bonding must also be changing. This will certainly affect the stiffness of the bonds to stretches and bends, and this should show up in the interatomic correlations that determine the deviations in the small r peak heights from a Debye model. Hopefully, it may be possible to use molecular dynamics calculations to estimate the force constants that produce these interatomic correlations and thereby learn something about the bonding in plutonium.[27] In addition to the issues raised above, the study of δ-phase plutonium alloy would be extremely interesting. First, the simpler structure may make it easier to derive the interatomic forces. Second, it is much softer than α-Pu, with a Debye temperature estimated at ≈130 K.[22] Finally, δ-Pu shows evidence of having a soft-mode instability and has the most anisotropic shear wave elastic constants [C_{44} is approximately 7 times greater than $C^* = 1/2(C_{11}-C_{12})$] known for any fcc metal,[28] even though its atomic volume (see Figure 1) would suggest that its $5f$ electrons are mostly localized. The elastic constants for the other phases are not known, but the greater participation of the $5f$ electrons in the bonding suggests that the other phases may be even more anisotropic, if they could be measured. Clearly effects like these should show up in the force constants as well.

As alluded to earlier, alloying plays an important role in stabilizing the δ-phase of plutonium. This brings up a whole new range of topics for future study centered on the role that the substituent plays in this stabilization and in the local structure of these alloys.

DIFFERENTIAL PAIR DISTRIBUTION FUNCTIONS OF YBa₂Cu₃O₆₊δ

The crystallography of $YBa_2Cu_3O_{6+\delta}$ has been thoroughly studied as a function of oxygen stoichiometry,[29,30] oxygen ordering[31,32] and temperature.[33,34] One of the first experiments that clearly showed an anomaly in the structure near T_c were the ion channeling experiments of Sharma et al.[35,36] In these experiments, ions channeled along the c-axes of single crystals of $YBa_2Cu_3O_{6+\delta}$ and $ErBa_2Cu_3O_{6+\delta}$ showed a broader angular width below T_c as the crystal axis is rotated, suggesting an ordering or correlation of the atomic displacements in the superconducting state. As is often the case, these changes were not observed in the crystallographic data.[33,34] A polarized XAFS study of the temperature dependence of magnetically aligned powders of $YBa_2Cu_3O_{6+\delta}$ showed an anomaly near T_c, which was interpreted as a split copper-axial oxygen bond length [Cu(1)-O(4)] whose splitting narrowed, or disappeared, near T_c.[37] Others have suggested the possibility of polaronic or bipolaronic transport. Subsequent work has been somewhat equivocal in the support of this finding. Stern et al.[38] have carried out similar experiments with several samples with a range of oxygen stoichiometries. They find that only the sample with the lowest T_c was consistent with a split Cu(1)-O(4) bond length; the sample with highest T_c could be fit with either a single site or a split site and there was no dependence on temperature. Recently, Booth et al.[39] reported careful polarized XAFS study of both single crystals and c-axis aligned thin films of $YBa_2Cu_3O_{6+\delta}$. The single crystal data can indeed be fit with a split site between 50 and 100 K but this splitting does not change near T_c. The thin film data can be fit well with a single site model, limiting any possible Cu(1)-O(4) bond length splitting to <0.09 Å. The general conclusion seems to be that samples with a low oxygen concentration show a Cu(1)-O(4) split bond length and that this splitting may be a static effect. Therefore, both the physical interpretation of this anomaly and its relationship to the superconductivity remains controversial.

The failure to provide more definitive evidence either way is disappointing in the context of other work that clearly established axial oxygen anomalies for other superconducting systems such as Tl-2212,[4] YBC-124,[6] and Bi-2212,[40] using pair distribution function analysis of neutron scattering data or XAFS. One of the problems is that the peak from the planar copper to axial oxygen [Cu(2)-O(4)] bond ca. 2.29Å is obscured by peaks from the many Y-O(2) and Y-O(3) bonds at 2.40 and 2.37 Å, respectively, and small changes in the correlations involving axial oxygen bonds are difficult to see.

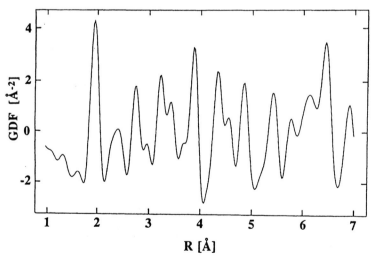

Figure 5. Measured GDF for $YBa_2Cu_3O_{6+\delta}$ showing the large number of peaks arising from the complex structure.

A number of years ago, we had access to relatively large (>10gms.) samples of $YBa_2Cu_3O_{6+\delta}$ that had been labeled to >99% with either ^{63}Cu or ^{65}Cu, and we decided to try a *differential* PDF (DPDF) study to try to resolve the axial oxygen anomaly. The labeled samples had been made 6-7 years earlier and had been stored in a humid climate. We made a few attempts to re-anneal the samples under identical conditions, but the T_c's and oxygen stoichiometries of the two samples were always somewhat different (89K and 6.90 for the ^{65}Cu sample versus 93K and 6.95, for the ^{63}Cu labeled sample). There was also a problem in that the ^{65}Cu labeled sample also contained a small amounta of Y_2BaCuO_5. However, since reprocessing these precious samples was difficult, they were used as re-annealed.

With scattering lengths of 6.43 fm and 10.61 fm for ^{63}Cu and ^{65}Cu, respectively, the contrast is substantial and all correlations between pairs that do not involve Cu should vanish in a DPDF. We collected neutron scattering data for both samples at four sets of temperatures, two each above T_c (97 and 107 K) and two below (87 and 77 K). As in the α-Pu studies, the data were collected on HIPD, for about 8 hrs at each temperature. Figure 5 shows the GDF for $YBa_2Cu_3O_{6+\delta}$ obtained from data collected at 107 K.* Many of the peaks at larger r contain contributions from a number of bonds and, as has been mentioned, the peak at ≈ 2.29 Å includes contributions from both the Y-O(2) anmd Y-O(3) bonds and the Cu(2)-O(4) bond.

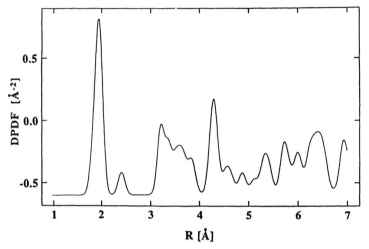

Figure 6. Model DPDFs for $YBa_2Cu_3O_{6+\delta}$ including only Cu correlations.

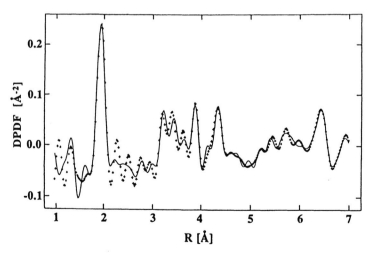

Figure 7. Measured DPDFs for $YBa_2Cu_3O_{6+\delta}$ above and below T_c. The solid line is the DPDF at 107 K, while the plus (+) signs denote the DPDF at 77 K.

---

*For Figures 5 and 10 in this article, we use $G(r)$ instead of $\rho(r)$, hence GDF instead of PDF, in order to accentuate the large r correlations. These functions are related by $G(r) = r[\rho(r) - \rho_0]$. In practice, we use PDFs and GDFs more or less interchangeably.

A model DPDF, corresponding to the crystal structure at 107 K is shown in Figure 6. All correlations that do not involve Cu are now absent. Clearly, the distribution function is considerably simpler: the peak at ≈2.3 Å, now consisting of only the Cu(2)-O(4) contribution, is much smaller and shifted slightly to smaller r; in addition, a number of peaks at larger r are now absent.

The measured DPDFs for 77 K and 107 K are shown in Figure 7. These DPDFs resemble the model shown in Figure 6, but major differences exist. Most importantly, the peak at 2.3 Å corresponding to the Cu(2)-O(4) bond is essentially *absent* at both temperatures. Unfortunately, the noise in the data is large, especially at 107 K. This may mean that the bond is indeed split and more difficult to see. However, given the quality of the data, this is probably all that can be said. Several other peaks, such as those at 3.8 Å and at 4.3 Å, are expected to be much smaller or larger, respectively; also the region between 5 and 6 Å, does not agree well. These deviations are important because the errors are much smaller at larger values of r and the difference between measurement and model becomes much more significant.

These initial experiments thus show some promise in giving a clearer view of the Cu(2)-O(4) correlation. However, both counting statistics and samples need to be improved. We are now making a new set of samples so that we can repeat these experiments in the near future.

APPLICATION OF LOCAL STRUCTURE TECHNIQUES TO BIOLOGICAL STRUCTURES

Traditionally, the structure of large biological molecules has been determined using x-ray absorption fine structure (XAFS) techniques or high-resolution nuclear magnetic resonance (NMR) techniques to provide information over relatively short length-scales or by growing the analogous single crystals and solving the full structure using x-ray crystallography. By using PDF techniques, we believe that we may be able to provide information over longer length scales for peptides, enzymes or proteins, without the need to grow and solve the structure of a single crystal. We can think of three different kinds of experiments in which these techniques may be applied, but all require that the many overlapping peaks arising from the complex structures be reduced in number. All of these should be adaptable to the study of biological molecules in *solution*.

(1) The first kind of experiment is aimed at the determination of the distance between heavy metal centers that depend on their proximity and cooperative action for their biological activity in peptides, enzymes, or proteins. These experiments would be done at very high x-ray energies so that the lighter atoms contribute only to the Compton scattering, which can be removed. Thus, we do not need to rely on DPDFs to reduce the number of pair correlations. This technique has recently been used to study the Peierls distortions in linear Pt-I chains, where the Pt ions exist in two valence states, and has been described by Billinge,[41] Egami[42] and Kycia.[43]

(2) The second kind of experiment is a DPDF technique that takes advantage of the substantial change in the x-ray scattering factor for a given atom at its absorption edge to give radial distribution functions about that atom. The only requirement is that there be a heavy metal center such as Mo, Ru, Ag or Cd, so that the K-edge is at sufficiently high energy. Since many biological systems of interest have such metal atoms, this is not much of a restriction. The resulting distribution functions will be much like those provided by XAFS, but since the scattering does not depend on the scattering of photoelectrons and their multiple scattering, the range of the technique is much greater than the ≈5 Å typically available. The DPDF data will be useful in determining the structure at much longer ranges and provide the complementary information needed for a more complete description of the structure. Thus this technique can be used to study the medium to long range structure near an active metal site in a peptide, enzyme or protein and to see how this structure changes with function.

(3) Finally we may be able to determine the tertiary structure of proteins from the local structure. Since the DPDFs will extend to large enough distances, the known amino acid sequence and the known covalent bond lengths and bond angles can be used together with an

efficient stochastic procedure to generate trial DPDFs corresponding to various possible conformations. These then can be tested against the measured DPDF until the correct *tertiary* structure is found. One possible difficulty is that each biological system will have to be treated separately; another is that in very large systems the bond lengths and angles may not be so rigidly constrained, especially with packing strains and hydrogen bonding. However, this technique has a good chance of working for smaller systems and should provide a new and powerful way to study biological structures in solution!

We now devote the remainder of this article to a description of our progress to date on experiments which would fit under the second category of experiments. Since Cd has a K-absorption edge at high enough energies (26.7 keV) to allow the collection of x-ray scattering data over a large range of momentum transfer, we concentrate our initial efforts on model compounds of the type $Cd(S-cys)_x(N-donor)_{4-x}$, e.g. cadmium bound to cysteine- and histidine-like residues, which may serve as analogs to a large number of Zn binding peptides and proteins, such as the phytochelatins and metallothioneins, and to the major class of "zinc finger" proteins. A number of model compounds that mimic the near neighbor Cd environment have recently been synthesized and their structures determined from single crystal x-ray crystallography.[44,45] We have been studying the compound $Cd(S-2,4,6-Pr^i_3C_6H_2)_2(bpy)$ and its Se analogue. The structure of this compound is shown in Figure 8.

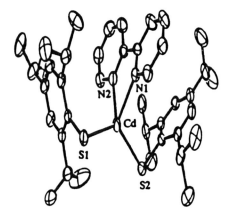

Figure 8. The structure of $Cd(S-2,4,6-Pr^i_3C_6H_2)_2(bpy)$ from single-crystal x-ray diffraction.

In this compound, the central Cd is coordinated to two sulfur ligands from the thiolate (cysteine-like) residues and two nitrogen ligands from the bipyridine (histidine-like) residue. The Cd-S and Cd-N bond lengths for these are approximately 2.46 and 2.28 Å, respectively.

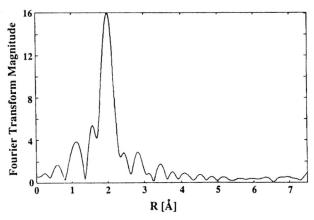

Figure 9. The radial distribution function of atoms around the central Cd obtained from Cd K-edge EXAFS data.

A radial distribution function obtained from room temperature Cd K-edge XAFS for this material (courtesy of Jane G. DeWitt of the San Francisco State University) is shown in Figure 9. The large peak at 2.00 Å is from the Cd-S bonds and is shifted down 0.46 Å from the known values of the Cd-S bond lengths (this "phase shift" is well known and can be accurately accounted for). The observed satellite peak ca. 1.58 Å may result from the Cd-N bonds; however, it is more likely arise from noise in the transformation which can be removed by optimizing the fit to the non-XAFS part of the absorption. Many of the peaks at larger values of r most likely are real, but they are difficult to distinguish from noise peaks in the same region. Thus in this case, XAFS provides information mostly about the nearest neighbor bond lengths and coordination, and little information about the structure at larger r.

X-ray diffraction data for a 50 mg powdered crystalline sample were collected using the six-circle Huber diffractometer on beam line 7-2 at the Stanford Synchrotron Radiation Laboratory (SSRL). The synchrotron ring was operating at 3.0 GeV and 100-50 mA and the wavelength was defined by a Si(220) double monochromator. The sample was mounted in a 2.0 mm thick sample holder with 0.0015" thick self-adhesive kapton windows and θ-2θ scans were carried out in transmission mode. Data were collected at 25 eV and 300 eV below the Cd K-edge which was taken to be 26711 eV at the midpoint of the absorption edge. Scattered signals were measured using an energy dispersive intrinsic Ge detector with three SCA windows set to collect the elastic scattering (including Compton scattering which was later subtracted), fluorescence, and the total signal; the incident intensity was monitored with a NaI(Tl) scintillation counter by measuring the scattering from a kapton film inserted upstream from the sample. We attempted to collect each data set in two scans $0.5{\leq}k{\leq}2.0$ Å^{-1} steps and $2.0{\leq}k{\leq}24.5$ Å^{-1} rocking the sample ±1° about θ in order to average over particle statistics (sample rocking was found to be absolutely essential to provide adequate powder averaging). This resulted in a count time of approximately 17.5 sec per point and excellent statistics in the signals. In the low k scans, Mo filters were inserted to keep the total count rate low enough so that there was no need to correct for photomultiplier and Ge detector dead time, even for the intense Bragg peaks. Separate scans of the empty sample cell (with only kapton windows) were also made at both energies and used later for background subtraction.

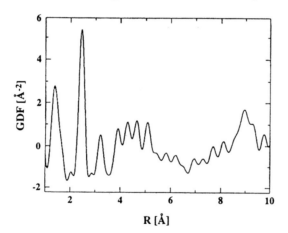

Figure 10. Measured GDF from data collected 300 eV below the Cd K-edge (see text for details).

A sample GDF obtained from the scan at 300eV below the Cd K-edge is shown in Figure 10. The first peak at 1.37 Å corresponds to a mixture of the C-C single bonds (1.54 Å) and double bonds (1.33 Å) in the Cd thiolate. With a little imagination, a break on the high r side of the peak is visible, corresponding to the two sets of peaks. The second peak centered at 2.45 Å corresponds to the Cd-S bonds. Smaller peaks corresponding to the Cd-N bonds may be hidden in the low r side of the Cd-S peaks. The very small peaks on either side of the 2.45 Å peak probably are noise peaks arising from ringing. Peaks at larger distances correspond to various Cd-C, S-S, N-N, C-C and C-N correlations. Some of the overlapping peaks, such as the C-C single and double-bonded peaks and the Cd-S and Cd-N peaks, may be better resolved at lower sample temperatures where the atomic displacement parameters are smaller. Thus in relatively simple molecular systems where there are not too many overlapping correlations, the PDF technique already offers advantages over XAFS in giving more accurate bond lengths at longer distances, better coordination numbers, and information about correlations with more distant neighbors.

In the experiment we performed, the complementary near-edge run (25 eV below the edge) did not have properly set windows and could not be used. Such as energy comparison would have been useful. In the meantime, the proper setting of windows and the treatment of Compton scattering has received much more attention[46] and can now be done in much better fashion. Thus our next attempt to repeat these experiments should work and should provide the DPDFs that can be compared with the known structure from single crystal x-ray diffraction.

Another challenge that remains is to carry out these experiments in solution, and preferably at low temperatures. Presumably in the liquid the intramolecular correlations will remain intact while the intermolecular correlations will be averaged out. The structure for the parts of the molecule that are "floppy" will also average out but these portions of the molecule are probably less important biologically. The interference from intramolecular contributions from the solvent and the necessary "anti-freeze" will each cancel out when taking differences. The structure of the Cd thiolate will most likely be somewhat different in solution and it should be possible to carry out real space refinements to determine the new structure.

In the future, there are many biological systems that contain Cd that would be interesting to look at. One type that comes to mind are the Cd-binding plant peptides (phytochelatins) that contain two to nine amino acids. These are found naturally in jimsun weeds and may be useful in the environmental clean-up of heavy metal contaminants. Another type consists of the animal analogs to the phytochelatins, the metallothioneins. These are small proteins that contain about twenty cysteines and their biological function is to bind and rid the body of heavy metal toxins such as Hg or Cd. There are also many Zn proteins have Cd analogs: this includes the Zn finger proteins that play an important role in RNA transcription. If we get away from Cd there are other heavy metals such as Mo which plays an important role in nitrogenase, and is responsible for nitrogen fixation. The list goes on. Much remains to be done!

CONCLUSION

Pair distribution function techniques have been extremely successful in revealing details of the local structure that are important to the behavior of the electronic oxides. Although we are still actively continuing our study of these systems, we are also beginning to broaden the scope of our work and apply these techniques to the study of other classes of materials. We have been very interested in using local structural techniques to study possible disorder in metallic alloys. As a start, we are studying the local structure of plutonium and its alloys in order to learn more about bonding in these systems; in particular, we hope to learn more about the role of the localization and itinerancy of the $5f$ electrons, about the forces between atoms, about possible disorder in the structure, and about the many phase transitions that take place. At the same time, we have also been interested in the extension of PDF techniques to include *differential* PDF techniques. We are hopeful that isotopic substitution neutron scattering or near-resonant x-ray diffraction will become just as useful a technique as the PDF technique has been and that it will help elucidate the local structure where interfering bonds obscure the correlations of interest or where the structure becomes so complex that the pair distribution function becomes too complicated for analysis. When the DPDF experiments described here are completed, we hope that they will serve as good examples of what can be done.

ACKNOWLEDGEMENTS

We have been fortunate to have many collaborators in our work. Barbara S. Cort, Raymond J. Martinez and Fidel A. Vigil prepared and encapsulated the α-plutonium sample used. The isotopically labeled $YBa_2Cu_3O_{6+\delta}$ samples were provided to us through the courtesy of Susumu Ikeda of KENS; and were prepared by Y. Tokura's group at the University of Tokyo. The cadmium thiolate complex was provided to us by Stephen Koch of the State University of New York at Stony Brook. The XAFS structure for the cadmium thiolate was determined by Jane G. DeWitt of the San Francisco State University. GHK thanks Corwin H. Booth for a critical reading of the manuscript. SJLB thanks the Alfred P. Sloan Foundation for support as a Research Fellow. The work at Los Alamos National Laboratory was done under the auspices of the U.S. Department of Energy (DOE). The neutron scattering experiments were done at the Manuel Lujan Jr. Neutron Scattering Center at Los Alamos, which is funded as a national user facility by the Division of Materials Sciences of the DOE. The synchrotron experiments were done at the Stanford Synchrotron Radiation Laboratory, which is funded by the Division of Chemical Sciences, DOE, and by the Biomedical Resource Technology Program, Division of Research Resources, National Institutes of Health.

REFERENCES

1. T. Egami, Materials Transactions **31**(3), 163 (1990).
2. B.H. Toby and T. Egami, Acta Crystallogr. **A48**, 336 (1992).
3. W. Dmowski, B.H. Toby, T. Egami and M.A. Subramaniam, J. Gopalakrishnan and A.W. Sleight, Phys. Rev. Lett. **61**, 2608 (1988).
4. B.H. Toby, T. Egami, J.D. Jorgensen and M.A. Subramaniam, Phys. Rev. Lett. **64**, 2414 (1990).
5. S.J.L. Billinge and T. Egami, Phys. Rev. **B47**, 14386 (1993).
6. T. Egami, T.R. Sendyka, W. Dmowski, D. Louca, H. Yamauchi, S. Tanaka and M. Arai, Physica **C235-240**, 1229 (1994).
7. H.D. Rosenfeld and T. Egami, Ferroelectrics **150**, 183 (1993).
8. G.H. Kwei, S.J.L. Billinge, S-W. Cheong and J.G. Saxton, Ferroelectrics **164**, 57 (1995).
9. S. Teslic, T. Egami and D. Veihland, J. Phys. Chem. Solids **57**, 1537 (1996).
10. S.J.L. Billinge, G.H. Kwei and H. Takagi, Physica **B199&200**, 244 (1994).
11. E. Bozin, S.J.L. Billinge and G.H. Kwei, Physica B, in press.
12. S.J.L. Billinge, G.H. Kwei and H. Takagi, Phys. Rev. Lett. **72**, 2282 (1994).
13. S.J.L. Billinge, G.H. Kwei and H. Takagi, Physica **C235-240**, 1281 (1994).
14. S.J.L. Billinge, R.G. diFrancesco, G.H. Kwei, J.J. Neumeier and J.D. Thompson, Phys. Rev. Lett. **77**, 715 (1996).
15. R.G. DiFranceso, S.J.L. Billinge and G.H. Kwei, Physica B, in press.
16. D. Louca, T. Egami, E.L. Brosha, H. Röder and A.R. Bishop, Phys. Rev **B56**, Rapid Communications, 1 (1997).
17. D. Louca and T. Egami, J. Appl. Phys. **81**, 5484 (1997).
18. D. Louca and T. Egami, Physica B, in press.
19. D.A. Young, *Phase Diagrams of the Elements*, University of California Press, Berkeley (1991), p. 225.
20. P. Söderlind, J.M. Wills, B. Johansson and O. Eriksson, Phys. Rev. **B55**, 1997 (1997).
21. W.F. Zachariasen and F.W. Ellinger, Acta Crystallogr. **16**, 780 (1963).
22. A.C. Lawson, private communication.
23. M.B. Brodsky, A.J. Arko, A.R. Harvey and W.J. Nellis, in *The Actinides: Electronic Structure and Related Properties*, ed. A.J. Freeman and J.B. Darby, Jr., Academic Press, New York (1974), p.185-264.
24. T.R. Loree and H.T. Pinnick, J. Nucl. Mater. **38**, 143 (1971).
25. A.C. Larson and R.B. VonDreele, Generalized Structure Analysis System (GSAS), LA-UR-86-748, (1987).
26. L.E Cox, R. Martinez, J.H. Nickel, S.D. Conradson and P.G. Allen, Phys. Rev. **B51**, 751 (1995).
27. M.F. Thorpe, "Advances in PDF Profile Fitting; Studying the PDFs of Alloys," this volume.
28. H.M. Ledbetter and R.M. Moment, Acta Metallurgica **24**, 891 (1976).
29. R.J. Cava, Science **247**, 656 (1990).
30. J.D. Jorgensen, B.W. Veal, A.P. Paulikas, L.J. Nowicki, G.W. Crabtree, H. Claus and W.K. Kwok, Phys. Rev. **B41**, 1863 (1990).
31. T. Zeiske, R.S.D. Hohlwein, N.H. Anderson and T. Wolf, Nature (London) **353**, 542 (1991)
32. T. Zeiske, R.S.D. Hohlwein, R. Sonntag, F. Kubanek and G. Collin, Z. Phys. **B86**, 11 (1992).
33. G.H. Kwei, A.C. Larson, W.L. Hults and J.L. Smith, Physica **C169**, 217 (1990).
34. R.P. Sharma, F.J. Rotella, J.D. Jorgensen and L.E. Rehn, Physica **C174**, 409 (1991).
35. R.P. Sharma, L.E. Rehn, P.M. Baldo and J.Z. Liu, Phys. Rev. Lett. **62**, 2869 (1989).
36. R.P. Sharma, L.E. Rehn, P.M. Baldo and J.Z. Liu, Phys. Rev. **B40**, 11396 (1989).
37. J. Mustre-de Leon, S.D. Conradson, I. Batistic, A.R. Bishop, I.D. Raistrick, M. Aronson and F.H. Garzon, Phys. Rev. **B45**, 2447 (1992) and references cited therein.
38. E.A. Stern, M. Qian, Y. Yacoby, S.M. Heald and H. Maeda, Physica **C209**, 331 (1993).
39. C.H. Booth, F. Bridges, J.B. Boyce, T. Claeson, B.M. Lairson, R. Liang and D.A. Bonn, Phys. Rev. **B54**, 9542 (1996) and references cited therein.

40. A. Bianconi, S. Della Longa, M. Missori, I. Pettiti and M. Pompa, in *Lattice Edffects in High T_c Superconductors*, ed. Y. Bar-Yam, T. Egami, J. Mustre-de Leon and A.R. Bishop, World Scientific, Singapore (1992), p. 65.
41. S.J.L. Billinge, "Real Space Rietveld," this volume.
42. T. Egami, "Dual-Space Analysis of Diffraction Data and Error Analysis in PDFs," this volume.
43. S. Kycia, "Use of High-Energy X-rays for Obtaining Accurate PDFs," this volume.
44. D.T. Corwin, Jr., E.S. Gruff and S.A. Koch, J. Chem. Soc., Chem. Commun., **1987**, 966.
45. R.A. Santos, E.S. Gruff, S.A. Koch and G.S. Harbison, J. Am. Chem. Soc. **112**, 9257 (1990).
46. S. Roorda *et al.*, unpublished work.

STUDIES OF LOCAL STRUCTURE IN POLYMERS USING X-RAY SCATTERING

Michael J. Winokur

Department of Physics
University of Wisconsin
Madison WI 53706

INTRODUCTION

The study of the intrachain and interchain structure in polymeric materials has been an exceedingly important research topic and techniques which can resolve these attributes remain one of the central research pillars used to characterize systems composed of these chain-like molecules. The enormous anisotropy intrinsic to the polymer chain structure is the defining materials property and, as a result, one is interested in structure and structural phase behavior at a large number of length scales. For investigations at the largest microscopic distances, ranging from 50 to 10,000 Å, small-angle neutron and x-ray scattering and light scattering techniques[1, 2] are often employed and these studies can yield a vast amount of detailed data. At shorter length-scales wide-angle x-ray, neutron and electron diffraction techniques all provide complementary information about crystalline, semi-crystalline and amorphous hosts[3, 4, 5].

In large part the most quantitatively accessible features are those generated by crystalline and semi-crystalline materials. In this case the elastic Bragg scattering, at wave vectors $q > 0.1$ Å^{-1} (as obtained from $q = 4\pi/\lambda \sin\theta$ where λ is the scatterer wavelength and 2θ is the angle between the incident beam and the scattered beam), provides direct information about the placement of the polymer chains within a periodic unit cell. Since the chain chemical architecture is generally a known attribute, it is often possible to ascertain the average geometric construction of the polymer host. Still it is important to emphasize that, even in the best-case scenario, this analysis is most sensitive to the interchain packing of the polymer chains and, in general, the unit cell contains a large amount of static and and dynamic disorder. Hence all of the Bragg scattering undergoes a rapid exponential falloff with increasing q so that by 4 Å^{-1} or so this scattering signal is no longer resolvable. Moreover most crystalline polymers contain a considerable fraction of amorphous material which generates a secondary diffuse scattering background. In semi-crystalline hosts the proportion of diffuse scattering signal is increased. Finally I note that many polymeric materials, including those in the melt state, exhibit only diffuse scattering signatures and thus are essentially amorphous or liquid-like. Obviously, in these cases, there can be no analysis of the Bragg scattering yet an understanding the atomic-length-scale structural organization is still desirable.

Local Structure from Diffraction
Edited by S.J.L Billinge and M.F. Thorpe, Plenum Press, New York, 1998

This diffuse scattering signal also contains a superposition of information concerning both the local interchain and intrachain structure of the polymer and, in many instances, it is this local structure which is the most relevant feature for understanding the physical properties of the polymer host. The recent and rapid development of molecular level materials engineering and new light sources[6] in combination with computer generated atomistic simulations of polymeric materials[7, 8, 9] has initiated a resurgence of interest in direct techniques which can adequately resolve the local chain structure. Simulations which do not accurately describe the local structure may not be expected to faithfully reproduce physical properties at larger length scales. Moreover there are a number of novel materials which are extraordinarily sensitive to the nature of the local structure. In conducting polymer hosts electronic charge transport requires motion of charge both along the backbone and between chains to create a three-dimensional conducting matrix. In these materials subtle variations in both the intra- and inter- chain organization can generate profound changes in the measured transport properties[10, 11, 12].

To adequately reconstruct structure at length scales ranging from 1 to 10 Å it is often possible to employ radial- (or pair-) distribution-function analysis (RDF or PDF respectively) techniques. PDF analysis can be successfully used to elucidate structure in a variety of structural "settings" ranging from crystalline to amorphous[13]. These techniques can even be exploited in polymer samples containing appreciable two phase mixtures of crystalline and amorphous components[14]. A number of excellent reviews of radial (or pair) distribution function analysis have appeared in the literature [3, 15, 16] and some these are even specific to polymers[17]. The overall goals in the text that follows are to briefly review a few of the most recent PDF studies of polymeric materials, to highlight the unique attributes of PDF analysis in polymers, and finally, to describe in modest detail some of the specialized difficulties and solutions for obtaining quantitative analysis of the local chain structure when using refinement of x-ray scattering data.

BASIC THEORY

PDF studies require data analysis and reduction techniques which are considerably different than those typically employed for analyses restricted to only the Bragg-like scattering features. In general it is necessary to acquire all scattering data, from moderately low to relatively high wave vector (or q), correct the experimental data for a variety of systematic effects[15] (e.g., geometry, absorption, multiple scattering[18], x-ray Compton scattering[19], backgrounds, etc.) and then perform a properly normalized Fourier transform. In theory this new spectrum contains a weighted distributional average over all atomic pairs. Because of the fundamental complexity in deciphering the superposition of all these distributions it is necessary to choose host system carefully so as to take best advantage of the resultant data. For polymeric materials it is desirable to choose model compounds containing a very limited number of monomer base units [e.g., polyethylene or poly(propylene)] or anomalous scattering scattering centers (e.g., large-Z cations or anion species in polyelectrolytes). In the latter case it then becomes necessary to employ differential anomalous scattering[20] methods.

A common starting point for this analysis is the Debye formula

$$I^{coh}(q) = \sum_{i=1}^{N} \sum_{j=1}^{N} f_i(q) f_j^*(q) \frac{\sin(qr_{ij})}{qr_{ij}}, \tag{1}$$

which describes the general form of the total coherent scattering intensity from a powder averaged system of N atoms where $q(= \frac{4\pi}{\lambda} \sin\theta)$ is the scattered photon momentum transfer, r_{ij} is i-j pair spacing and f_i is the ith atom scattering factor. With neutrons the atom scattering factors are essentially q-independent so that the coherent scattering signal remains strong even at high q. This expression must be recast as

$$I^{coh}(q) = N \sum_{i=1}^{N} |f_i(q)|^2 + \sum_{i,j}^{(i \neq j)} f_i(q) f_j^*(q) \frac{\sin(q r_{ij})}{q r_{ij}}, \tag{2}$$

and then normalized to give

$$i^{coh}(q) = \sum_{i}^{N} |f_i(q)|^2 + \frac{1}{N} \sum_{i,j}^{(i \neq j)} f_i(q) f_j^*(q) \frac{\sin(q r_{ij})}{q r_{ij}}, \tag{3}$$

and rewritten in terms of the structure function

$$H(q) = \frac{[i^{coh} - \langle f^2 \rangle]}{\langle f \rangle^2} = \frac{1}{\langle f \rangle^2} \sum_{i'}^{m} \sum_{j'}^{n} c_{i'} c_{j'} f_{i'}(q) f_{j'}^*(q) e_{i'j'}(q) \tag{4}$$

where

$$\langle f^2 \rangle = \sum_{i'=1}^{n} c_{i'} |f_{i'}(q)|^2 \qquad \text{and} \qquad \langle f \rangle = \sum_{i'=1}^{n} c_{i'} |f_{i'}(q)| \tag{5}$$

with $\langle f^2 \rangle$ representing the self-scattering term, $\langle f \rangle$ a mean scatterer, $c_{i'}$ an element specific fractional concentration (and n or m are the number of elemental constituents) and $e_{i'j'}(q)$ is an element specific partial structure function. Comprehensive discussions are given elsewhere for both conventional and anomalous scattering methods. Equation (4) may be rewritten in terms of a pair of Fourier transforms:

$$G(r) = 1 + \frac{1}{2\pi r \rho_0} \int_0^\infty q H(q) \sin(q r) dq \qquad \text{and}$$

$$q H(q) = \int_0^\infty 4\pi r^2 \rho_0 [G(r) - 1] \frac{\sin(q r)}{r} dr + \int_0^\infty 4\pi r^2 \rho_0 \frac{\sin(q r)}{r} dr \tag{6}$$

where $G(r)$ is the total weighted pair correlation function and ρ_0 is the average scatterer number density. The last integral term represents scattering from the entire sample having a uniform density and is experimentally unobservable. Beyond this point everything becomes problem specific so that one must invoke experiment or model specific details. With x-rays the atom scattering factors are both strongly q and atom dependent[21] so that if one wishes to evaluate the $\int \ldots dq$ integral expression (in order to obtain the total $G(r)$ from various partial $e_{i'j'}(q)$'s) one must use the approximation that all atom types, $f_{i'}(q)$, can be replaced by a single average scatterer $\langle f(q) \rangle$. Is it important to note that atomistic simulations avoid this difficultly since the second term in Eqn. 3 may be evaluated directly. Anomalous differential[20] and compositional difference[22] measurements require significantly more extensive treatments.

POLYMER SPECIFIC ISSUES

The central goal is to develop quantitative methods for resolving both the interchain and intrachain structure in polymeric compounds using either x-ray or neutron based scattering methods. Because there exists an enormous difference between these two structural components, polymer PDF profiles manifest a rather unique line profile. In the case of amorphous polyaniline[23], a relatively rigid rod-like polymer given by

$$([(-C_6H_4-NH-C_6H_4-NH-)_{1-x}] [(-C_6H_4-N=C_6H_4=N-)_x])_n$$

with $0 \leq x \leq 1$, the short-range intrachain pair correlations are extremely well-defined and superimposed on a nearly featureless interchain $G(r)$[24]. Figure 1 shows a representative calculation of the various components in real space. This also has implications for the scattering data in momentum (or q) space. Figure 2 shows a single-chain $q\, H(q)$ model calculation in superposition with experimental data. At low q (under 4 Å^{-1}) the scattering is dominated by interchain pair correlations while at higher wave vectors the oscillatory behavior of the structure function is primarily derived from intrachain pair correlations. If an accurate single chain model is available, then the calculated difference curve, given by subtracting the model $qH(q)$ from the experimental profile, yields a profile containing a superposition of the residual experimental errors and the interchain structure function. From this profile it then is possible to independently reconstruct the interchain $G(r)$. In the case of more flexible polymers (e.g., polyethylene) the interchain/intrachain segregation is markedly less pronounced but still significant.

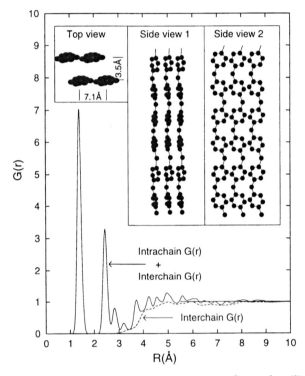

Figure 1: A representative example of a polymer (i.e. polyaniline) G(r) profile explicitly identifying the interchain and intrachain contributions using a nominal crystal structure depicted in the three inset panels assuming a ~ 10 times larger root-mean-square displacement of the interchain atom pairs. Reprinted with permission from ref. [24]. Copyright © 1995 American Chemical Society.

In all situations scattering data out to very high momentum transfer is a prerequisite for a comprehensive analysis. For neutrons, with their q independent atom scattering factors, this does not present an undue complication. X-rays are more problematic. By 20 Å^{-1} the carbon structure function has diminished by almost 95% and is now superimposed on a Compton scattering background almost nine times larger. A typical intensity spectrum highlighting this loss of scattering intensity is shown in Fig. 3. Moreover the intrachain scattering is only a small fraction of the total coherent

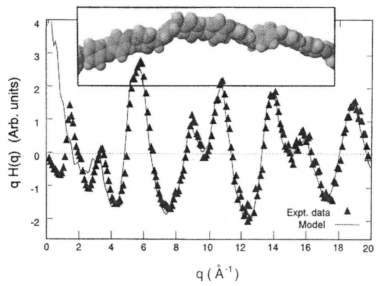

Figure 2: A comparison between the experimentally derived $qH(q)$ (filled triangles) and that obtained from a simple single chain model (thin line) as shown in the inset for an as-cast film of emeraldine base ($x=0.5$).

scattering signal. Ideally a large aperture detector (to increase the effective count rate) combined with exceptional energy resolution (to fully isolate the coherent scattering) would be best. The limited energy resolution of current detector technologies allows for only a partial separation although, as the inset of Fig. 3 shows, the Compton profile can be adequately resolved at higher q-values. The actual shapes of the both the elastic and Compton scattering profiles can be ascertained so that one only need vary the intensities of the two constituents. By fitting all energy dispersive profiles at the higher q-values the ratio of these two components can be experimentally deduced. To reduce the $\sqrt{N}$ statistical noise still present in the extracted Compton profiles this data is typically passed through a low-pass Fourier filter[25]. At some intermediate q value the coherent scattering profile must then be suitably spliced to the existing low-q experimental data. In this way quantitative x-ray data is obtained out to reasonably high q values. For x-ray studies, as compared to those using neutrons, there is an addition advantage in that the hydrogen atom scattering is essentially negligible at q values beyond 10 Å^{-1} so that the structure function is *dominated* by only the major skeletal atomic constituents.

An absolute measure of the coherent/Compton ratio has an additional benefit which is of some consequence. Often the sample geometry (i.e., reflection, transmission) in combination with absorption effects requires pronounced corrections of the profile shape function. These corrections can be very sensitive to sample alignment and measurements of the absorption coefficient. Moreover polymer samples are often macroscopically inhomogeneous; a property which further impacts these corrections. By obtaining this direct measurement of the relative coherent and Compton fractions one can, after appropriately scaling the experimental ratio to match the theoretical curve, immediately execute a Krough-Moe normalization[26] of the data without resorting to many of these aforementioned corrections.

Even after a rigorous attempt is made to apply all of the generally accepted curve correction schemes[15], these extended q profiles can still display small systematic deviations which both hamper direct refinements of the structure function and create unphysical artifacts at low r. Although a thorough investigation establishing the true

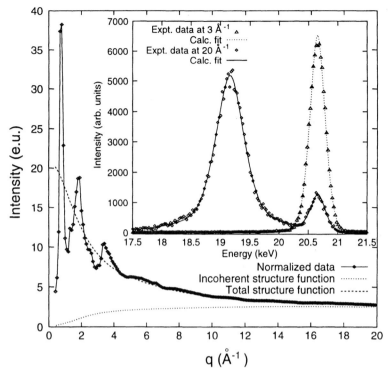

Figure 3: An example of the various contributions to the x-ray scattering intensity for a typical polymer sample. Inset: Two representative MCA curves obtained at high- and low-q.

origin of these discrepancies has not been attempted it may well be that the derivations of the theoretical atomic scattering curves themselves are not sufficiently accurate. Despite these limitations there are a few *ad hoc* procedures which can "correct" the data without imparting significant new systemic errors in the spectral regions of interest. One correction scheme is to identify nodes of the structure function (where $H(q) = 0$) from a representative model and perform a series of Krough-Moe normalizations. This information is then used to generate a correction curve to the nominal full-range normalization procedure. To supplement this approach it is also possible to effect a low r (typically at 1.2 Å and less) correction of G(r) by simply using the known profile as generated by a representative model and replace the low r portion of G(r). This is then Fourier back transformed to yield an additionally modified $H(q)$. Figure 4 depicts the sequential variations in the structure function as each indicated modification is implemented.

There is also the ubiquitous problem of artifacts generated in the Fourier transform procedure itself. As Fig. 2 clearly demonstrates, significant oscillatory behavior exists in the structure function well beyond the highest q-values accessed. As such, a simple transform of data is expected to produce a variety of truncation effects including a broadening of the G(r) peaks and the so-called "ringing" artifacts (which arise from the Fourier transform of a step function). One seemingly effective solution, proposed by Mitchell and Lovell[27], is to execute a 'sampled transform' in which a series of fast Fourier transforms are overlaid using the same $H(q)$ data set having fewer and fewer high q data points. In the way the ringing occurs at different frequencies so that, on average, this undesirable feature is minimized. Moreover, the low q data, which typically has less statistical error, is weighted more heavily. The Fig. 4 inset shows a typical real space profile that is obtained after this scheme has been employed. Other

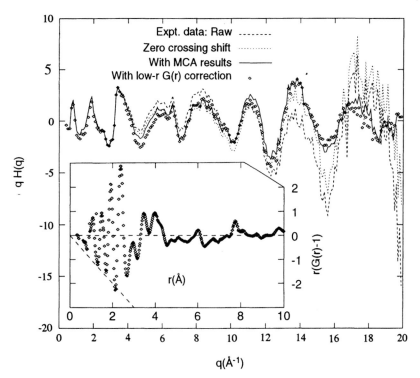

Figure 4: Variations in $H(q)$ as various corrections to the data are made to data from a poly(alkylsilane) sample. Inset: The final real space profile.

procedures, such as artificial damping and/or extension of the $H(q)$[28], have also been applied with varying degrees of success.

EXPERIMENTAL DATA AND MODELING

Ultimately the experimentally derived correlation functions and structure functions need to be rigorously compared against a physically realistic model. Since the actual chemical organization is predetermined, this gives a well-defined starting point for undertaking these comparisons. Still there is a formidable number of pair correlations even in the most basic of polymer model compounds [e.g., polyethylene or poly(tetrafluoroethylene)]. In the simplest setting, only the nearest-neighbor, next nearest-neighbor and greater distances are used assuming a nearly isotropic root-mean-square (*rms*) distribution of the atomic positions and a fixed chain geometry[28]. In this case the $e_{ij}(q)$ term in Eqn. 5 is approximated by

$$e_{ij}(q) = w_{ij} \exp(-\ell_{ij}^2 q^2/2) \frac{\sin qr}{qr} \qquad (7)$$

where r is the i-j atom separation, ℓ_{ij} is the *rms* deviation and w_{ij} is the relative number of i, j atom pairs. For polymer chains, which are locally linked by strong chemical bonds, this is clearly only a first-order approximation because there are highly correlated motions of neighboring atoms.

Detailed knowledge of the local pair, bond-angle, and torsion-angle potentials enables a far more sophisticated modeling approach. Since these atomistic models contain all atom locations, including disorder, the need for employing the exponential term in Eqn. 5 becomes unnecessary for evaluating $qH(q)$. In the case of hydrocarbons and fluorocarbons a number of full-fledged molecular simulations, using either molecular dynamics or molecular mechanics, have been performed and the computationally generated curve profiles compared to experimental data[29, 30, 31]. As the recent work of Londono et al.[31] in Fig. 5 shows there is relatively good agreement between x-ray experiments and theory for a "simple" hydrocarbon system, in this case for isotactic poly(propylene) in the melt. Neutron based experiments, which allow for much higher q ranges, of polyethylene and poly(tetrafluoroethylene) melts have also demonstrated the utility of this approach.

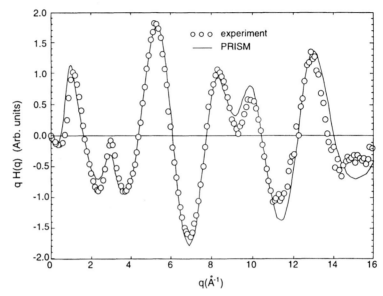

Figure 5: A comparison of the total $qH(q)$ for isotactic polypropylene with results from the integral equation theory, PRISM . Reprinted with permission from ref. [31]. Copyright © 1997 John Wiley & Sons, Inc.

An alternate scheme, also appearing in ref. [29], is to employ a reverse Monte Carlo algorithm. In this approach the atoms forming the polymer backbone are simply moved at random so that all configurations which generate acceptable structure functions are sampled. In this case the Monte Carlo partition is governed by the chi-square error function arising from the differences between the experimental and model-derived structure functions. For polyethylene the best-fit torsion-angle distributions generated by RMC compare very favorably with those obtained from molecular simulations.

While there is considerable merit in all these approaches, continued progress requires implementation of more tightly nested modeling schemes. For a true refinement of the experimental data an indirect comparison with molecular simulation has its limitations. Often there is an incomplete knowledge of the molecular force-fields, particularly with respect to torsion potentials. Ideally the potentials used by an appropriate simulation could be modified in an iterative process using direct and immediate comparisons to the experimental PDF data. In this way the simulation algorithms could be optimized by integrating them into a comprehensive refinement scheme. Moreover it is important to recognize that there still exists an extensive range of polymeric materials which are not easily mimicked using the current generation of computer simulations. In particular there is considerable difficulty in obtaining physically correct solutions for polymers having shallow minima in their torsional potentials (e.g., the polysilanes), extensive π-conjugation along the backbone (as in the case of conducting polymers), or polymer hosts having ionic interactions (as in the of polyelectrolytes). Semi-crystalline and liquid crystalline polymers are also problematic because of the intrinsically slow kinetics and the subtle competing interactions which often exist at the shorter length scales.

An unrestricted RMC approach may be problematical as well. Since, *a priori*, no atom interactions are used, unphysical and energetically unfavorable configurations may be extensively sampled and thus leading to a maximal set of potential solutions. In many instances the local correlated motions of nearest-neighbors are well understood. A hybrid scheme in which these known attributes are incorporated into the RMC would then lead to a more tightly constrained set of solutions.

The choice of which data representation to use is also an important consideration. As is the case for other classes of materials, a combined real-space and momentum space refinement of the data can be advantageous since the two representations emphasize different attributes. As noted previous, the $G(r)$ profiles derived from x-ray scattering often contain unphysical artifacts at low r. These artifacts appear across the entire $S(q)$ profiles and complicate refinements which utilize the structure function. By weighting the refinement towards $G(r)$ data points at larger r's one can minimize their impact. On the other hand, the intrachain $G(r)$ is necessarily superimposed on the interchain $G(r)$ and refinement of this data requires a prior knowledge of the interchain pair correlation function. In contrast this interchain feature is limited to the low q range of the structure function so that restricting a refinement to the higher q's guarantees that only intrachain components are used when optimizing an intrachain model. Thus a carefully orchestrated refinement of both sets of data can lead to a more rapid convergence towards a physically appropriate model. In many instances local chain polymers structures can be found which work well in either r-space or q-space but not both.

A MODEL HOST POLYMER FAMILY

As a model test system, we have recently begun a series of experiments to ascertain the local intrachain and interchain structure in various polysilane derivatives. Polysilanes are well-known polymers ($[-SiRR'-]_n$) comprised of only σ-bonded silicon atoms along the polymer backbone and short alkyl and/or alkoxy segments as the side-chains. From a structural perspective there is an intriguing UV absorption feature[32, 33], due to a $\sigma-\sigma^*$ transition along the Si-atom backbone, which has been shown to be extremely sensitive to the choice of side chain constituents, sample temperature and processing history. In the limiting case of an all-*trans* main-chain conformation, typically seen in some symmetric alkylsilane samples[34, 35] such as poly(di-hexylsilane), the peak in the UV absorption is centered near 370 nm. Helical and other, at present, unknown chain conformations exhibit transitions which are "blue"-shifted to shorter wavelengths. One long-standing question is simply how small changes in the local main-chain and side-chain conformations influence the nature of this transition. A proper interpretation requires a quantitative assessment of the local main-chain and side-chain structure in a variety of settings. Direct molecular simulations are of limited use at present because *ab initio* calculations of the Si-Si-Si-Si torsion potential obtain only slight energy differences between *trans, gauche* and various intermediate conformations[36]. Hence there is a strong impetus to use direct probes of local structure to gain further insight.

From the perspective of technique development, the polysilane polymer family also appears to have a significant advantage over conventional polymeric materials. The presence of an all-silicon backbone significantly enhances the strength of the x-ray scattering signal as compared to lower Z hosts. Thus the PDF profile will be strongly dominated by scattering from the well-defined (CH_x-Si-CH_x) core units. Moreover the nearest-neighbor Si-Si pair distance of 2.4 Å is well separated from the nominal 1.9 Å and 1.5 Å pair distances of C-Si and C-C, respectively, so that at intermediate length scales (from 2 to 5 Å) the pertinent information is better differentiated. For refinements against the structure function, $H(q)$, the larger repeat unit also guarantees that more relevant scattering features are concentrated in the 20-25 Å^{-1} q-range now easily accessible with existing light sources and instrumentation.

The smallest poly(alkyl) derivatives exhibiting thermochromism are the symmetric poly(di-ethylsilane) (PdeS) and the asymmetric poly(methy-n-propylsilane) (PmpS). At room temperature PdeS is crystalline with an all-*trans* Si-backbone conformation while PmpS is found to be semi-crystalline with a monoclinic approximate having lattice parameters of **a**=8.38 Å, **b**=10.13 Å, **c**=3.92 Å and γ=66° where c is along the Si backbone direction[37]. Nominally an all-*trans* conformation is consistent with the Bragg scattering analysis but the UV absorption maximum is near 320 nm which is much more suggestive of a non-planar arrangement. Slightly longer methyl-n-alkyl polysilanes exhibit a somewhat different thermal behavior (with thermochromism occurring at reduced temperatures) and with chain conformations which are, at present, unknown. To be entirely fair there are some drawbacks. For instances, existing synthetic routes produce materials which are atactic so that isotactic and syndiotactic dyads occur at random and in equal proportions. Modeling of an atactic system is, necessarily, more complicated than for more regular chain structures. Polysilanes are also found to be radiation sensitive. Exposure to the x-ray beam for over 24 hours caused no discernible differences in the scattering profiles despite a certain drop in the molecular weights.

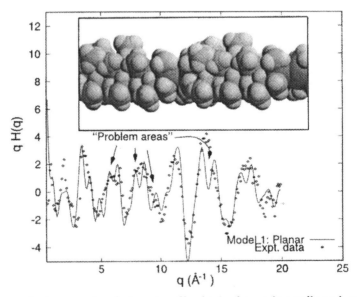

Figure 6: A comparison between profile obtained experimentally and generated by the poly(methyl-n-propylsilane) chain model shown in the inset.

Our present modeling approach employs a direct refinement of the various bond angles, bond lengths and dihedral angles within a single "atactic" eight Si unit oligomer model having periodic boundary conditions. Hard core packing constraints prevent overlap of neighboring atoms. This scheme represents an attempt to find a single conformational setting which adequately reproduces the scattering data. As such it is quite limited but it is still illustrative for demonstrating the sensitivity of PDF analysis in studies of local polymer structure. Figure 6 compares a q-weighted H(q) from a room temperature PmpS sample and a model restricted to a trans-planar configuration with refinement of only the alkyl side chains. As denoted by the arrows, a fixed planar backbone structure is clearly an unphysical description of the local chain structure. An equally as poor $G(r)$ comparison is also obtained (not shown). By allowing for variations in both the Si-Si-Si bond angles and the Si-Si-Si-Si dihedral angles substantial improvements are obtained. The current "best-fit" model is shown in Fig. 7 using both q- and r-space representations. On average there appears to be a 20° deviation from planarity although it can exceed 40° across at least one of the eight Si-Si-Si-Si linkages. The presences of these large twists, in combination with random disorder, is consistent with the shorter wavelength UV absorption maximum found experimentally. At present there also appears to be strong cross-correlations between the local Si bond and dihedral angles. With continued improvements in the modeling algorithms it may be possible to quantitatively specify this behavior.

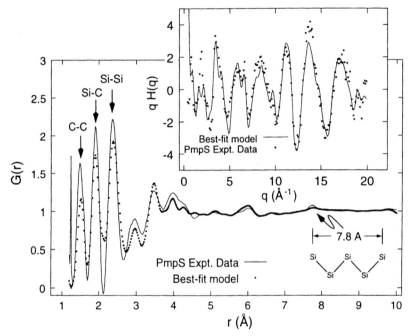

Figure 7: A comparison between $G(r)$ and $qH(q)$ profiles obtained experimentally and those generated by a poly(methyl-n-propylsilane) model in which all Si bond angles and dihedral angles are allowed to fluctuate independently.

Finally, in Figure 8, preliminary $G(r)$ (as $r[G(r) - 1]$) profiles are shown for a poly(methyl-n-hexylsilane) sample observed at temperatures clearly above and below the reported thermo-chromic transition point. The actual UV absorption spectra exhibit a clear isosbestic point and are indicative of a distinct two-phase coexistence over the transition region. The two displayed $G(r)$ profiles exhibit a number of differences suggestive of distinct changes in the chain structure. In the 3 to 5 Å region the peaks and valleys appear measurably sharper in the low temperature profile. In addition, pair correlation peaks centered near 5.9 Å and 7.8 Å become more pronounced in the low temperature curve. These two latter features may be tentatively assigned to the Si-Si-Si-Si and Si-Si-Si-Si-Si pair distances and, if this is valid, then imply a transition to a more planar main chain conformation. Once again improvement modeling will be required before a proper assessment of this structural transformation can be reached. Despite the tentative nature of the polysilane data and the various interpretations, these spectra hopefully demonstrate the structural information which can be extracted from PDF analysis of polymer systems.

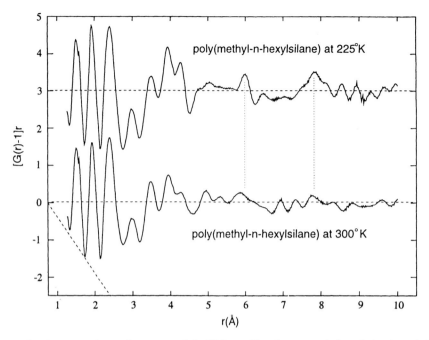

Figure 8: A comparison of two rescaled $G(r)$ profiles from a poly(methyl-n-hexylsilane) sample at temperatures above and below the thermo-chromic transition region.

CONCLUSIONS

Existing polymer PDF studies have, on the whole, only touched upon a few of the possible host systems which could benefit from this methodology. The continuing development of computer modeling algoritms will rapidly advance the quantitative analysis capabilities of the technique. With the ongoing development of dedicated x-ray and neutron spectrometers, an even larger range of structural studies will be available in the future. The high-q ranges now becoming accessible should motivate a reevaluation of the theoretical elastic and Compton scattering profiles. Better knowledge of these attributes will reduce the guess work now required to appropriately scale experimental spectra.

It is important to note that PDF studies of oriented polymer samples are also possible. This would allow the anisotropy in the local structure to be better reconstructed thus yielding less uncertainty in the structures obtained during the refinement process. However this would significantly increase the number of calculations necessary for these refinements.

In addition to the conventional x-ray scattering results briefly reviewed here, there also are an expanding number of polymer PDF studies employing anomalous scattering techniques. These include NiBr-doped polyelectrolytes[38, 39], HBr-doped polyaniline[40] and, most recently, Li_6 and Li_7 substitutions in polyethylene oxide[41]. These studies are potentially more powerful than the much more commonly used EXAFS technique because anomalous PDF methods can resolve structural features beyond 5 Å and, in the case of neutron scattering, investigate a large number of low Z materials.

There are many avenues for the application of PDF techniques in the study of local structure in polymeric materials.

ACKNOWLEDGMENTS

This support of this work by a NSF grant (DMR-9631575, M.J.W.) is gratefully acknowledged. I also wish to acknowledge fruitful discussions and collaborations with G.R. Mitchell, B.R. Mattes, Robert West and J.R. Koe.

References

[1] *Small Angle X-ray Scattering*, edited by O. Glatter and O. Kratky (Academic Press, London, New York, 1982).

[2] D. I. Svergun, *Structure Analysis by Small-angle X-ray and Neutron Scattering* (Plenum Press, New York, 1987).

[3] B. E. Warren, *X-ray Diffraction* (Addison-Wesley, Reading, Mass., 1969).

[4] L. E. Alexander, *X-ray Diffraction Methods in Polymer Science* (Wiley, Interscience, New York, 1969).

[5] J. S. Higgins, *Polymers and Neutron Scattering* (Clarendon Press, Oxford University Press, New York, 1994).

[6] For a recent review see: *Science*, 277:1213 (1997).

[7] *Monte Carlo Methods in Statisitcal Physics*, edited by K. Binder (Springer, Berlin, 1979).

[8] E. Leontidis, J. J. de Pablo, M. Laso, and U. W. Suter, *Advances in Polymer Science* (Springer-Verlag, Belin Heidelberg, 1994), Vol. 116, p. 283.

[9] K. Binder and W. Paul, *J. Polym. Sci., Polym. Physics* 35:1 (1997).

[10] A. J. Epstein, J. M. Ginder, F. Zuo, R. W. Bigelow, H. S. Woo, D. B. Tanner, A. F. Richter, W. S. Huang, and A. G. MacDiarmid, *Synth. Met.* 18:303 (1987).

[11] A. J. Heeger, J. R. Schriefer, and W.-P. Su, *Rev. Mod. Phys.* 40:3439 (1988).

[12] M. Reghu, Y. Cao, D. Moses, and A. J. Heeger, *Phys. Rev. B* 47:1758 (1993).

[13] A. H. Narten, *J. Chem. Phys.* 90:5857 (1989).

[14] A. H. Narten, A. Habenschuss, and A. Xenopoulos, *Polymer* 32:1923 (1991).

[15] H. P. Klug and L. E. Alexander, *X-ray Diffraction Procedures for Polycrystalline and Amorphous Materials* (Wiley, Interscience, New York, 1974).

[16] T. Egami, W. Dmowski, J. D. Jorgensen, D. G. Hinks, D. W. C. II, C. U. Segre, and K. Zhang, *Rev. Sol. St. Sci.* 1:247 (1987).

[17] G. R. Mitchell, in *Comprehensive Polymer Science*, edited by G. Allen, J. C. Bevington, C. Booth, and C. Price (Pergamon, Oxford, 1989), Vol. 1, p. 687.

[18] C. W. Dwiggins, *Acta Cystallogr., Sect. A* 28:1580 (1972).

[19] D. T. Cromer and J. B. Mann, *J. Chem Phys.* 47:1893 (1967).

[20] Y. Waseda, *Novel Application of Anomalous X-ray Scattering for Structural Characterization of Disordered Materials* (Springer-Verlag, Belin Heidelberg, 1984), Vol. 204.

[21] *International Tables for X-ray Crystallography* (Kynoch Press, Birmingham, 1974), Vol. IV.

[22] B. K. Annis, A. H. Narten, A. G. MacDiarmid, and A. F. Richter, *Synth. Met.* 22:191 (1988).

[23] A. G. MccDiarmid and A. J. Epstein, *Science and Applications of Conducting Polymers* (Adam Hilger, Bristol, England, 1990), p. 141.

[24] J. Maron, M. J. Winokur, and B. R. Mattes, *Macromolecules* 28:4475 (1995).

[25] W. H. Press, B. P. Flannery, S. A. Teukolsky, and W. T. Vetterling, *Numerical Recipes: The Art of Scientific Computing* (Cambridge University Press, Cambridge, 1986).

[26] J. Krogh-Moe, *Acta Cystallogr.* 9:951 (1956).

[27] G. R. Mitchell and R. Lovell, *Acta Crystallogr., Sect. A* 37:135 (1981).

[28] A. H. Narten and A. Habenschuss, *J. Chem. Phys.* 92:5692 (1990).

[29] B. Rosi-Schwartz and G. R. Mitchell, *Polymer* 35:5398 (1994).

[30] B. Rosi-Schwartz and G. R. Mitchell, *Polymer* 35:3139 (1994).

[31] J. D. Londono, A. Habenschuss,J. G. Curro and J. J. Rajasekaren, *J. Polym. Sci., Polym. Phys.*, in press.

[32] R. Miller and J. Michl, *Chem. Rev.* 89:1390 (1989).

[33] C. Yuan and R. West, *Macromolecules* 27:629 (1994).

[34] F. C. Schilling, A. J. Lovinger, J. M. Zeigler, D. D. Davis, and F. A. Bovey, *Macromolecules* 22:3055 (1989).

[35] E. Karikari, A. Greso, B. Farmer, R. Miller, and J. Rabolt, *Macromolecules* 26:3937 (1993).

[36] S. H. Tersigni, P. Ritter, and W. J. Welsh, *J. Inorg. and Org. Polym.* 1:377 (1991).

[37] M. J. Winokur and J. Koe and R. West, *Polym. Prepr., Div. Polym. Chem., ACS* (1997), in press.

[38] H. Cai, R. Hu, T. Egami, and G. C. Farrington, *Sol. State Ionics* 52:333 (1992).

[39] J. R. Fishburn and S. W. Barton, *Macromolecules* 28:1903 (1995).

[40] M. J. Winokur and B. R. Mattes, *Synth. Met.* 84:725 (1997).

[41] B. Annis, private communication.

DIFFUSE X-RAY AND NEUTRON REFLECTION FROM SURFACES AND INTERFACES

Sunil K. Sinha[1] and Roger Pynn[2]

[1]Experimental Facilities Division
Argonne National Laboratory
Argonne, IL 60439
[2]Los Alamos Neutron Science Center
Los Alamos National Laboratory
Los Alamos, NM 87545

INTRODUCTION

Neutron and X-ray reflectometry were initially conceived as tools for studying the average structure of materials in a direction perpendicular to planar surfaces or interfaces. As such, they probed the variation of the intensity of specular scattering at grazing incidence as the momentum transfer perpendicular to the surface was changed. Any off-specular or diffuse scattering observed in these experiments was regarded as a nuisance, i. e. background that simply had to be subtracted to obtain reliable values for the specular reflectivity. More recently, both neutrons and X-rays have been used to study diffuse scattering from surfaces and interfaces in order to learn about lateral inhomogeneities within the reflecting surface. A variety of phenomena ranging from capillary waves on liquids to roughness correlations in the growth of thin films have been studied in this way. Because diffuse scattering is generally weak, the intense beams of X-rays available from modern synchrotron sources have been more widely used in the study of diffuse surface scattering than neutrons. Nevertheless, in certain cases, such as the study of magnetic systems or certain polymer problems, neutrons still have a role to play.

In all cases, diffuse scattering observed in reflectometry experiments arises from inhomogeneities in the reflecting medium, and by far the most common cause of such inhomogeneity is surface roughness. Rough surfaces give rise to diffuse scattering when waves (i.e. neutrons or x-rays) scattered from "valleys" and "peaks" of the surface are significantly out of phase with one another. A calculation of the difference in path length for such waves quickly leads to the conclusion[1] that a surface will appear smooth when

$q_z \sigma \ll 1$, where q_z is the wavevector transfer of the radiation perpendicular to the surface (the z direction) and σ is a measure of the amplitude of surface roughness. This so-called Rayleigh criterion is easy to violate in reflectometry experiments: a value of $q_z \sim$ 0.02 Å^{-1} corresponds to a typical critical wavevector (i.e. the wavevector below which the smooth surfaces of materials reflect completely), and almost all experiments probe values of q_z out to at least 0.1 Å^{-1}. In such situations, roughness amplitudes of a few Angstroms can give rise to easily measurable diffuse scattering.

Although investigations of surface roughness using neutrons and X-rays have only become common recently, there is a considerable literature from the past 50 years or so in which the scattering of radio waves by rough surfaces has been studied in connection with phenomena such as "sea clutter" in radar images or tropospheric scatter of radio waves[1]. This literature makes it clear that exact calculations of scattering by rough surfaces are possible only in a few well-defined situations. In general, approximations have to be made, and the most appropriate of these depend on the surface morphology. For example, if a surface is comprised of smooth, misoriented facets each larger than a Fresnel zone, scattering from the surface can be calculated, by summing specular scattering from each of the facets. In this case, provided one ignores multiple scattering and shadowing of one facet by another, the total reflectivity of the surface is given by the expression

$$R = R_o \exp\left(-q_z^2 \sigma^2\right) \tag{1.1}$$

if the height (z) distribution of the surface is a Gaussian of standard deviation σ and R_o is the reflectivity of an ideal smooth surface. In the case of X-ray or neutron scattering from a surface, the longitudinal dimension of the first Fresnel zone is of order of a few microns, so Eq. (1.1) only applies if surface facets of this size scale or larger are present. In many interesting cases, the lateral scale of surface roughness - i.e. the distance over which a surface departs from flatness - is much smaller than this.

X-ray specular reflectivity and off-specular diffuse scattering experiments are usually carried-out with a monochromatic beam incident on a sample surface mounted on a 2-axis or 4-axis x-ray diffractometer programmed to carry out the scans in reciprocal space indicated in Fig. 1. (In the case of liquid surfaces, a special liquid diffractometer has

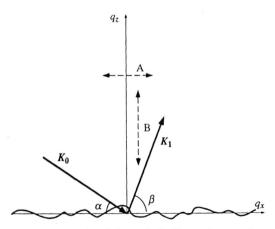

Fig. 1 Schematic for X-ray surface scattering , k_0, k_1 are the incident and scattered wavevectors respectively, making grazing angles α, β respectively with average surface. $q = k_1 - k_0$. Scans A & B represent transverse and longitudinal diffuse scans respectively.

to be used which tilts the incident beam down onto the liquid surface and which allows the scattered beam to be detected emerging at an upward angle from the surface in the plane of scattering, or in the case of grazing incidence scattering geometry by rotating the sample detector plane out of the plane of scattering,). To measure the "true" specular reflectivity for a surface, often the diffuse scattering close to the specular and reciprocal space is subtracted off point by point, by carrying-out "longitudinal diffuse" scattering scans parallel to but slightly offset from the specular ridge (scans "B" in Fig. 1). Sometimes, for high resolution, an analyzer crystal is employed in front of the detector. Specular reflectivity measurements from solid surfaces can usually be carried out on tube x-ray sources, whereas measurements of the much weaker diffuse scattering is usually carried out using rotating anode or synchrotron radiation sources. Specular reflectivities have often been measured in this way down to values as low as 10^{-8} or 10^{-9}.

Neutron reflection experiments can be done in two different ways: either by using a monochromatic beam and varying the angle of incidence or by using a pulsed, polychromatic beam and recording neutron time of flight at constant incident angle. Since the former method is very similar to that used with x-rays we will not describe it separately here. Reflectometers installed at pulsed spallation neutron sources are among those spectrometers that compete well with similar instruments at reactor sources, because the inherently large dynamic range which they offer is often useful. The natural coordinates in which to plot data obtained with a time-of-flight reflectometer are neutron wavelength and scattering angle from the surface, as shown in Figure 2. Although this space is not the most natural one in which to describe the physics of the reflecting sample it does have some important uses. For example, loci of points at which the outgoing beam

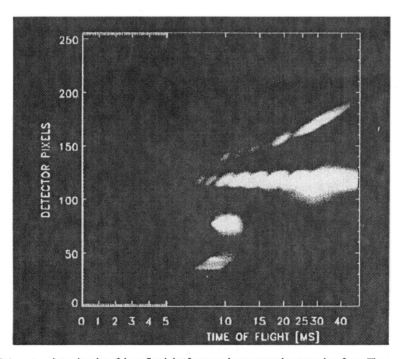

Fig. 2 A neutron intensity plot of the reflectivity from a polymer-coated patterned surface. The specular streak falls near pixel 120 (~1 degree for the angle of incidence). The time of flight is proportional to the neutron wavelength ($\lambda \sim TOF/3$)

emerges at the critical angle for that wavelength, where the Fresnel transmission function has a sharp cusp are straight lines through the origin in lambda-theta space. Furthermore, the resolution of the spectrometer needs to be understood and deconvoluted in this space, before a transformation can be made to the more physical q_x - q_z space.

Because neutron sources are much weaker than modern synchrotrons (the most powerful CW neutron sources provide about the same flux as a rotating anode), neutron reflectometers, designed primarily to measure specular scattering, attempt to overcome their flux disadvantage by using incident beams with a relatively large incident divergence. A position sensitive neutron detector defines the scattered angle with an accuracy of better than 1 mrad, and the fact that incident and scattered angles are equal for specular scattering can then be used to identify the angle of incidence of each neutron with the same accuracy. Of course, this exact correlation between incident and scattered angle does not apply to diffuse scattering from surfaces, so neutron reflectometers generally have poor resolution for such scattering. Until now, no real attempt has been made to deconvolute resolution from diffuse scattering data obtained at a pulsed spallation source, so there are no quantitative results available.

Before leaving this section, it is useful to establish magnitudes of lateral correlation lengths that can be probed by surface diffuse x-ray or neutron scattering. The wavevector resolution in the direction parallel to the surface in the plane of scattering is given (around the specular position) by roughly $(q_z \Delta\alpha)$ where q_z is the wavevector transfer normal to the surface and $\Delta\alpha$ the in-plane angular divergence contribution from the incident and outgoing beams. Since both q_z and $\Delta\alpha$ are small quantities, this typically yields a value of $\sim 10^{-4}$ Å^{-1} corresponding to an accessible length scale of ~ 10 microns! The resolution parallel to the surface but normal to this scattering plane on the other hand is of order $(k_o \Delta\beta)$ where $\Delta\beta$ is the out-of-plane angular divergence and k_o is the incident wavevector of this radiation, and is much larger. However, the accessible range of wavevectors is greatly increased in this direction.

ROUGHNESS AT A SINGLE INTERFACE

Let us consider a surface centered on the plane $z = 0$, but with fluctuations $\delta z(x,y)$ as a function of the lateral coordinates (x,y). Such fluctuations may often have a statistical distribution which is Gaussian, which we shall assume to be the case here. (The case of non-Gaussian roughness will be discussed later). An important quantity is the mean-square height-deviation function $g(\mathbf{R})$ ($\mathbf{R} \equiv x,y$), defined by

$$g(\bar{R}) \equiv \left\langle \left[\delta z(\bar{r}) - \delta z(\bar{r} + \bar{R}) \right]^2 \right\rangle \tag{2.1}$$

representing a statistical average over position $\bar{r}$ on the surface. A surface which is self-affine has the property that

$$g(R) \sim R^{2k} \tag{2.2}$$

where h (known as the roughness exponent) has a value between 0 and 1. The morphology of the roughness depends sensitively on h. For small h, the surface is sharp

354

and jagged, while as h approaches 1, the surface becomes more gently rounded. $h=1/2$ corresponds to the case of random-walk fluctuations. In order to be realistic at large values of R, we may introduce a cut-off length for the roughness x which makes g(R) saturate as $R \to \infty$, i.e., we write[2]

$$g(R) = 2\sigma^2 \left(1 - e^{-(R/\xi)^{2h}} \right)$$

(2.3)

Since $\left\langle \left[\delta z(\bar{r}) \right]^2 \right\rangle = \sigma^2$, Eqs. (2.1) and (2.3) imply that

$$C(\bar{R}) \equiv \left\langle \delta z(\bar{r}) \delta z(\bar{r}) \right\rangle = \sigma^2 e^{-(R/\xi)^{2h}}$$

(2.4)

C(**R**) is referred to as the height-height correlation function. The justification of Eqs (2.3) and (2.4) may to some extent be found in the equations which govern the growth of deposited films, such as the Kardar-Parisi-Zhang (KPZ) theory.[3] For such surfaces, the height function H(**r**,t) as a function of **r** and time is governed by an equation of the form

$$\frac{\partial H}{\partial t} = \gamma \nabla^2 H + \frac{\lambda}{2} (\nabla H)^2 + \eta(\bar{r}, t)$$

(2.5)

where γ, λ are constants and η (**r**,t) is a white-noise random function. Analysis of the solutions of this equation, either numerically or using renormalization group methods, reveal that the corresponding width function,) obeys the following scaling form

$$g(\bar{R}, t) = R^{2h} G(t / R^z)$$

(2.6)

where G(x) is a scaling function, and h,z are scaling exponents. The scaling function G(x) must have the properties that

$$G(x) \to \text{constant for } x \gg 1$$
$$G(x) \to x^{2\beta} \text{ for } x \ll 1$$

(2.7)

A function which satisfies these conditions is

$$G(x) = C_1 x^{2\beta} \left(1 - e^{-\left(\frac{1}{x}\right)^{2\beta}} \right)$$

(2.8)

where C_1 is a constant. Substituting Eq. (2.8) into Eq. (2.6), it may be verified that we recover the form of Eq. (2.3) if

$$\sigma^2 = \left(\frac{1}{2} C_1 \right) t^{2\beta}$$

(2.9)

and

$$\xi = t^{\frac{1}{z}}$$

(2.10)

where β, h, and z satisfy the relation

$$z = h / \beta \tag{2.11}$$

Thus, we conclude that a self-affine surface with a finite cut-off length for the roughness is consistent with growth models if we assume that it corresponds to the surface which results after <u>finite</u> growth time. Eqs. (2.9) and (2.10) are specific predictions for such growth models. Regardless of the validity of growth models for surfaces, Eq. (2.4) seems to work remarkably well in describing a wide variety of rough surfaces encountered in nature, as deduced from both scattering experiments and direct imaging probes used to profile the surface.[4] Its convenience is that it characterizes the surface roughness mathematically in terms of three parameters, the root-mean-square roughness (σ), the roughness exponent (h) and the roughness cutoff length (ξ). Other forms have been proposed for the height-height correlation function (see for instance Palasantzas and Krim in Ref. 5), but most of them do not show any preferred advantages. One slightly modified form which does have certain advantages was first proposed by Church[6] and later discussed by Palasantzas[7] and de Boer[8] is

$$C(\bar{r}) = \frac{2h\sigma^2}{\Gamma(1+h)} \left(\frac{r\sqrt{2h}}{2\xi} \right) K_h \left(\frac{r\sqrt{2h}}{\xi} \right) \tag{2.12}$$

where $\Gamma(x)$ is the Gamma function, and $K_h(x)$ is the modified Bessel function of non-integral order. This has the advantage that it does yield an analytic expression for the Fourier transform of $C(\mathbf{R})$ which is simply the noise spectral function $\langle |\delta z(\mathbf{q})|^2 \rangle$, i.e., it yields

$$\langle |\delta z(\bar{q})|^2 \rangle = \frac{A}{(2\pi)^5} \frac{\sigma^2 \xi^2}{\left(1 + uq^2\xi^2 \right)^{1+h}} \tag{2.13}$$

where A, u are constants, and $\mathbf{q}$ is a two-dimensional wavevector in the (x,y) plane. Eqs. (2.12) and (2.13) also have the advantage that they show that $C(R) \sim \ln R$ as $h \to 0$, making this a special case of the general form. (A logarithmic height-height correlation function is characteristic of liquid surfaces having capillary wave fluctuations[9], or a surface undergoing a roughening transition[10] and will be discussed in more detail below).

SCATTERING BY A SINGLE INTERFACE

Let us assume, for the moment, that the surface represents an interface between two media with uniform scattering length densities ρ_1, and ρ_2. Let $\Delta\rho = (\rho_1 - \rho_2)$. (For X-ray scattering, ρ_1 is simply the Compton scattering length (e^2 / mc^2) times the electron number density, while for neutrons it is a weighted average of the coherent nuclear scattering length times the nuclear number density, averaged over all types of nuclei). For scattering experiments, (see Fig. 1) where the magnitude of the wavevector transfer $\mathbf{q}$ (defined as $\mathbf{k}_1 - \mathbf{k}_0$, where $\mathbf{k}_0, \mathbf{k}_1$ are the incident and scattered wavevectors of the radiation, making grazing angles α, β respectively with the average surface) is small compared to the

356

inverse of the interatomic distances, we may neglect the crystallinity of each medium. For experiments at small grazing angles, we may also neglect polarization effects in the x-ray scattering and consider for simplicity the case of TE polarization so that we may use a scalar wave equation. Then the Born approximation for scattering yields

$$S(\bar{q}) = (\Delta\rho)^2 \iint d\bar{r} d\bar{r}' e^{i\bar{q}(\bar{r}-\bar{r}')}$$

(3.1)

where the integral is over the volume on one side of the surface. (We may use periodic boundary conditions and a small absorption in the lower medium to ignore all surfaces except the one shown in Fig. 1, such as the surface at $z \to \infty$, etc.except the one shown in Fig. 1).

The integration over the z-coordinates may then be carried out in the above integral, yielding

$$S(\bar{q}) = \frac{(\Delta\rho)^2}{q_z^2} \iiiint dx dy dx' dy' e^{i\left[q_z\left(\delta z(x,y) - \delta z(x',y')\right)\right]} e^{i\left[q_x(x-x') + q_y(y-y')\right]}$$

(3.2)

Since δz is a Gaussian random variable, carrying out a statistical average yields an integrand which depends only on the <u>relative</u> separation (X,Y) of the coordinates (x,y), (x', y'), and we obtain:

$$S(\bar{q}) = A \frac{(\Delta\rho)^2}{q_z^2} \iint dX dY e^{-\frac{1}{2}q_z^2 g(R)} e^{-i(q_x X + q_y Y)}$$

(3.3)

where g(R) is defined by Eq. (2.1), and A is the surface area. Eq. (3.3) may also be rewritten by using the relation between g(R) and the height-height correlation function (Eq. 2.3) as

$$S(\bar{q}) = \frac{(\Delta\rho)^2}{q_z^2} A e^{-q_z^2 \sigma^2} \iint dX dY e^{q_z^2 C(R)} e^{-i(q_x X + q_y Y)}$$

(3.4)

Since $C(R) \to 0$ as $R \to \infty$ the integral in Eq. (3.4) contains a delta function which yields the specular reflectivity and we may write

$$S(\bar{q}) = S_{sp}(\bar{q}) + S_{diff}(\bar{q})$$

(3.5)

where

$$S_{sp}(\bar{q}) = 16\pi^2 A \frac{(\Delta\rho)^2}{q_z^2} e^{-q_z^2 \sigma^2} \delta(q_x)\delta(q_y)$$

(3.6)

and

$$S_{diff}(\bar{q}) = A \frac{(\Delta\rho)^2}{q_z^2} e^{-q_z^2 \sigma^2} \iint dX dY \left[e^{q_z^2 C(R)} - 1\right] e^{-i(q_x X + q_y Y)}$$

(3.7)

The specular part may be converted into an expression for the specular <u>reflectivity</u> by integrating over the detector solid angle and dividing by the incident beam intensity (see Ref. 2 for details) to yield

$$R(q_z) = R_F(q_z)e^{-q_z{}^2\sigma^2}$$

(3.8)

where $R_F(q_z)$ is identical to the limiting case of the Fresnel Reflectivity from a smooth surface at large q_z,

$$R_F(q_z) = 16\pi^2(\Delta\rho)^2 / q_z{}^4$$

(3.9)

and Eq. (3.8) modifies it with a Debye-Waller-like factor due to the roughness "smearing" of the average surface. Eq. (3.7) yields the off-specular or diffuse scattering in this approximation, which can only be written down as an analytical function of $\mathbf{q}$ for special cases[2], but may be calculated quite generally if the height-height correlation function is known. If sufficiently accurate data can be taken over a wide enough range of q_x, q_y (e.g. with a linear or 2D position-sensitive detector) it may be possible to Fourier transform $S_{diff}(\mathbf{q})$ and thus invert Eq. (3.7) to yield g(R) directly. Such experiments have been done recently[11] and the g(R) obtained is consistent with the self-affine form given in Eq. (2.3). In other cases, forms such as given by Eqs. (2.3) or (2.4) have been used to fit scattering data, and compared with the corresponding statistical quantities derived by digitizing STM or AFM data taken on the same surfaces. These methods are now yielding consistent results.[4]

The Born approximation results are only valid if the scattering is weak, which will be the case when α or β (Fig. 1) are small or close to the critical angle for total reflection. A slightly better approximation in this case is to use the so-called distorted wave Born approximation (DWBA). In this case, instead of using plane waves to calculate the matrix elements of the scattering as in the Born approximation, one uses the actual wave functions, which closely approximate the actual system, i.e., one may use the true wave functions for reflection and transmission at the corresponding smooth surface. In this case, the main effect on $S_{diff}(q)$ as given by Eq. (3.7) is to modify it to [2,12]

$$S_{diff}(\bar{q}) = |T(\alpha)|^2 |T(\beta)|^2 A \frac{(\Delta\rho)^2}{|\bar{q}_z|^2} e^{-\frac{1}{2}\left(\bar{q}_z^2 + \bar{q}_z^{*2}\right)\sigma^2} \iint dXdY \left[e^{|\bar{q}_z^2|C(R)} - 1 \right] e^{-i(q_xX + q_yY)}$$

(3.10)

where $T(\alpha)$ is the Fresnel transmission coefficient of the average interface for grazing angle of incidence α, and q_z is the z-component of the wave vector transfer in the medium under the surface (which may be complex or even purely imaginary for evanescent waves in the case of total external reflection). The main effect of this is to produce side-peaks in the transverse diffuse scans (rocking curves) when α or β is equal to the critical angle, since at that point $T(\alpha)$ reaches a maximum value of 2, the incident and specularly reflected waves being in phase so that the field at the surface is at a maximum. Such side-peaks are known as "Yoneda wings"[13] (see Fig. 3). For large α, β, and hence large q_z, Eq. (3.7), is a good approximation, and we may neglect the difference between q_z and $\bar{q}_z$. For large q_z but with $q_x = q_y \cong 0$, it may be shown that S_{diff} has the asymptotic form[2]

358

$$S_{diff}(\bar{q}) \cong q_z^{-(2+2/h)} \qquad (3.11)$$

so that the exponent h may be found from such asymptotic power laws. For many experiments, the instrumental resolution in the direction of q out of the scattering plane is kept rather loose, i.e., q_y is effectively integrated over (if the plane of scattering is defined as the x-z plane). Then Eq. (3.7) shows that what is measured is:

$$I(q_x, q_z) = 2\pi A \frac{(\Delta\rho)^2}{q_z^2} e^{-q_z^2\sigma^2} \int dX \left[e^{q_z^2 C(X)} - 1 \right] e^{-iq_x X} \qquad (3.12)$$

i.e., a one-dimensional, rather than a two-dimensional Fourier transform. For isotropic rough surfaces, this yields the same information but may be misleading if the surface is anisotropic (e.g., miscut single crystal surfaces with steps). The asymptotic form in this case which corresponds to Eq. (3.11), is

$$I(q_x \sim 0, q_z) \sim q_z^{-(2+1/h)} \qquad (3.13)$$

The specular reflectivity is also modified in the DWBA from the simple expression Eq. (3.8) and is replaced by the form[2,12]

$$R(q_z) = R_F(q_z) e^{-q_z \bar{q}_z \sigma^2} \qquad (3.14)$$

which was first derived by Nevot and Croce.[14] The reflectivity deduced from the DWBA may be checked against the exact solution for a surface whose density profile follows a hyperbolic tangent function. In this case the DWBA gives a value for the reflectance which differs from the exact answer only by a phase factor, so the reflectivity given by the approximation (3.14) is the same as the exact result.

Another approximation made in the DWBA is that replacing the x-ray or neutron wave function across the rough surface by an analytic continuation of the wave function below the surface. If the amplitude of the surface roughness is too large this approximation breaks down. In practice this limits the DWBA to values of $\sigma < 10$ Å.

The conditions under which the DWBA is expected to be applicable have been carefully analyzed by de Boer. His analysis shows that if $(k_z^2 \xi / k_0) < 1$ (where k_0, k_z are respectively the magnitude and the z-component of the incident wavevector in free space, and ξ is the roughness correlation length) then the DWBA is applicable and the Nevot-Croce formula for the specular reflectivity (given by Eq. 3.14) applies. If on the other hand, $(k_z^2 \xi / k_0) \gg 1$, then the DWBA breaks down, and the specular reflectivity is better given by the Rayleigh formula (Eq. 3.8). This limits the applicability of the DWBA to accurately describe diffuse scattering to cases where ξ is less than 10 microns or so.

LIQUID SURFACES

In the case of liquid surfaces, surface roughness is due to capillary wave fluctuations. There are some problems in connection with a truly first-principle

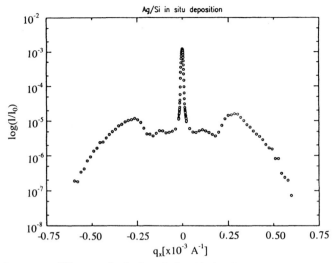

Fig. 3 Typical transverse diffuse scan for Ag deposited on Si, showing the specular peak, diffuse scattering and Yoneda wings referred to in the text.

calculation of such fluctuations (for a recent discussion, see Ref. 15), but one may write down at least a phenomenological expression for the surface free energy of the liquid and derive from it the spectral function for surface height fluctuations in the form:

$$\left\langle \left| \delta z(q) \right|^2 \right\rangle = \frac{kT}{\gamma \left(q^2 + \kappa^2 \right)} \tag{4.1}$$

where γ is the surface (or interface) tension, and κ is the inverse of the capillary length defined by $\kappa^{-1} = (\Delta \rho_0) g / \gamma$, $\Delta \rho_0$ being the mass-density difference between the fluids on either side of the interface. For bulk liquids κ is typically of the order of 10 cm^{-1}. Fourier transformation of Eq. (4.1) leads to a form for the height-height correlation function

$$C(r) = -\frac{1}{2} B K_0(\kappa r) \tag{4.2}$$

where

$$B = kT / (\pi \gamma) \tag{4.3}$$

and $K_0(x)$ is the modified Bessel function. At length scales $\ll \kappa^{-1}$ (which are, in practice, those relevant for scattering experiments), the Bessel function may be replaced by a logarithm, and, to prevent short (molecular) length scale problems, we may also introduce a lower length scale cutoff.[16] Thus, we finally write

$$C(r) = -\frac{1}{2} B \ln \left[\kappa \left(r^2 + r_0^2 \right)^{\frac{1}{2}} \right] \tag{4.4}$$

where r_0 is defined to give the correct lateral surface roughness given by the integral of Eq. (4.1) (r_0 turns out to be the inverse of the upper q cutoff for the capillary waves, q_u defined below).

From Eq. (4.1) we see by integration over q that the true mean square roughness due to surface capillary waves is given by

$$\sigma^2 = \frac{1}{4} B \ln\left[\left(q_u^2 + \kappa^2\right)/\kappa^2\right] + \sigma_o^2$$

(4.5)

where σ_o is an "intrinsic roughness" due to the size of the molecules at the surface, and q_u is an upper cut-off for the capillary wavevectors introduced to make the integral converge. It is $(1/r_0)$ where r_0 is the cut-off introduced in Eq. (4.4). Since κ is in general $<< q_u$, Eq (4.5) may be written as

$$\sigma^2 = \frac{1}{2} B \ln(q_u / \kappa) + \sigma_o^2$$

(4.6)

Substituting this in Eq. (3.10), we may calculate the scattering in the DWBA after folding with the resolution function. If the latter is approximated by a Gaussian, Sanyal et al[17], have derived the form for the scattered intensity at q_x, q_z (with q_y integrated over as before)

(4.7)

$$I = I_0 \frac{q_c^4}{16} \frac{1}{q_z^3} \left(\frac{1}{2k_0 \sin\alpha}\right) \exp\left(-q_z^2 \sigma_{eff}^2\right) \frac{1}{\sqrt{\pi}} \Gamma\left[\frac{1-\eta}{2}\right]_1 F_1\left[\frac{1-\eta}{2}; \frac{1}{2}; \frac{q_x^2 L^2}{4\pi^2}\right] |T(\alpha)|^2 |T(\beta)^2|$$

where I_0 is the incident beam intensity, k_0 the incident wave vector, q_c the wave vector corresponding to the critical angle of incidence, $\Gamma(x)$ is the gamma function, $_1F_1(x;y;z)$ is the Kummer function,

$$\eta = \frac{1}{2} B q_z^2$$

(4.8)

and

$$\sigma_{eff}^2 = \sigma^2 + \frac{1}{2}(0.5772)B - \frac{1}{2} B \ln(2\pi/\kappa L)$$

(4.9)

L is the coherence length of the beam along the surface or the inverse of the resolution width in q_x space. For $q_x <$ the resolution function width, this saturates and merges into the nominal "specular" reflectivity. For larger q_x, this has the asymptotic form

$$I(q_x, q_z) \sim q_x^{-(1-\eta(q_z))}$$

(4.10)

This is analogous to the algebraic decay $q^{(-2-\eta)}$ of S(q) in a 2D <u>crystal</u> for which the displacement correlations possess logarithmic correlation functions (q being the lateral

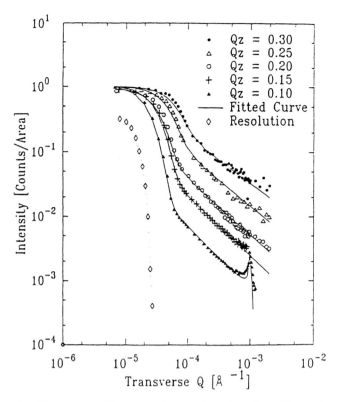

Fig. 4 Log-log plot of the transverse diffuse scattering scans from the surface of liquid ethanol normalized to unity at $q_x = 0$ for different values of q_z. Backgrounds have been subtracted and corrections for variation of illuminated area made. The solid curves represent the calculated scattering. The dashed curve represents the main beam profile converted to an effective transverse resolution at $q_z = 0.1$ Å^{-1} (From Ref. [17]).

component of wavevector transfer, and the $(1-\eta)$ rather than $(2-\eta)$ arises in Eq. (4.10) from integrating over q_y). In this sense the diffuse scattering around the "specular ridge" in the case of surface scattering is the analogue of the diffuse scattering around the Bragg rods in a 2D crystal. However, in the present case, the exponent η is a continuous function of q_z, being given by Eq. (4.8), which can be calculated knowing the surface tension. Experiments carried out with X-ray synchrotron radiation on the surface of liquid ethanol show excellent agreement with the above predictions.[17] (See Figs. 4,5).

By Eqs. (4.6 and 4.9) the effective roughness σ_{eff} measured in a reflectivity experiment is given by

$$\sigma_{eff}^2 = \frac{1}{2}B\ln(q_u/\Delta q) + \sigma_o^2$$

(4.11)

where Δq is the instrumental resolution width $(2\pi/L)$.[18] The so-called "specular" reflectivity from a liquid will be governed by the Debye-Waller-like factor $\exp(-q_z^2 \sigma_{eff}^2)$ rather than $\exp(-q_z^2 \sigma^2)$. The fact that the effective roughness measured for a liquid surface by scattering is less than the true roughness has been known for some time[18] and is due to the unavoidable inclusion of some capillary-wave diffuse scattering inside the

362

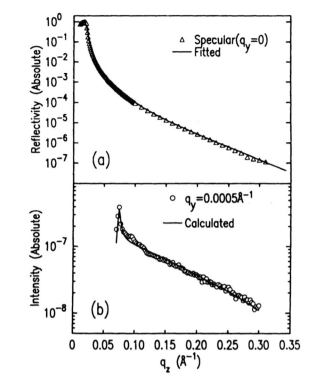

Fig. 5 (a) Measured and fitted "specular" reflectivity for liquid ethanol in an absolute scale (b) Measured and calculated longitudinal diffuse scattering (normalized to direct beam) for $q_x = 5 \times 10^{-4} \, \text{Å}^{-1}$. The background has been subtracted form both spectra.

resolution broadening of the specular peak. A measurement of σ^2_{eff} from specular x-ray reflectivity measurements on liquid alkanes of different chain lengths at different temperatures by Ocko et al[19] (see Figs. 6) yields consistent values for q_u, which appears to scale inversely with the chain length. This leads to the reasonable speculation that the short wavelength cut-off for capillary waves is at a length scale corresponding to the intermolecular spacing, as in the Debye cut-off for phonons in crystals, although a rigorous proof is lacking. Mode-coupling theory yields such a cut-off naturally by introducing a q^4 term in the denominator of Eq. (4.1), which yields an effective cut-off which is of the same order of magnitude as measured from experiments. We note that such a q^4 term in the denominator of Eq. (4.1) also occurs naturally from a curvature-resisting term in the surface free energy as for a surfactant-covered surface and has been used to fit x-ray scattering from such surfaces.[20]

For thin liquid films, the Van-der-Waals interaction with the substrate can enormously increase the value of κ defined in Eq. (4.1), which is now given by

$$\kappa^2 = A/4\pi\gamma d^4 \tag{4.12}$$

where A is the Hamaker constant for the Van der Waals interaction and d is the film thickness. In such cases, κ may actually become larger then the resolution-width Δq and a distinct shoulder is seen in the capillary wave diffuse scattering at a value of $q_x \sim \kappa$. This

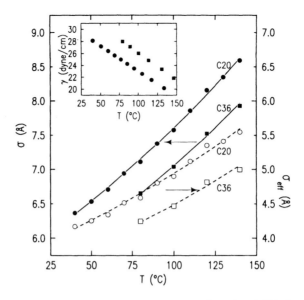

Fig. 6 The "true" roughness σ and the "effective roughness" σ_{eff} for C20 and C36 alkane chains as a function of temperature. The lines are corresponding values calculated with $\sigma_o = 1.1\text{Å}$, $q_u = 0.44\text{Å}^{-1}$ for C20 and $q_u = 0.27$ Å^{-1} for C36. This inset shows the measured surface tension γ of C20 (circles) and C36 (squares).

is seen in Fig. 7, which shows diffuse scattering (transverse diffuse scans as a function of q_x) from a thin polystyrene film on a silicon substrate.[21]

SCATTERING BY MULTIPLE INTERFACES

When one has a thin film on a substrate or a multilayer, the roughness at the various interfaces may be correlated. This is indicated in Fig. 8, which shows clearly in a sectioned TEM micrograph of a multilayer how interface fluctuations propagate from the substrate to each deposited interface. The roughness is said to be "conformal" in such cases. We may discuss the scattering from such interfaces in terms of the height-height correlation function between different interfaces, i.e. we generalize Eq. (2.4) to define

$$C_{ij}(R) = \left\langle \delta z_i(r)\delta z_j(r+R) \right\rangle \tag{5.1}$$

$z_i(r)$, $z_j(r+R)$ are now height fluctuations of the i-th and j-th interfaces. The generalization of Eq. (3.7) for the diffuse scattering in the Born approximation is [12,22-26]

$$S_{\text{diff}}(q) = \left(A/q_z^2\right)\sum_{i,j=1}^{N} e^{-\frac{1}{2}q_z^2\left(\sigma_i^2+\sigma_j^2+\delta^2\lfloor i-j\rfloor\right)}\Delta\rho_i\Delta\rho_j e^{iq_z\left(z_i-z_j\right)}E_{ij}(\bar{q}) \tag{5.2}$$

where

$$E_{ij}(\bar{q}) = \iint dXdY\left[e^{q_z^2 C_{ij}(R)} - 1\right]e^{-i\left(q_x X + q_y Y\right)} \tag{5.3}$$

364

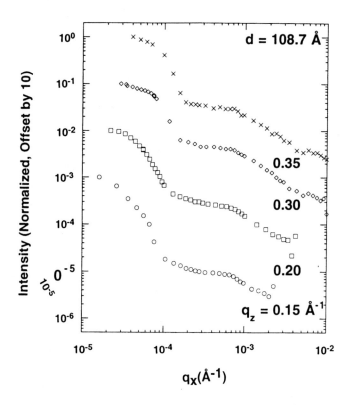

Fig. 7 Transverse diffuse scans measured for a 108.7 Å film of polystyrene on a silicon substrate for different values of q_z showing the finite thickness cut-off outside of the instrument resolution which is the width of the specular peak.

and σ_i is the rms roughness of the i-th interface, $\Delta\rho_i$ is the scattering length density contrast across it, z_i is its average height, and δ is the rms deposition error in the layer spacing, which is cumulative from layer to layer. If there is no correlation between the interfaces, $C_{ij}(R)=0$ for $i \neq j$ and Eq. (5.2) reduces to the sum of the diffuse scattering from the individual interfaces. However, in general $C_{ij}(R) \neq 0$, and the phase factors $\exp(iq_z(z_i-z_j))$ in Eq. (5.2) will cause the diffuse scattering to peak in ridges of constant q_z at the q_z values corresponding to the maxima in the specular reflectivity, i.e. at the Kiessig-fringe maxima in the case of a thin film on a substrate, or at the positions of the multilayer Bragg peaks in the case of multilayers.[22-26] This is illustrated in Figs. 9(a) and 9(b) for a single thin film of water wetting a glass substrate,[27] and for a multilayer film, respectively.[24] The fringes and peaks due to conformal roughness between the interfaces is quite evident. In the case of the multilayer, an excellent fit was obtained to the data using the expression in Eq. (5.2) and assuming perfect conformality between all the interfaces, i.e. $C_{ij}(R)$ independent of i, j.[24] Most thin films, unless extremely rough, show a degree of conformality with the substrate over often surprisingly large thicknesses. For multilayers, Spiller et al.[28] have developed a theoretical model for the propagation of conformal height fluctuations through a multilayer. They approach it from the noise spectral function, or the Fourier transform of C(R), since the conformality is obviously a function of the lateral Fourier component of the fluctuations, i.e. large q (rapid)

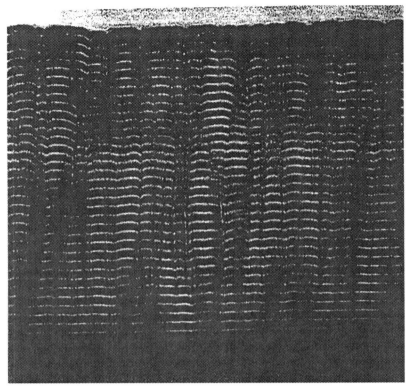

Fig. 8 Micrograph of a Nb/Si multilayer (after sectioning) showing the conformal roughness (plus independent random fluctuations) of the interfaces. (From E.B. Fullerton et al., Phys. Rev. B **48** (1993) 17432).

fluctuations will be more likely to be uncorrelated than small q (long wavelength) fluctuations.

For the case of liquid films wetting a rough substrate, Robbins et al.[29] have calculated theoretically the amplitude of the fluctuations of the upper surface of the liquid film in terms of those of the substrate and shown that the coefficient of proportionality depends on the lateral wavevector q of the fluctuation, the liquid/vapor surface tension, and the interaction between the liquid and the substrate. Recent X-ray scattering experiments on thin liquid[30] and polymer[31] films have been used to check the predictions of these types of theories, with on the whole satisfactory results. The substrates used were patterned as in a one dimensional grating with the grating being of order 1 micron. On this structure polymer films of various thicknesses were deposited, which were subsequently annealed in the molten state. Analysis of the modulations along q_z on the specular rod and also the rods at multiples of the lateral wavevector corresponding to the grating periodicity yielded the degree of the conformality between the grating's height-profile and that of the overlying film. Very similar experiments have recently been carried-out with neutron scattering by A. Mayes and her collaborators.[37] This experiment illustrates a potential strength of the neutron scattering method as it is applied at pulsed spallation sources, where data are naturally obtained simultaneously for a range of values of q_x and q_z. The data obtained in this case are shown in Figure 2, which is a gray-scale intensity plot in the space of detector pixel versus neutron time of flight. The former is

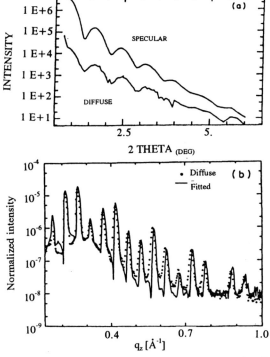

Fig. 9 (a) Specular reflectivity and longitudinal diffuse scattering from a wetting film of water on a glass substrate, showing conformal roughness as evidenced by Kiessig fringes in the diffuse scattering which mirrored those in the specular (From Ref. [27]). (b) Longitudinal diffuse scattering from a GaAs/AlAs multilayer showing conformal roughness, as evidenced by the "quasi-Bragg" peaks. The solid line is a fit to the data assuming perfect conformal roughness. (From Ref. [26])

simply a direct measure of the scattering angle from the surface β, while the latter is proportional to neutron wavelength. To obtain the data in Figure 2, Mayes and her collaborators used a silicon substrate which had on its surface a square array of silicon oxide cylinders about 1500 Å in diameter and 300 Å tall. The spacing between adjacent cylinders was about 4000 Å. The patterned silicon wafer was coated by a film of deuterated poly (methyl methacrylate) about 2000 Å thick. During their experiment, Mayes et al used various thicknesses of polymer and repeated their measurements with polymers of different molecular weight.

The data in Figure 2 show a number of qualitative features that are expected. Firstly, the polymer film gives rise to modulated specular scattering that appears in the figure as a series of diagonal "blobs" at a constant value of the scattering angle beta (at around pixel number 120), equal to the incident angle of the neutron beam. In addition, the Figure shows streaks of diffuse scattering that result from the lateral periodicity of the surface; these are the same fringes, (or rods of scattering as discussed above) at regular intervals in q_x, that one would observe in scattering from an optical grating. They are present even in the absence of the polymer film although, in that case, there is no intensity modulation along each fringe. In the Figure, the intensities of the diffuse fringes are modulated along their length in a way that reflects the thickness of the polymer film. To the extent that the air/polymer interface is conformal with the patterned substrate, one would expect the maxima of these modulations to appear at the same value of q_z along each constant-q_x fringe, as discussed elsewhere in this article. From a series of

367

experiments and an analysis of the way in which the intensity of the modulated fringes changes, one should be able to learn about the ability of various polymers to smooth out surface irregularities of the size scale of the islands etched on the silicon wafer. Unfortunately, the analysis of the data displayed in Figure 2 has proved to be less straightforward as one might have hoped - even the deconvolution of the instrumental resolution has not yet been done satisfactorily. Nevertheless, one hopes that, when the analysis is complete, a new tool for studying surface roughness using neutron scattering will emerge.

For multiple interfaces, going beyond the Born approximation becomes very complicated[25,32]. We cannot go into the details here, but simply point out that in the vicinity of the critical angles for total reflection, as well as angles of incidence for multilayer Bragg reflections, the use of the DWBA, or the "dynamical theory" (where one uses the true eigenfunctions)[33] yields additional sharp structure in the diffuse scattering due to multiple wave interferences. These are the generalizations of the "Yoneda wings" seen for single interfaces and discussed previously. A detailed discussion of diffuse X-ray scattering from multilayers is presented in this volume in the article by R. Paniago.

SCATTERING BY NON-GAUSSIAN SURFACE FLUCTUATIONS

Many kinds of surfaces have specific surface features that cannot be discussed within the random Gaussian Self-Affine model of roughness discussed in Section 2. Examples are surfaces with islands of fixed or variable heights above the reference surface, surfaces with pits (as in the case of corrosion), surfaces with steps, etc. In order to discuss the scattering from such surfaces in the Born Approximation, we must go back to the basic formula for S(q) given in Eq. (3.2). Let us for the moment imagine that the height function $z(x,y)$ for the interface has a bimodal distribution, being 0 with probability $P_1(x,y)$ and Δ with probability $P_2(x,y)= 1 - P_1(x,y)$. (This corresponds to islands of fixed height Δ across the interface). Then Eq. (3.2) may be written as

$$S(q) = \frac{A(\Delta\rho)^2}{q_z^2} \iint dxdy \left\{ \begin{bmatrix} P_1(0,0)P_1(x,y) + P_2(0,0)P_2(x,y) \end{bmatrix} + P_1(0,0)P_2(x,y)e^{-iq_z\Delta} \\ + P_2(0,0)P_1(x,y)e^{iq_z\Delta} \right\} e^{-i(q_x x + q_y y)}$$

(6.1)

Writing ϕ for the fractional coverage of islands, we may express the above in terms of the two-dimensional analogue $\gamma_0(x,y)$ of the Debye correlation function related to the probability of crossing over from no island to an island within a relative separation of (x,y), and obtain

$$S(q) = S_{spec}(q) + S_{diff}(q)$$

(6.2)

where

$$S_{spec}(q) = \frac{A4\pi^2(\Delta\rho)^2}{q_z^2}\delta(q_x)\delta(q_y)\left[1 - 4\phi(1-\phi)\sin^2(q_z\Delta/2)\right]$$

(6.3)

368

and

$$S_{diff}(q) = \frac{A(\Delta\rho)^2}{q_z^2}\phi(1-\phi)Sin^2(q_z\Delta/2)\iint dxdy\gamma_0(x,y)e^{-i(q_xx+q_yy)}$$

(6.4)

Note that the longitudinal diffuse scattering ($q_x, q_y \cong 0$) has a modulation along q_z with period ($2\pi/\Delta$) which is exactly out of phase with a similar modulation in the specular reflectivity. This is in contrast to the case of conformal roughness, where the Kiessig fringes in the specular and the diffuse are in phase. The above theory can be easily generalized to the case of a film with islands deposited on a substrate, and to include roughness fluctuations as well. The expression for the specular reflectivity may be written as

$$R(q_z) = \frac{16\pi^2}{q_z^4}\left\{\rho_i^2 e^{-q_z^2\sigma_i^2}\left[1-4\phi(1-\phi)Sin^2(q_z\Delta/2)\right]+(\rho_2-\rho_1)^2 e^{-q_z^2\sigma_2^2}\right.$$
$$\left.+2\rho_1(\rho_1-\rho_2)e^{-\frac{1}{2}q_z^2(\sigma_1^2+\sigma_2^2)}\left[\phi Cos(q_z(t+\Delta))+(1-\phi)Cos(q_zt)\right]\right\}$$

(6.5)

where t is the total film thickness, ρ_1 is the film electron density, ρ_2 that of the substrate and σ_1, σ_2 are respectively the roughness values at the film/air and film/substrate interfaces. This reflectivity expression yields both the rapid Kiessig fringes, as well as modulations due to the islands on the surface. Fig. 10 shows the specular reflectivity and longitudinal diffuse scattering from a polymer film decorated with such islands, where both the Kiessig fringes (which appear in phase in the specular and the diffuse scattering due to conformal roughness of the film and substrate) and the "island modulations "

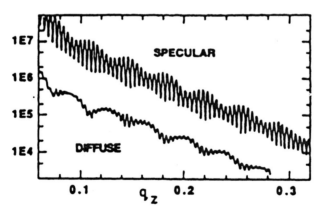

Fig. 10 Specular and longitudinal diffuse scattering as a function of q_z for a polystyrene/PMMA film decorated with islands on the surface. The fitted curves are not shown for clarity as they are indistinguishable from the experimental curves.
(From Ref. 34]).

369

(which are out of phase in the specular and in the diffuse) are observed.[34] Fig. 11 shows transverse diffuse scans (rocking curves) obtained for another system studied, namely a copper film in contact with an electrolyte in an electrochemical cell to which a negative (oxidizing) voltage is applied, as a function of the time for the applied voltage.[35] It may be seen that side peaks grow in the diffuse scattering on either side of the specular reflection. These are a consequence of the pit correlations, which are reflected in a peak in the Fourier transform of the γ_o (x,y) function.

Steps on a surface can result from a miscut of a single crystal surface relative to high symmetry crystallographic planes, or facets, and such steps lead to a roughness which is very anisotropic. The steps can be quasi-periodic and of uniform height, resulting in satellite peaks about the specular position[36], or they can meander in a disordered way and give rise to a characteristic diffuse scattering which has been discussed elsewhere[26]. At the so-called "roughening transition"[10], the height-height correlation function between the steps become logarithmic, as for a liquid surface with capillary wave fluctuations and results in power-law tails in the transverse diffuse scans.

CONCLUSION

In conclusion, we have seen that, with the rapid growth of both experiments and theoretical analyses, surface diffuse scattering, which was first simply a nuisance in the measurement of specular scattering has become a popular tool for the study of lateral

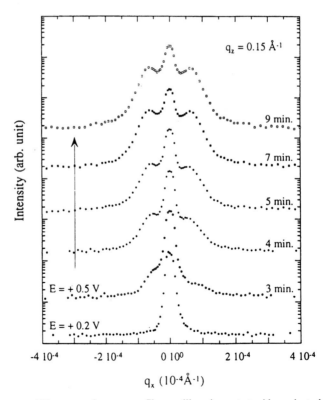

Fig. 11 Transverse diffuse scans for a copper film on silicon in contact with an electrolytic solution at a voltage of + 0.5V for various times of application of the voltage. (From Ref. [35]).

defects and fluctuations at surfaces and interfaces. While not providing real space images as in complementary tools such as AFM and STM, these types of experiments provide global statistical information about the whole surface "in one shot," and have the additional advantage of being able to probe buried interfaces. While X-rays are currently the probe of choice for surface diffuse scattering rather than neutrons, owing to the much higher brightness available from synchrotron sources, we may anticipate increasing neutron scattering studies of diffuse scattering from magnetic and polymeric interfaces in the future.

ACKNOWLEDGMENTS

SKS wishes to thank many of his colleagues for their essential collaborations on much of the work described here. They include M.K. Sanyal, B.M. Ocko, Y.P. Feng, J. Wang, C. Melendres, S.K. Satija, T.P. Russell, E.B. Sirota, M. Tolan, M. Rafailovich, J. Sokolov, E.B. Fullerton, X.Z. Wu, C. Thompson, J. Krim, S. Garoff and M. Deutsch. RP wishes to acknowledge helpful discussions with G. Smith, M. Fitzsimmons and J-P. Schlomka.

This work has been supported by the U.S. Dept. of Energy Basic Energy Sciences under contract numbers W-31-109-Eng-38 and W-7405-Eng-36.

REFERENCES

[1]   P. Beckman and A. Spizzichino, *The Scattering of Electromagnetic Waves from Rough Surfaces*, Pergamon, New York, 1963.

[2]   S.K. Sinha, E.B. Sirota, S. Garoff and H.B. Stanley, *Phys. Rev.* B 38:2297 (1988); P.Z. Wong and A. Bray, *Phys. Rev.* B 37:7751 (1989) .

[3]   M. Kardar, G. Parisi and Y. Zhang, *Phys. Rev. Lett.* 56:889 (1986).

[4]   W. Press et al. *Physics* B 198:42 (1994); M. Wormington, *Philos. Mag. Lett.* 74:211 (1996); see also Ref. [11].

[5]   G. Palazantzas and J. Krim, *Phys. Rev.* B 48:2873 (1993).

[6]   E.L. Church and P.Z. Takacs, *SPIE* 645:107 (1986).

[7]   G. Palasantzas, *Phys. Rev.* E 49:1740 (1994).

[8]   D.K.G. de Boer, A.J.G. Leenaers, and W.W. van den Hoogenhof, *J. de Physique* III France 4:1559 (1994); D.K.G. de Boer, *Phys. Rev.* B 53:6048 (1996).

[9]   H.T. Davis in *Waves on Fluid Interfaces*, (Ed. R.E. Meyer) Pp. 123-150 (Academic Press New York, 1983); C.A. Croxton, *Statistical Mechanics of the Liquid Surface* (Wiley, New York, 1980).

[10]   J.D. Weeks in *Ordering in Strongly Fluctuating Condensed Matter Systems* (T. Riste, Ed.), Plenum, New York, 293 (1960).

[11]   T. Salditt, T.H. Metzger and J. Peisl *Phys. Rev. Lett.* 73:2228 (1994).

[12]   R. Pynn, *Phys. Rev.* B 45:602 (1992); R. Pynn and S. Baker, *Physica* B 198:1 (1994).

[13]   Y. Yoneda, Phys. Rev. 131:2010 (1963); A.N. Nigam, *Phys. Rev.* A 4:1189 (1965).

[14]   L. Nevot and P. Croce, Rev. *Phys. Appl.* 15:761 (1980).

[15] M. Napiorkowski and S. Dietrich, *Phys. Rev.* E 47:1836 (1993).

[16] B.R. McClain, D.D. Lee, B.L. Carvalho, S.G.J. Mochrie, S.H. Chen and J.D. Litster, Phys. *Rev. Lett.* 72:246 (1994).

[17] M.K. Sanyal, S.K. Sinha, K.G. Huang, and B.M. Ocko, *Phys. Rev. Lett.* 66:628 (1991).

[18] A. Braslau, P.S. Pershan, G. Swizlow, B.M. Ocko, and J.Als-Nielsen, *Phys. Rev.* A 38:2457 (1988); J. Meunier and D. Langevin, *J. Phys.* (Paris) *Lett.* 43:L185 (1982).

[19] B.M. Ocko, X.Z. Wu, E.B. Sirota, S.K. Sinha and M. Deutsch, *Phys. Rev. Lett.* 72:242 (1994).

[20] C. Fradin, J. Daillant, A. Braslau, D. Luzet, C. Gourier, M. Alba, G. Grubel, G. Vignaud, J.F. Legrand, J. Lal, J.M. Petit, and F. Rieutard, Proceedings of 5th International Conference on Surface X-ray and Neutron Scattering (SXNS-5), J. Penfold and D. Norman, Eds., *Physica.* B (1997) (to be published).

[21] J. Wang, M. Tolan, A.K. Sood, X-Z. Wu, Z. Li, O. Bahr, M.H. Rafailovich, J. Sokolov and S.K. Sinha, Proceedings of 5th International Conference on Surface X-ray and Neutron Scattering (SXNS-5), J. Penfold and D. Norman, Eds., *Physica.* B (1997) (to be published).

[22] J. Daillant and O. Bélorgey, *J. Chem. Phys.* 97:5824 (1992); ibid. 97:5837 (1992).

[23] D.E. Savage, J. Leiner, N. Schimke, Y.H. Phang, T. Jankowski, J. Jacobs, R. Kariotis and M.G. Lagally, *J. Appl. Phys.* 69:1411 (1991); D.E. Savage, N. Schimke, Y.-N. Phang and M.G. Lagally, *J. Appl. Phys.* 71:3282 (1992).

[24] M.K. Sanyal, S.K. Sinha, A. Gibaud, S.K. Satija, C.F. Majkrzak, and H. Homma, *Surface X-ray and Neutron Scattering*, H. Zabel and I.K. Robinson, Eds. (Springer-Verlag, Berlin, Heidelberg, 91 (1992); M.K. Sanyal et al in K. Liang, M.P. Anderson, R.F. Bruinsma and G. Scoles, Eds., *Mat. Res. Symp. Proc.* 237:393 (1992).

[25] S.K. Sinha, *J. Phys.* III, France 4:1543 (1994).

[26] S.K. Sinha et al, *Physica* B 198:72 (1994).

[27] S. Garoff, E.B. Sirota, S.K. Sinha, and H.B. Stanley, *J. Chem. Phys.* 90:7507 (1989).

[28] E. Spiller, D. Stearns and M. Krumrey, *J. Appl. Phys.* 74:107 (1993).

[29] M.O. Robbins, D. Andelmann, J.F. Joanny, *Phys. Rev.* A 42:4344 (1991).

[30] I.M. Tidswell, T.A. Rabedeau, P.S. Pershan and S.D. Kosowskiy, *Phys. Rev. Lett.* 66:2108 (1991).

[31] M. Tolan, G. Vacca, S.K. Sinha, Z. Li, M. Rafailovich, J. Sokolov, H. Lorenz, and J.P. Kotthaus, J. Phys. D: *Appl. Phys.* 28:A231 (1995) and to be published.

[32] V. Holy and T. Baumbach, *Phys. Rev.* B 49:10688 (1994).

[33] S.K. Sinha in Neutron Scattering in Materials Science II, (D.A. Neumann, T.P. Russell, B.J. Wuensch, Eds), MRS Symposium Proceedings No. <u>376</u> (Materials Research Society, Boston, 1995); V.M. Kaganer, S.A. Stepanov and R. Kohler, *Phys. Rev* B 52:16369 (1995).

[34] S.K. Sinha, Y.P. Feng, C.A. Melendres, D.D. Lee, T.P. Russell, S.K. Satija, E.B. Sirota and M.K. Sanyal, *Physica* A 231:99 (1996).

[35] C.A. Melendres, Y.P. Feng, D.D. Lee and S.K. Sinha, J. Electrochem. Soc. 142:L119, (1995); Y.P. Feng, S.K. Sinha, C.A. Melendres and D.D. Lee, *Physica* B 221:251 (1996).

[36]   K.S. Liang, E.B. Sirota, K.L. D'Amico, G.J. Hughes and S.K. Sinha, *Phys. Rev. Lett.* 59:2447 (1987).

[37]   A. Mayes (to be published).

STUDYING GROWTH KINETICS OF METALLIC MULTILAYERS USING ELASTIC X-RAY DIFFUSE SCATTERING

Rogerio Paniago

Department of Physics, University of Houston, Houston, TX 77204 USA
and Sektion Physik der Ludwig-Maximilians Universität München,
Geschwister-Scholl-Platz 1, 80539 München, Germany[*]

INTRODUCTION

The field of surface roughening during growth has been extremely active in the past years, due to its clear importance for artificially grown thin films[1-5]. The most important consequence of the fluctuation of the height of the surface (due to the build-up of roughness during deposition) is its influence on the thickness of the film. Some of the properties of thin films are intrinsically related to their thickness. Furthermore, in the case of multilayers, any surface roughness present during growth results in a smearing of the interfaces between the two components of the films, and this may change some of their properties. In the case of giant magneto-resistive metallic multilayers, it has been know for a few years that roughness plays a positive role in increasing the scattering of polarized electrons at the interfaces[6,7], which in turn increases the magneto-resistance. However, although the presence of interfacial roughness is crucial for the performance of these films, in case the height fluctuations are too large, they may induce a change in the thickness of the individual bilayers, especially if the roughness is not conformal. This may result in changes in the interaction between neighboring magnetic layers, which is essential to the performance of the metallic thin film as a magnetoresistive device[8].

In this review, we show how X-ray non-specular reflectivity can provide us with a comprehensive picture of the interfacial morphology of metallic multilayers. It is our objective here to determine how roughness increases as a function of length scale and time, and to relate this behavior to a specific growth model. The dynamic information about the physical mechanisms present during growth is usually lost after the film deposition, but it may be retrieved from the contrast of the interfaces of the multilayers. A combination of the static and dynamic properties of these films allows us to predict what is the shape of the growth front, after the film has reached a stationary mode of growth. A number of excellent reviews on surface roughening of simple thin films is already available[1-5], and here we will only focus on metallic multilayers. We will show how one can determine the lineshape of the correlation functions associated with the roughness, and how more subtle interfacial parameters (for instance, the roughness cutoff length) can be determined from this lineshape. We will also determine some of the dynamic scaling properties of these films.

---

[*] Present Address

SURFACE ROUGHNESS FORMALISM

Our objective here is to determine the type of growth of metallic multilayers by comparing the roughness exponents[4], obtained by using X-ray diffuse scattering, to the available local models of growth[5]. In reality, several different physical mechanisms may be present at the same time during growth. For example, atomic adsorption, surface migration (lateral diffusion), desorption, lattice incorporation of atoms to an existing layer, nucleation of two or more atoms, shadowing of regions of the surface by neighboring hills and deposition noise are typical processes that occur during the film formation. However, some of these processes are not easily modeled by an equation of growth, since they are usually non-local phenomena, which involve several atomic sites at the same time. A simpler approach is to relate the increase of the surface to its local morphology. However, only effects that can be related to a local model can be identified using the following roughness formalism.

Roughness exponents allow us to classify the growth of a surface according to a specific universality class[5]. They describe the evolution of a surface as a function of time and length scale. In figure 1 we show the length scales associated with the growth of a thin film, as well as a typical shape of this surface for several times of deposition. The quantities shown here help us to describe the evolution of the morphology of the surface, which ultimately should be associated with one particular model of growth. This occurs since, although several physical mechanisms may be present during the growth, usually only a few of them are important for the final shape of the surface.

We start by describing the height of the surface by a variable $H(x,t)$, which is the deviation of the surface from its average. The most important parameter is the interfacial width $\sigma(R,t)$, which is equivalent to the height difference function

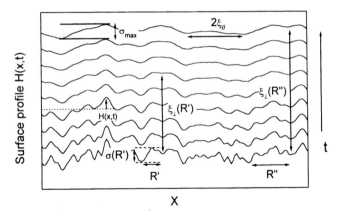

Figure 1. Schematic representation of the relevant parameters during surface growth. The interfacial profile is described by $H(x,t)$. The interfacial width $\sigma(R)$ is a function of the in-plane length R, the in-plane cutoff length $\xi_{//}(t)$ is a measure of the length for which the interfacial width ceases to increase (equivalent to the lateral distance between one valley and one peak in the figure). Both parameters are expected to increase as a function of time. The out-of-plane correlation length $\xi_{\perp}(R)$ is proportional to the saturation time of replication of interfacial features of size R (see text).

$\sigma(R,t) \equiv \{<[H(x,t)-H(0,t)]^2>_R\}^{0.5}$, where $<>_R$ denotes an average over the length R. The interfacial width is therefore a function of the microscopic length scale R with which one examines the surface and is also a function of the time of deposition t. In case the surface is self-affine, its width is expected to scale according to[5]

$$\sigma(R,t) = R^\alpha f\left(\frac{t}{R^z}\right), \quad \text{with} \quad f(y) \propto y^\beta \tag{1}$$

where α, β and z are different roughness exponents. As a result of this scaling property, the interfacial width is expected to increase as a function of length scale R following $\sigma(R) \propto R^\alpha$, until it saturates ($\sigma(R) = \sigma_{max}$) for $R \gg \xi_{//}$, where $\xi_{//}$ is the in-plane cutoff length[9]. The interfacial width (or height difference function) must therefore be modelled by a function that exhibits these two asymptotic regimes[9], e.g., $<[H(R)-H(0)]^2> = 2\sigma_{max}^2[1-\exp(-(R/\xi_{//})^{2\alpha}]$. In an X-ray scattering measurement we usually determine the height-height correlation function $<H(R)H(0)>$, and this can be related to the height difference function using $<[H(R)-H(0)]^2> = 2\sigma_{max}^2 - 2<H(R)H(0)>$.

The maximum interfacial width also increases as a function of time according to $\sigma_{max}(t) \propto t^\beta$, until it becomes saturated. Although not explicitly seen in equation (1) the cutoff length $\xi_{//}$ increases as a function of time following $\xi_{//} \propto t^{1/z}$. The roughness exponents are related to each other ($z = \alpha/\beta$), and it is therefore not necessary to determine all three exponents. One must, however, obtain at least two of them, and especially the static exponent α. Other scaling properties can also be associated with equation (1). For instance, the saturation time scales according to[10] $t_{sat}(R) \propto R^z$. The saturation time $t_{sat}(R)$ is the time it takes for a feature of size R to disappear during growth. This scaling law reveals that larger features tend to be better replicated than smaller ones. The small features usually disappear faster, either due to the action of interfacial tension (which favours the predominance of larger wavelengths) or the deposition noise, which is usually uncorrelated and is added to the local roughness height fluctuations.

In case the deposition rate is constant during film growth, this saturation time can be translated into a correlation length $\xi_\perp(R)$. This correlation length, therefore, also scales according to[11] $\xi_\perp(R) \propto R^z$ It turns out that in homoepitaxy the roughness is hidden after new layers are deposited, and usually one does not have access to $\xi_\perp$. In the case of multilayers, however, this becomes possible due to the contrast of the interfaces, as will be shown later.

GROWTH MODELS

Local growth models determine the variation of height of a surface against time as a function of its local morphology. Therefore, only physical effects that can be associated with some function (or derivative) of the surface profile may be modelled. The final identification of the type of growth is performed by a comparison between the roughness exponents predicted by these models with the measurements of the interfacial roughness using, for instance, X-ray diffuse scattering. These terms of the growth models, however, must obey some simple basic symmetry rules : invariance under translation in time and space (both parallel and perpendicular to the growth direction), as well as rotation and inversion about the axis to the surface. In the case of the growth of crystalline structure, rotation symmetry may not be imposed. The replication of roughness, for instance, may be dependent on the crystallographic direction.

One can then describe the variation of the local height of the surface $\partial H(\mathbf{x},t)/\partial t$ as a function of several derivatives of the surface profile. The most general equation that respects all symmetries is

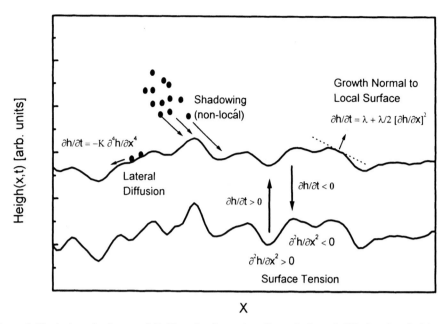

Figure 2. Physical mechanisms modelled by a local equation of growth. Lateral diffusion, interfacial tension and growth normal to the surface can be translated into an equation which relates the variation of the local height H(x,t) to the shape of the surface. Also shown is a non-local effect (shadowing), where a part of the surface is not exposed to the deposited atoms. This effect cannot be modelled by a local equation.

$$\frac{\partial H(\mathbf{x},t)}{\partial t} = F + \nu\nabla^2 H - K\nabla^4 H + (\lambda/2)[\nabla H]^2 + c_n\nabla^{2n}H + d_{m,n}\nabla^{2m}H(\nabla H)^{2n} + \eta(x,t) \qquad (2)$$

where we separated the terms that can be directly identified with one mechanism. F is the net flux of atoms that are deposited on the surface and $\nu\nabla^2 H$ can be related the local surface tension[12]. The surface tension is responsible for the smoothing of the surface for $\nu > 0$ (see figure 2) and counteracts the action of the deposition noise (which is mainly responsible for the increase of the interfacial width). The interfacial tension is proportional to the local curvature of the surface. The term $(\lambda/2)[\nabla H]^2$ is the result of the assumption that the growth is locally normal to the surface[13] with a speed λ, yielding a sideways component in the direction of growth. The term $-K\nabla^4 H$ can be associated with the presence of surface lateral diffusion, which is driven by a gradient of the local chemical potential. Finally $\eta(\mathbf{x},t)$ is an uncorrelated (non-conservative) noise term, where $<\eta(\mathbf{x},t)\ \eta(\mathbf{x}',t')> = 2D\ \delta(\mathbf{x}-\mathbf{x}')\ \delta(t-t')$.

One of the important factors that must be taken into account is that usually only the lower order terms determine the final morphology of the surface. For instance, in case both lateral diffusion ($-K\nabla^4 H$) and interfacial tension ($\nu\nabla^2 H$) are present, the lowest term ($\nu\nabla^2 H$) is predominant and the roughness exponents of a growth model including these two terms would correspond to the values predicted by an equation of growth with only the interfacial tension.

Only these simple mechanisms that may exist during film deposition can be modelled, and several other mechanisms are unfortunately left out. For instance, in figure 2 we show the effect of shadowing, where a region of the surface is not exposed to the deposited atoms, in case the angle of incidence of the atoms is too shallow. In order to include these local models one must take into account the effect of several neighbouring sites of the position where the growth is studied[14].

Table 1. Local models of growth with their respective roughness exponents for 2-dimensional surfaces. The variation of the height of the surface is a function of the local morphology. In the last set of equation the two variables refer to the height of the surface and the width of the surfactant layer.

	$\sigma \propto R^{\alpha}$	$\sigma \propto t^{\beta}$	$\xi_{//} \propto t^{1/Z}$; $t_{sat} \propto R^{Z}$
Edward –Wilkinson[12]			
$\dfrac{\partial h}{\partial t} = \nu \nabla^{2}h + \eta(x,t)$	$\alpha = 0$ $\sigma \propto \ln(R)$	$\beta = 0$ $\sigma \propto \ln(t)$	$Z = 2$
Kardar-Parizi-Zhang[13]			
$\dfrac{\partial h}{\partial t} = \nu \nabla^{2}h + \lambda/2(\nabla h)^{2} + \eta(x,t)$	$\alpha = 0.385$	$\beta = 0.240$	$Z = 1.61$
Diffusion in controlled growth (linear)			
$\dfrac{\partial h}{\partial t} = -K\nabla^{4}h + \eta(x,t)$	$\alpha = 1$	$\beta = 1/4$	$Z = 4$
Diffusion (non-linear)			
$\dfrac{\partial h}{\partial t} = -K\nabla^{4}h + \lambda\nabla^{2}(\nabla h)^{2} + \eta(x,t)$	$\alpha = 2/3$	$\beta = 1/5$	$Z = 10/3$
Surfactant-Mediated Growth[15]			
$\dfrac{\partial h}{\partial t} = \nu \nabla^{2}h + \lambda/2(\nabla h)^{2} + \gamma_{1}(\nabla v)^{2} + \eta_{1}(x,t)$ $\dfrac{\partial v}{\partial t} = -K\nabla^{4}v + \gamma_{2}\nabla^{2}[(\nabla h)(\nabla v)] + \eta_{2}(x,t)$	$\alpha \leq 0$ (flat surface)	-	$Z \geq 8/3$

The most general and physically motivated growth models are shown in Table 1. They incorporate the simplest mechanism. The exponents for the Kardar-Parisi-Zhang (KPZ) model were obtained using growth simulations[16]. We also include in Table 1 a more sophisticated model of growth, which assumes that the evolution of the surface is mediated by a segregated layer on top of the surface[15]. This model predicts a value for the roughness exponent of $\alpha < 0$, meaning that the interfacial width does not increase as a function of R. Thus local growth models may also predict that surfaces can be completely flat. This is not surprising, since a segregated layer is usually well spread on top of the film surface, naturally resulting in a flat interface. This effect is particularly important in case one of the two components of a multilayer segregates during the film growth, and this must be taken into account.

For instance, this effect is very important for the growth of Fe/Au multilayers. It is well known that Au acts as a surfactant during the growth of one Fe monolayer[17]. This could lead to the segregation of Au during the initial stages of deposition of Fe on Au, inducing the formation of flat interfaces. However, it is not completely clear if this may occur, since surface lateral diffusion of Fe (allowing the atoms to move freely on Au) is also an important factor for the formation of the Au/Fe interface. This atomic mobility is present only during the growth of the first monolayer of Fe on Au, however, and it is not a factor for the growth of subsequent layers of Fe, as well as during the growth of Au. Fe/Au multilayers do exhibit a change in the roughness replication as a function of temperature[18]. It has been shown that the interdiffusion of Fe and Au starting at 300^{0}C leads to a very poor replication of the roughness, in contrast with lower temperatures, where the features are very well reproduced from layer to layer.

379

X-RAY SCATTERING FROM ROUGH INTERFACES

X-ray diffuse scattering provides us the Fourier transform of the height-height correlation between the interfaces of a multilayer. The initial observation of diffuse intensity coming from rough surfaces was usually done using rocking scans[9], in which both specular as well as diffuse intensity were observed. This scan was performed in the plane of specular scattering, and it had inherently a very good momentum resolution. The momentum transfer associated with the angles in this geometry is $q_y = (2\pi/\lambda)$ (cos α_i - cos α_f), where α_i and α_f are the incident and exit angles of the X-ray beam with respect to the surface of the sample. This geometry, however, restricted the achievable momentum transfer to very small values. It was particularly useful in the case of semiconductor films, where the cutoff length is usually very large, and a very good momentum resolution is needed.

More recently, an alternative geometry of scattering has been introduced[19], in which the sample is kept fixed and is illuminated by a highly collimated (both in the scattering plane as well as out-of-plane) X-ray beam at a grazing angle α_i with respect to the surface. As seen in figure 3, the X-ray scattering is then collected by an image detector and recorded as a function of three momentum coordinates; $q_x = (2\pi/\lambda)$ sin 2θ cos α_f, $q_y = (2\pi/\lambda)$ (cos α_i - cos α_f) and $q_z = (2\pi/\lambda)$ (sinr α_i + sinr α_f), where sinr $\alpha \equiv (\sin^2\alpha - 2\delta + 2i\beta)^{1/2}$. $n = 1 - \delta + i\beta$ is the average index of refraction of the sample. Except for $\alpha_i = \alpha_f$ and $2\theta_{//} = 0$ (specular reflected beam), all collected intensity is due to the diffuse scattering coming from the interfaces. The measured pattern is therefore essentially a reciprocal space mapping of the diffuse intensity as a function of one out-of-plane (q_z) and one in-plane (q_x) momentum coordinate (with respect to the sample surface). Although providing a poorer momentum resolution, this geometry allows us to achieve a much higher momentum transfer. Another advantage is that all diffuse intensity is collected for the same incident and exit angles α_i and α_f. Therefore, there is no influence of the Fresnel transmission function (discussed below) on the diffuse profile[19,20].

We also show in figure 4 an alternative geometry, using a Position-Sensitive Detector (PSD), placed parallel to the surface of the sample. The geometry using image plates suffers from a distortion in the q_y direction, corresponding to the α_f-scan also shown in the figure. A scan using the PSD in a straight line, parallel to the q_z-axis, with a fixed non-zero value for q_x (avoiding the specular truncation rod), would avoid this distortion. Although not presented here, the results obtained from such a corrected geometry would provide us with a perfect mapping of the diffuse sheets. In case the cutoff length is short (meaning that the diffuse sheets are sharp), the higher order satellites are not collected by the image plate (see dashed line). This occurs since the value of q_y exceeds $2\pi/4\xi_{//}$, and thus the image plate leaves the region of intense diffuse scattering. α_f-scans, therefore, do not completely allow us to determine the interfacial roughness parameters associated with the direction of growth.

The X-ray diffuse intensity obtained from such an experiment can be modelled by the Distorted-Wave Born Approximation, in which the diffuse scattering intensity is given by[21,22]

$$I(q_x q_z) = \frac{I_0 A k^4}{4\pi^2} \frac{|T(\alpha_i)|^2 |T(\alpha_f)|^2}{q_z^2} e^{-q_z^2\sigma^2} \sum_{j,k=0}^{2N} (\delta_{j+1} - \delta_j)(\delta_{k+1} - \delta_k) e^{iq_z(z_j - z_k)} S_{j,k}(q_x, q_z) \quad (3)$$

where $k = 2\pi/\lambda$, A is the area of the sample surface, $T(\alpha)$ the Fresnel transmission function, z_j the average height of interface j and N the total number of bilayers. Here we have simplified the correct expression and taken into account only the Fresnel transmission function for the whole multilayer.

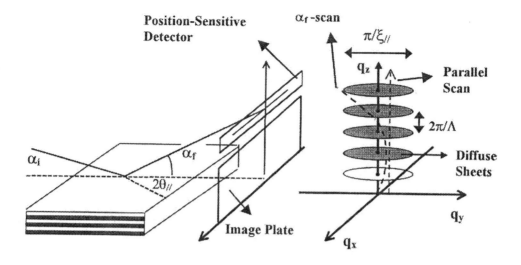

Multilayer   **Reciprocal Space**

Figure 3. Out-of-plane geometry used to collect X-ray diffuse scattering over a large momentum transfer range[19]. The image plate is positioned behind the sample, and collects diffuse intensity as a function of q_x and q_z. We also show an alternative geometry, where the PSD is placed parallel to the surface of the multilayer, in order to scan along a straight line parallel to q_z. This provides us with a perfect mapping of the diffuse intensity.

The roughness structure factor

$$S_{j,k}(q_x, q_z) = \int_0^\infty \left[e^{q_z^2 \langle H_j(R) H_k(0) \rangle} - 1 \right] J_0(q_x R) R dR \qquad (4)$$

contains the information about the replication of roughness for two interfaces (through the in-plane height-height correlation function $\langle H_j(R) H_k(0) \rangle$ of interfaces j and k) as well as the in-plane correlation of the roughness. We may assume a perfect roughness replication from layer to layer, i.e., $\langle H_j(R) H_k(0) \rangle = \langle H(R) H(0) \rangle$ (conformal roughness), which is appropriate only in the small q_x (large R) regime. The comparison between such a simulation using the model given by equations (3-4) with an α_f-scan thus determines the deviation of the interfacial roughness from conformal behavior[23].

In figure 4 we present a typical diffuse pattern obtained with the image plate for a Fe/Au multilayer. We observe a series of extended diffuse satellites starting at $q_z = 0.32 Å^{-1}$, showing that the roughness is very correlated from one interface to another. The diffuse satellites are parallel to the surface of the multilayer. One may notice, however, that the center of the satellites is shifted to one side, especially for higher order satellites. As pointed out recently[24], the direction of replication of roughness does not necessarily coincide with the direction of growth, and this results in a significant shift of the satellites, as a function of the momentum q_z. Assuming that the direction of replication of roughness makes an angle χ with q_z, one may rewrite the correlation function $\langle H_j(R) H_k(0) \rangle$ as

$$\langle H_j(R) H_k(0) \rangle = \langle H_k[R - (z_k - z_j)\tan\chi] H_k(0) \rangle = \langle H[R - (z_k - z_j)\tan\chi] H(0) \rangle \qquad (5)$$

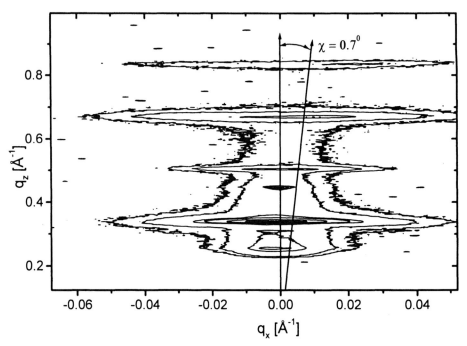

Figure 4. Mapping of the X-ray diffuse intensity in the q_x-q_z plane from a Fe/Au multilayer grown by dc sputtering (T = 40°C). The incident angle was α_i = 2.25°, and therefore the first satellite was hidden below the horizon of the sample. Note the shift of the center of the satellites from q_z = 0, especially for higher order satellites. This effect occurs since the direction of replication of roughness is not parallel to the growth direction (see text).

This model simply assumes that a roughness feature at a position R of interface j will be displaced by a lateral distance $[(z_k - z_j) \tan\chi]$ at interface k, corresponding to the displacement associated with the direction of replication. This assumption leads to a modified form for the diffuse profile[24]

$$I(q_x, q_z) = \frac{I_0 A k^4}{4\pi^2} \frac{|T(\alpha_i)|^2 |T(\alpha_f)|^2}{q_z^2} e^{-q_z^2 \sigma^2} \sum_{j,k=0}^{2N} (\delta_{j+1} - \delta_j)(\delta_{k+1} - \delta_k) e^{i(q_z - q_x \tan\chi)(z_j - z_k)} S(q_x, q_z) \quad (6)$$

As a result of this transverse component of the replication of roughness, the intensity maxima of the diffuse satellites lie on a line making an angle χ with the q_z axis. The analysis of the diffuse peaks from figure 4 yielded that the angle between the direction of replication and the surface normal is χ = 0.7. The last satellite seen in the mapping, however, seems to deviate even more than this angle of 0.7°. We also point out that we did not know the direction of replication of roughness in the q_x-q_y plane. Thus the value of χ obtained here is simply the angle between the surface normal and the projection of the roughness replication vector *in the q_x direction defined by the experiment*.

INTERFACIAL MORPHOLOGY OF METALLIC MULTILAYERS

We turn now to the determination of the correlation functions of a series of metallic multilayers. In our experiments several metallic thin films were grown by sputtering by S.S.P. Parkin at the IBM-Almaden Research Center at different substrate temperatures and a 3.0-5.0 mTorr argon pressure. The deposition rate was typically ~2.0Å/s, considerably faster than in MBE techniques. The growth mode of the film is presumably affected by the sputtering deposition, particularly in view of the very short time for accommodation of each atomic layer.

The samples were taken to the Australian National Beamline Facility (ANBF) at the Photon Factory in Tsukuba, Japan. This facility is a synchrotron X-ray scattering station optimised for both 2-circle diffractometry and small angle scattering. The beamline optics consist of a Ge(111) double crystal monochromator. To increase the signal to noise ratio, the diffractometer and the detector are enclosed in a cylindrical low vacuum chamber. This is particular important to avoid the contribution from air scattering, which would introduce a very high background level in the forward direction.

In figure 5 we present α_f-scans ($\alpha_i = 2.8^0$) of the diffuse intensity coming from $Ni_{0.81}Fe_{0.19}$ [45.2Å] /Au[10.3Å] multilayers with 50 bilayers[20]. We observe a series of very strong diffuse satellites up to the ninth order, showing that the interfacial roughness of these metallic multilayers is very correlated. The simulation was performed using the formalism given in eq. (2), assuming perfect conformal roughness and using the interfacial parameters determined by the specular reflectivity. This shows that the smearing of the interfaces of these samples is completely related to interfacial roughening, and not interdiffusion.

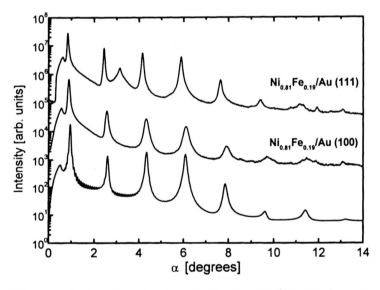

Figure 5. Diffuse intensity along the exit angle α_f for $Ni_{0.81}Fe_{0.19}$(45.2Å)/Au(10.3Å) multilayers with 50 bilayers and two different orientations, showing the periodicity of the replication of roughness. Note the specular peak at $\alpha_f \sim 3.0^0$, which is simply the intersection of the Ewald sphere with the [111] truncation rod. The lower curve is a simulation using the nominal parameters of the multilayers and the theory given in the text. From ref. [20].

The real importance of the out-of-plane scattering geometry can be understood from figure 6. We show here the diffuse scattering parallel to the surface of the same films, and we clearly observe two regimes of scattering. Starting at large momentum transfers, the diffuse intensity exhibits an asymptotic decay, following[11] $I(q_x) \propto q_x^{-2-2\alpha}$. For $q_x < 0.08\text{Å}^{-1}$, however, the diffuse intensity starts to levels off. This lineshape is simply the result of the Fourier transformation of the height-height correlation function, which also has shows two regimes of scaling. A fit of the diffuse profile using the model function $\langle H(R)H(0)\rangle = \sigma^2[\exp(-(R/\xi_{//})^{2\alpha}]$ yielded the roughness parameters[20] $\xi_{//} = 32\text{-}36\text{Å}$ and $\alpha = 0.90\text{-}0.92$.

This plot is a clear example where the determination of the cutoff length would only be possible using such a scattering geometry. For the incident angle used here of $\alpha_i = 2.8^0$ the maximum achievable momentum transfer in a rocking scan would be $2\pi/\lambda(1 - \cos \alpha_i) = 4.3 \times 10^{-3}\text{Å}^{-1}$ (corresponding to a length scale of ~1500Å) and this clearly would not be enough to determine $\xi_{//}$. This out-of-plane geometry was necessary to achieve the very large momentum transfer corresponding to $q_x > 2\pi/4\xi_{//} = 0.05\text{Å}^{-1}$, and also to observe the asymptotic decay which occurs at even larger momentum transfers.

The surface diffusion growth models[5], which yield values for α in between 0.66 and 1.0 (see Table 1), seem to be appropriate for the type of growth of these NiFe/Au multilayers. This would be supported by the high substrate temperature (T = 500^0C) during the deposition of the film. However, it was not possible for these multilayers to determine a dynamic roughness exponent, since the growth correlation length was limited to the total thickness of the film (50 bilayers). The mosaic quality of these samples was also considerably poor (about $3\text{-}5^0$) and this raised questions about the influence of the lattice mismatch on the interfacial formation[20].

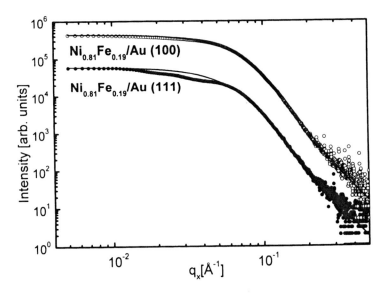

Figure 6. Diffuse scattering parallel to the surface sample for $Ni_{0.81}Fe_{0.19}$ multilayers. The fits to the profiles according to a self-afine roughness model yields very short cutoff length of ~ 30Å. This diffuse profile is a typical example where only the out-of-plane geometry provides us with the correct interfacial parameters, in particular the cutoff length. From ref [20].

Another set of Fe[15Å]/Au[21Å] multilayers grown at T = 200°C, with a much better crystalline quality, was also examined, and here a much more comprehensive understanding was achieved[23]. The measurements of X-ray diffuse scattering were performed in a similar way as with the other multilayers.

In figure 7 we present the diffuse profiles of three Fe/Au multilayers of increasing thickness, which were fitted using a different height correlation function[25] $\langle H(R)H(0)\rangle = \sigma^2\{1-[1-\exp(-(R/\xi_{//})^2)]^\alpha\}$. The fits yielded an increasing cutoff length as a function of film thickness. We obtained $\alpha = 0.36 \pm 0.02$ and $\xi_{//} = 160.0 \pm 20.0$ Å for one Fe/Au bilayer (95Å), $\alpha = 0.49 \pm 0.02$ and $\xi_{//} = 320.0 \pm 10.0$ Å for 40 bilayers (1400Å) and $\alpha = 0.45 \pm 0.02$ and $\xi_{//} = 500.0 \pm 10.0$ Å for 100 bilayers (3500Å). The values of $\xi_{//}$ could also be determined approximately from the position of the characteristic "knee" of the in-plane diffuse scattering (see arrow in fig.7).

These "knees" correspond to points where the roughness correlation has an insignificant effect on the diffuse intensity[26], i.e., approximately at $q_x = 2\pi/4\xi_{//}$. We clearly observe that the positions of the knees (where the diffuse intensity levels off) are shifted to lower momentum transfer for thicker films, meaning longer cutoff lengths, and they agree with the values we have obtained for $\xi_{//}$. The increase of $\xi_{//}$ for thicker films indicates that the evolution of roughness is established through the predominance of longer features over the shorter ones. In other words, the length at which the in-plane roughness is correlated increases as a function of time, as should be expected from the power-law $\xi_{//} \propto t^{1/z}$. Notice also that all films exhibit the same asymptotic decay for higher momentum transfer, yielding similar values of the static exponent α. As an average value for all films we obtained $\alpha = 0.43 \pm 0.05$. The values of the static roughness exponent are all in the range of $\alpha = 0.38$ predicted for the KPZ equation[23].

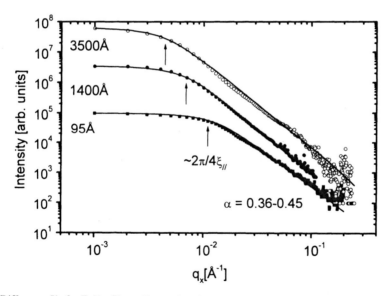

Figure 7. Diffuse profile for Fe/Au films of increasing thickness. Note the shift of the point where the diffuse intensity starts to decay, corresponding to longer cutoff lengths for thicker films. The asymptotic decay of the diffuse intensity for $q_x > 2\pi/4\xi_{//}$ follows $I(q_x) \propto q^{-2-2\alpha}$ and for these films it yields values of α predicted by the Kardar-Parisi-Zhang model. From ref [23].

These static results suggest that the Kardar-Parisi-Zhang model of growth should be appropriate to describe the growth of these multilayers. On must, however, determine also the dynamic scaling properties of growth. In the case of multilayers it is possible to determine the roughness exponent z from the diffuse X-ray pattern. This can be done using the scaling law $t_{sat} \propto \xi_\perp(R) \propto R^z$. One must first translate the relationship $\xi_\perp(R) \propto R^z$ into a reciprocal space language appropriate for X-ray scattering. In reciprocal space the width of the diffuse satellite sheets along the growth direction q_z corresponds to $\Delta q_z \propto \xi_\perp^{-1}$. The in-plane momentum transfer q_x can be associated to the feature size R by $q_x = 2\pi/R$. One then obtains a direct relationship between the width Δq_z and the in-plane momentum transfer q_x, i.e., $\Delta q_z \propto \xi_\perp^{-1} \propto R^{-z} \propto q_x^z$.

Thus one may determine the growth correlation length $\xi_\perp(R)$ as well as the length scale $R = 2\pi/q_x$ associated with it from one single diffuse satellite obtained with the image plate[11]. In the data analysis, however, one must also consider the effect of the resolution function in the q_z direction. We have observed that the diffuse satellites from Fe/Au multilayers are Lorentzian-shaped, and they were convoluted by a Lorentzian resolution function. In this case, the width of the observed convoluted peaks is given by the linear sum of the width of the resolution function and the width of the diffuse satellite (Δq_z).

In figure 8 we present the increase of $\xi_\perp(R)$ as a function of R, taken from the second diffuse satellite from a Fe/Au multilayer with 100 bilayers grown at $T = 40^0C$. A fit of this curve to the power-law $\xi_\perp \propto R^z$ yielded $z = 1.61 \pm 0.14$, in very good agreement with the value of $z = 1.58$-1.61 predicted by the KPZ equation. To our knowledge, this is the first example where a measurement of both static and dynamic roughness exponents, yields values in agreement with the KPZ equation[23]

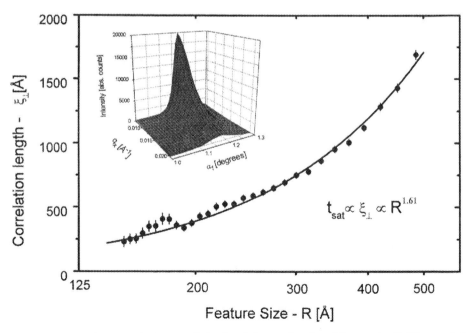

Figure 8. Increase of the out-of-plane correlation length $\xi_\perp$ as a function of feature size R. The correlation length is assumed to be proportional to the saturation time $t_{sat}(R)$, which scales following $t_{sat}(R) \propto R^z$. The fit yields $z = 1.61 \pm 0.14$, in accordance with $z = 1.58$-1.61 for the KPZ model. The inset shows the region of the broadening of the second satellite from which the correlation length was determined. From ref. [23].

It is also possible to verify, at least qualitatively, the prediction that sputtered Fe/Au multilayers grow according to the KPZ model. We start with the noiseless KPZ equation, where the noise term $\eta(x,t)$ is set to zero. Although this equation is not completely realistic (it predicts that the interfacial width decreases as a function of deposition time), it contains the basic physical mechanisms present during growth. The result of an integration of this equation should provide us with the shape of the growth front.

In the deterministic (or noiseless) KPZ model the height profile $H(x,t)$ is solely determined by the initial profile $H_0(\mathbf{x}) \equiv H(\mathbf{x},0)$ and the constants ν and λ. This equation is exactly solvable[13], and therefore one can predict what is the shape of the surface. What is most important for the surface profile, however, is the format of the interfaces after long deposition times. After the initial (uncorrelated) height fluctuations due to the substrate are smoothed out, the growing surface is composed of a number of paraboloids. The asymptotic form of one of these paraboloids is given by[13] $H(x,t) = C - (x^2 + y^2)/(2\lambda t)$, where λ is the speed of growth normal to the local surface, as given before. The paraboloids tend to coalesce and increase in lateral size, and their radii correspond exactly to the in-plane cutoff length $\xi_{//}$ (which increases as a function of time as described above)[18].

This prediction may be analyzed with a surface imaging technique. For instance, in figures 9-10 we show a comparison between the growth front, obtained from the deterministic KPZ equation, to an Atomic Force Microscopy scan of the surface of 3500Å thick Fe/Au multilayer. As is seen in figure 10, the surface of the multilayer is composed of a number of paraboloid "hills", which suggests that this asymptotic form should be correct[18].

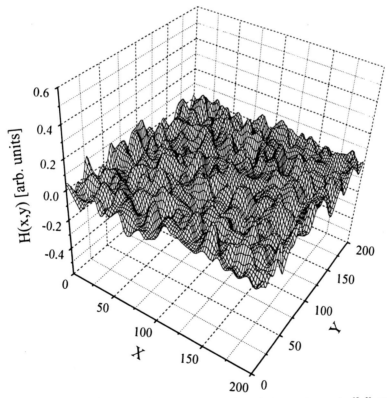

Figure 9. Asymptotic shape of a growth front according to the noiseless KPZ equation[18, 23]. The front is composed of paraboloids of a radius equivalent to the in-plane cutoff length $\xi_{//}$. Since the cutoff length increases as a function of time, the paraboloids tend to be enlarged for later times of growth[23].

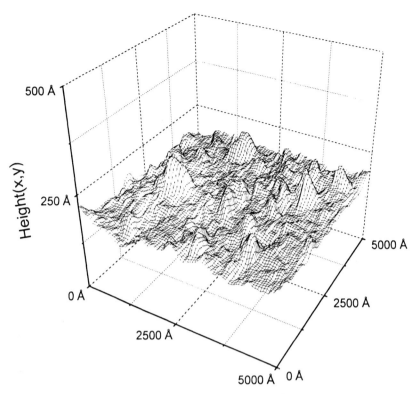

Figure 10. Atomic-force microcopy scan of the surface of a 3500Å thick Fe/Au multilayer grown by sputtering (T = 40°C)[18]. Although the paraboloid hills vary significantly in height, their widths are all in the range of 500Å, corresponding to the cutoff length measured for this multilayer using X-ray diffuse scattering. From ref. [18].

We observe, however, a much higher fluctuation in the heights of the paraboloids, in comparison with the growth front obtained analytically. The uniformity of the paraboloids in the simulation is probably due to the "well behaved" rough substrate, in which the initial fluctuations where assumed to be completely uncorrelated, and their distribution corresponded to a normal function. Nevertheless, we observe in figure 10 that all hills have about the same radii, and this radius is equivalent to the cutoff length $\xi_{//} = 500$Å, determined from the fit of the upper curve of figure 8.

This model seems therefore to be quite realistic in describing the growth of sputtered Fe/Au multilayers. We point out that, although surface tension (present in the KPZ equation) seems to be the most important factor for the surface formation, our results do not exclude the possibility of the presence of lateral diffusion. They only show that lateral diffusion is not the *dominant* factor for the interfacial morphology. Furthermore, at this time it is not clear if a similar structure grown by a different technique (for example, Molecular Beam Epitaxy) would yield similar results. As is known, the deposition rate of MBE is considerably slower, and a layer-by-layer growth is much more likely to occur. The very short time for accommodation of the atoms in sputter-deposition, due to the fast deposition rate (~1 monolayer/second), is an important factor in the build-up of roughness in our multilayers.

This same structural roughness formalism may be used to study magnetic roughness. This is a crucial information for the understanding of scattering of conduction electrons at the interfaces. In fact, recent results[27] show that magnetic correlation lengths are slightly longer than their structural counterpart, meaning that the interfacial magnetization varies more smoothly that the interfaces themselves.

388

In conclusion, in this review we have shown how X-ray diffuse scattering can provide us with a comprehensive understanding of the interfacial morphology of metallic multilayers. The determination of the very short cutoff lengths of these films was only possible using a geometry where the diffuse intensity is collected out of the plane of the specular reflected beam, and in the plane of the film. The comparison of the obtained interfacial roughness exponents with several growth models showed that, at least in one case, the growth is compatible with one of the most general and physically sensible growth models (KPZ). A subsequent comparison between a simulation of a growth front, predicted by the noiseless KPZ equation, with Atomic Force Microscopy results showed that the interfaces are composed of a number or paraboloids, with a radius corresponding to the cutoff length. These paraboloids tend to increase in lateral size (following the cutoff length) and coalesce, forming even larger structures. Further research should be performed trying to relate the structural roughness studied here with the magnetic roughness, which is essential for the understanding of the magnetoresistance of metallic multilayers.

ACKNOWLEDGEMENTS

The author wishes to thank the group of researchers with whom this study has been done over the past few years: Drs. Paul Chow and Hitoshi Homma, Rebecca Forrest and especially Prof. Simon C. Moss, from the University of Houston, Dr. Stuart Parkin, from IBM-Almaden (San Jose, CA) for supplying the samples, Dr. David Cookson for help at the ANBF beamline and Prof. Zwi Barnea, from the University of Melbourne (Australia). This research has been supported by grants from the National Science Foundation U.S.-Australia Cooperative Program INT-9114683 and DMR-9208450. My graduate studies in Houston were financed by the "Conselho Nacional de Desenvolvimento Científico e Tecnológico" (Brazil). This work was also supported in part by the MRSEC program of the NSF, DMR-9632667.

REFERENCES

1. See, for example, J. Krug and H. Spohn, in *Solids Far From equilibrium: Growth, Morphology and Defects*, ed. C. Godréche (Cambridge Univ. Press, Cambridge, 1991) and the following 4 references.
2. H.-N. Yang, G.-C. Wang and T.-M. Lu, *Diffraction from Rough Surfaces and Dynamic Growth Fronts* (World Scientific, Singapore, 1993).
3. P. Meakin, Phys. Rep. **235**, 189 (1994)
4. J. Krim and G. Palasantzas, Int. J. Mod. Phys. B **9**, 599 (1995).
5. A.-L. Barabási and H. E. Stanley, *Fractal Concepts in Surface Growth* (Cambridge Univ. Press, Cambridge, 1995).
6. E.E. Fullerton, D.M. Kelly, J. Guimpel and I.K. Schuller, Phys. Rev. Lett. **68**, 859 (1992).
7. S.S.P. Parkin, Phys. Rev. Lett. **71**, 1641 (1993).
8. S.S.P. Parkin, R.F.C. Farrow, R.F. Marks, A. Cebollada, G.R. Harp and R.J. Saviy, Phys. Rev. Lett. **72**, 3718 (1994).
9. S.K. Sinha, E. B. Sirota, S. Garoff and H. B. Stanley, Phys. Rev. B **38**, 2297 (1988).
10. J. Krug and H. Spohn, Phys. Rev. B **38**, 4271 (1988).
11. T. Salditt, D. Lott, T. H. Metzger, J. Peisl, G. Vignaud, P. Høghøj, O. Schärpf, P. Hinze and R. Lauer, Phys. Rev. B **54**, 5860 (1996).
12. S.F. Edward and D.R. Wilkinson, Proc. R. Soc. Lond. A **381**, 17 (1982).
13. M. Kardar, G. Parisi and Y.-C. Zhang, Phys. Rev. Lett. **56**, 889 (1986).
14. See, for example, J. Krug and P. Meakin, Phys. Rev. B, **47** R17 (1993).
15. A.-L. Barabási, Phys. Rev. Lett. **70**, 4102 (1993).
16. J.G. Amar anf F. Family, Phys. Rev. B **41**, 3399 (1990).
17. A.M. Begley, S.K. Kim, J. Quinn and F. Jona, Phys. Rev. B **48**, 1779 (1993).
18. R. Paniago, P.C. Chow, R. Forrest and, S.C. Moss, Physica B (in press).
19. T. Salditt, T. H. Metzger and J. Peisl, Phys. Rev. Lett. **73**, 2228 (1994).

20. R. Paniago, H. Homma, P. C. Chow, S. C. Moss, Z. Barnea, S. S. P. Parkin and D. Cookson, Phys. Rev. B **52**, R17052 (1995).
21. S. Dietrich and A. Haase, Phys. Rep. **260**, 1 (1995).
22. V. Holý and T. Baumbach, Phys. Rev. B **49**, 10668 (1994).
23. R. Paniago, R. Forrest, P.C. Chow, S.C. Moss, S.S.P. Parkin and D. Cookson (submitted to Phys. Rev. B).
24. V. Holý, C. Gianinni, L. Tapfer, T. Marschner and W. Stols, Phys. Rev. B **55**, 9960 (1997).
25. G. Palasantzas, Phys. Rev. B **48**, 14472 (1993).
26. G. Palasantzas and J. Krim, Phys. Rev. B **48**, 2873, (1993).
27. J.F. Mackay, C. Teichert, D.E. Savage and M.G. Lagally, Phys. Rev. Lett. **77**, 3925 (1996).

PARTICIPANTS

Madhav Acharya
University of Delaware
Department of Chemical Engineering
347 Colburn Laboratory
Newark DE 19716
USA
acharya@che.udel.edu

Rozaliya Barabash
National Technical University of Ukraine
Apt. 25, Street Proreznaya 4
Kiev 252003
UKRAINE
barabash@amath.pp.kiev.ua

Simon Billinge
Michigan State University
Physics and Astronomy Department
East Lansing MI 48824-1116
USA
billinge@pa.msu.edu

Emil Bozin
Michigan State University
Physics and Astronomy Department
East Lansing MI 48824-1116
USA
bozin@pa.msu.edu

Jean Chung
Chungbuk National University
Department of Physics
Cheongju, Chungbuk 361-763
REPUBLIC OF KOREA
chung@cbucc.chungbuk.ac.kr

Remo DiFrancesco
Michigan State University
Physics and Astronomy Department
East Lansing MI 48824-1116
USA
difrance@pa.msu.edu

Martin Dove
University of Cambridge
Department of Earth Sciences
Downing Street
Cambridge CB2 3EQ
UNITED KINGDOM
martin@minp.esc.cam.ac.uk

Takeshi Egami
University of Pennsylvania
Department of Materials Science and
Engineering
223 LRSM
Philadelphia PA 19104
USA
egami@pdfvax.lrsm.upenn.edu

Friedrich Frey
Institut für Kristallographie, LMU
Am Coulombwall 1
Garching D-85748
GERMANY
frey@yoda.kri.physik.uni-muenchen.de

Brent Fultz
California Institute of Technology
Keck Lab of Engineering, 138-78
Pasadena CA 91125
USA
btf@hyperfine.caltech.edu

Cybele Hijar
Los Alamos National Laboratory
MS J514
Los Alamos NM 87545
USA
chijar@lanl.gov

David Keen
Rutherford Appleton Laboratory
ISIS Science Division
Chilton, DIDCOT
Oxon OX11 0QX
UNITED KINGDOM
D.A.Keen@rl.ac.uk

George H. Kwei
Los Alamos National Laboratory
Cond. Matter and Thermal Physics Group
MST-10, MS K764
Los Alamos NM 87545
USA
ghk@lanl.gov

Stefan Kycia
CHESS, Cornell University
Wilson Lab
Ithaca NY 14853
USA
sk85@cornell.edu

Simon Moss
University of Houston
Department of Physics
4800 Calhoun
Houston TX 77204-5506
USA
smoss@uh.edu

Victoria Nield
University of Kent, Canterbury
School of Physical Sciences
Canterbury, Kent CT2 7NR
UNITED KINGDOM
v.m.nield@ukc.ac.uk

Rogerio Paniago
c/o Prof. Dr. J. Peisl
LMU München-Sektion Physik
Geschwister-Scholl Platz 1
München 80539
GERMANY
paniago@lspserver.roentgen.physik.uni-muenchen.de

Sidhartha Pattanaik
Georgia Institute of Technology
Dept. of Chemistry and Biochemistry
Atlanta GA 30332-0400
USA
sidhartha.pattanaik@gatech.edu

David Price
Argonne National Laboratory
Building 223
Argonne IL 60439
USA
price@anpns1.pns.anl.gov

Roger Pynn
Los Alamos National Laboratory
LER-PO, MS H845
Los Alamos NM 87545
USA
pynn@lanl.gov

Lee Robertson
Oak Ridge National Laboratory
P.O. Box 2008, MS 6393
Oak Ridge TN 37831-6393
USA
robertsonjl@ornl.gov

Sjoerd Roorda
University of Montreal
Gr. Rech. Tech. Couches Minces
Laboratoire Physique Nucleaire, Rm 202
Montreal PQ H3C 3J7
CANADA
roorda@lps.umontreal.ca

Marie-Louise Saboungi
Argonne National Laboratory
9700 South Cass Avenue
Room MSD 223 D233
Argonne IL 60439
USA
saboungi@anpns1.pns.anl.gov

Werner Schweika
IFF
Forschungszentrum Jülich
Jülich D-52425
GERMANY
w.schweika@fz-juelich.de

Sunil Sinha
Argonne National Laboratory
9700 South Cass Avenue
Building 401
Argonne IL 60439
USA
sksinha@aps.anl.gov

Alan Soper
Rutherford Appleton Laboratory
ISIS Facility
Chilton, DIDCOT
Oxon OX11 0QX
UNITED KINGDOM
aks@isise.rl.ac.uk

Michael F. Thorpe
Michigan State University
Physics and Astronomy Department
East Lansing MI 48824-1116
USA
thorpe@pa.msu.edu

Igor Tsatskis
University of Cambridge
Department of Earth Sciences
Downing Street
Cambridge CB2 3EQ
UNITED KINGDOM
it10001@esc.cam.ac.uk

T. Richard Welberry
Australian National University
Research School of Chemistry
Canberra City ACT 0200
AUSTRALIA
welberry@rsc.anu.edu.au

Angus Wilkinson
Georgia Institute of Technology
School of Chemistry and Biochemistry
Atlanta GA 30332
USA
angus.wilkinson@chemistry.gatech.edu

Michael Winokur
University of Wisonsin
Department of Physics
1150 University Avenue
Madison WI 53706
USA
winokur@ewald.physics.wisc.edu

Christopher Wolverton
National Renewable Energy Laboratory
Solid State Theory Group
1617 Cole Blvd.
Golden CO 80401
USA
cmw@sst.nrel.gov

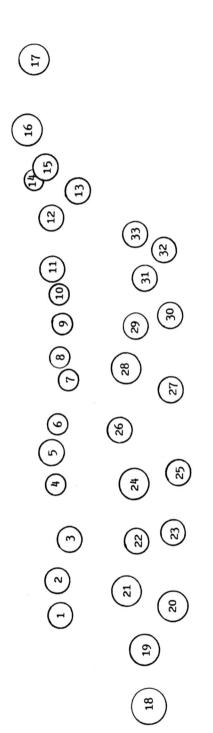

[1] R. DiFrancesco, [2] C. Wolverton, [3] S. Moss, [4] D. Keen, [5] A. Wilkinson, [6] S. Kycia, [7] D. Price, [8] S. Roorda, [9] S. Sinha, [10] L. Robertson, [11] R. Pynn, [12] T. R. Welberry, [13] J. King, [14] M. Winokur (half-hidden), [15] J. Chung, [16] M. F. Thorpe, [17] A. Soper, [18] L. Neuman [19] E. Bozin, [20] S. Billinge, [21] S. Pattanaik, [22] M. Dove, [23] M. Acharya, [24] B. Fultz, [25] C. Hijar, [26] T. Egami, [27] V. Nield, [28] G. Kwei, [29] R. Paniago, [30] R. Barabash, [31] I. Tsatskis, [32] W. Schweika, [33] F. Frey

Not in photograph: M-L. Saboungi

INDEX

AgBr, 122, 127
Alloys, 175, 189
  semiconductor, 157
Alpha-plutonium, 324
Amorphous silica, 76
Anharmonic vibrations Co_3V, 290
Anomalous dispersion, 181
Anomalous scattering, 23
Atomic arrangements, 158
Atomic force microscopy, 387
Atomic motion
  correlated, 141, 144
AXS, 23

Binary alloy, 208
Bond angle, 78
Born approximation, 88
Bragg diffraction, 158
Bragg peaks, 95

C_{60}, 122, 126
Charge density wave, 17
Clebsch-Gordan coefficients, 70
Cobalt, 313
Competing interactions, 200
Compounds, 17
Continuous random network, 106
Convolution, 142, 143
Correlation function
  density-density, 160
Coulomb, 233
Cristobalite, 111, 113, 257
Cu_3Au, 195, 289
Cu-Al, 197
Cu-Pd, 199
Cuprates, 16

Debye-Waller factor, 98, 233
Debye-Waller theorem, 158
Defects, 233
Depletion layer, 243
Difference structure factor, 25
Diffraction, 3
Diffraction theory, 300
  domain structures, 303

Diffuse scattering
  background, 158
  computer simulation, 39
  diffraction equations, 36
  experimental measurement, 37
  experimental methods, 181
  force models, 46
  interatomic, 46
  modulation wave, 55
  modulation wave method, 43
  theory, 177
Diffuse scattering, 35, 116, 121, 137, 175, 189,
  190, 207, 233, 239, 248, 262, 295, 351, 375
Disorder, 85, 101
Disordered alloys, 199, 207, 211
Disordered materials, 23, 59
Distorted wave Born approximation, 358
Distribution functions, 60
Domain structures, 295
  lamellar, 308, 313
Dynamic scaling, 386
Dynamical matrix, 160
Dyson equation, 218

Effective pair interactions, 95
Electronic theory, 211
Empirical potential, 73
Enstatite, 315

FeCr, 182
Fermi surface, 189, 199, 225
Fourier series, 191
Friedel oscillations, 192

$Ga_{0.5}In_{0.5}As$, 168
Germanate glasses, 27
Germania, 27
Glass, 111

Harmonic potential, 157
Height-deviation function, 354
Height-height correlation function, 381, 384
Hydrides, 85

Ice Ih, 122, 126

Impurity atoms, 233
InAs, 165
Incommensurate phase transitions, 264
Indium arsenide, 144
Interfaces
    roughness, 354, 356, 380
    scattering, 364
Ising model, 217

Kanzaki force models, 94
Keating model, 159
Kinetics, 375
Kirkwood model, 159
Kohn anomaly, 194

$La_{1-x}A_x MnO_3$, 14
La_2CuO_4, 148
$La_{2-x}(Sr,Ba)_xCuO_4$, 149
Lattice dynamics, 10
    Born-von Kármán model, 282
Layer-layer correlation function, 6
Liquids, 59
Locator, 212

Many body correlations, 99
Many body problems, 60
Markov chain, 105
Maximum Entropy, 64
Mean-field approximation, 208
Metallic alloys, 192, 323
Minimum Noise, 60, 66
Molecular Dynamics, 71, 107
Monte Carlo, 67, 71, 85, 190
Monte Carlo
    displacements, 45
    inverse, 87, 89
    Metropolis, 87
    occupancy, 42
    reverse, 87, 101, 108
    vacancy ordering, 52
Multicomponent systems, 62, 161
Multilayers
    metallic, 375, 382

Nanoclusters, 29
National Synchrotron Light Source, 27
Neutron, 3
Neutron scattering
    inelastic, 284
Nickel, 146, 163

Oxides, 85

Pair correlation function, 208
Pair correlation function
    orientational, 69
Pair distribution function, 1, 3, 62, 137, 138, 158, 337,
Pair distribution function
    accuracy, 4
    calculating, 139
    differentials, 329
    element specificity, 9

Fourier transformation, 138
    higher dimensions, 5
    layer correlation function, 6
    partial wave analysis, 8
    two-dimensional, 8
Partial structure factor, 29, 63
PDF
Perovskites, 96
Phonon density of states, 286
Phonon modes, 99
Point defects, 233
Polymers, 337
Polymers
    interchain pair correlations, 339, 340
    modeling, 343
Polysilane, 345
Potential
    Born-Meyer, 76
Precipitates, 239, 241
Propagator, 212
Proteins, 323, 331
Pyroxenes, 316
PZ, 11
PZT, 11

Quality factor, 64
Quartz, 111
Quasi-elastic broadening, 99

Radial distribution function, 62
Real-space Rietveld, 137, 139
Reflection
    neutron, 351, 353
    x-ray, 351
Reflectivity
    off-specular, 352
    specular, 352
Reflectometry, 351
Reverse Monte Carlo, 121, 261
    codes, 134
    constraints, 133
    measuring and preparing the data, 127
    moves, 130, 132
    single crystal, 123
    super-cell, 129
Rietveld profile refinement, 30, 60
Rigid unit mode, 253
Roughness
    conformal, 364

Self-energy, 216, 218
Semiconductor crystals, 158
Short-range order parameters, 96, 178
Short-range order, 207, 215
Silica, 101, 106, 109
Silicates, 254, 256
Single crystal, 121
Size-effect scattering, 180
Solid solutions, 175
Spectrometer
    backscattering, 97
Specular reflectivity, 359
Spherical harmonic expansion, 69

Spherical model, 208, 221
Structure factor, 60, 104
  disordered system, 61
Surface
  fluctuations, 368
  growth, 375
  liquid, 359
  non-Gaussian, 368
  roughness, 361
  self-affine, 354

Thermal diffuse scattering, 144
Transition metal aluminides, 286
Tridymite, 111, 262

Vanadium hydride, 94
Vegard, 157

Vibrational entropy, 273, 276
  quantum statistical mechanics, 278
Virtual crystal, 159, 208

Wüstite, 91

X-ray, 3
X-ray absorption fine structure (XAFS), 3

$YBa_2Cu_3O_{6+\delta}$, 329
Yoneda wings, 358, 368

Zeolites, 29, 265
Zinc-blende, 158, 313
Zirconia, 49

Breinigsville, PA USA
30 November 2009
228083BV00002BA/1/A